POLYFACE DESIGNS

A Comprehensive Construction Guide
for Scalable Farming Infrastructure

Joel Salatin & Chris Slattery

Foreword by Justin Rhodes

Polyface, Inc.
Swoope, Virginia

© 2020 Joel Salatin and Chris Slattery.
All rights reserved.
ISBN: 978-1-7336866-1-7
Library of Congress Control Number: 2020911639

First Edition, 2020
Edited by Jennifer Dehoff, Joel Salatin, Chris Slattery, Debbie Slattery

Cover Design© Jennifer Dehoff.
Cover Photo by Jessa Howdyshell.
Joel's Caricature by Rachel Salatin.
All rights reserved.
Book design/desktop publishing by Jennifer Dehoff Design.

Printed and bound in the USA.

This book is dedicated to farm crafters who do more than talk.

photo courtesy of Jessa Howdyshell

POLYFACE DESIGNS

Table of Contents

CHICKENS

COWS

PIGS

RABBITS

TURKEYS

WINTER HOOP HOUSES

MATERIALS HANDLING

APPENDIX

photo courtesy of Justin Rhodes

It was a destitute time. The Great Depression. A little girl named Mafry belonged to the family just up the holler. With her dad in jail and no government assistance, they were hungry. In those days, if you didn't grow it, you didn't eat. They simply couldn't grow it.

Fortunately for this destitute family, they had some good-hearted neighbors who had not only grown enough food for their family but had enough to share. Mafry, tears in her eyes, 70 plus years later, still recalls the day she saw those heads bobbing over the horizon of the hill in her direction.

Those neighbors were my grandparents and their children (including my dad) bringing Mafry and her family a bushel of apples, thus beginning a life-saving ministry.

Decades later, times are easier and my Dad makes sense of leaving the farm lifestyle to pursue a life of sales totally unrelated to farming. (Joel would call this a "town job"). Sure, Pop kept a dozen or so head of cattle throughout my formative years, but I think it was more about nostalgia and a tax write off than a way to provide food for our family or the community. My father represents so many of that "grandfather generation" who forsook the farm for a different kind of life.

And if you listen to the ol' timers stories, it makes perfect sense. Farming evolved into something that didn't pay well, was unpredictable at the market, required large amounts of land and it literally stunk from the large scale, close proximity factory farming style.

More decades pass and I'm of age and beginning to explore my own stories and convictions. I find myself living on Dad's land that fed two generations of ancestors before me. Like my father, I had a "Town Job." But after a series of events my wife and I begin to explore a healthier way of living. First, we replaced our conventional milk with organic and soon after, all our food followed down this (expensive) path. Our food bill skyrocketed and Tightwad Justin was in a panic. Just before financial collapse and panic attacks, a mentor (Steve) told me about how he'd buy a steer in the spring, graze it all summer and sell it in the fall. He'd keep a quarter for himself and sell the remainder to other families. Viola, he'd have food for himself and a profitable side hustle. Then, Steve told me about Joel Salatin's book, *"Salad Bar Beef"* and said it was all I needed to know for raising grass-fed beef.

Guidebook in hand, it was go time! I went

by the book, which included setting up temporary paddocks and lanes throughout the pasture with multiple electric fence lines. My dad must have thought I had gone looney because he always called them my "crazy wires." In my Dad's day, cows were just left to graze a single pasture and maybe moved to another one once or twice a year. But, I had learned from Joel that if you'll just move your cows every day into smaller paddocks, giving each paddock adequate rest to recover, you have the potential to double your edibles and halve your weeds in a very short amount of time. Given that I didn't have much land and might want to expand my business, I was game.

I knew Dad had done things differently but what Joel was saying made a lot of sense so I stuck with it for years despite feeling unsupported, and even slightly mocked by my Dad. In my heart, it must have threatened everything he knew from life on the farm (which was an intimate, and pleasant part of his life). I didn't argue or plead my case. Unsure of how all this would turn out myself, I just kept at it.

Time would tell.

In the end, I was extremely successful and able to grow enough grass-fed beef for our family, and profit a nice $1,000 per steer. Oh, and the pastures? I thought they were starting to look greener, less weedy, and more productive.

Several years later my family and I set out on the Great American Farm Tour highlighting America's greatest farms. We couldn't help but notice as we traveled just how many farmers were inspired by Joel and implementing his methods. I noticed that the closer folks followed his instructions, the more successful they were. In the meantime, Dad was back home watching every show. He got to see us explore America, visit amazing farms and even got to hear Joel explain the mob stocking herbivorous solar conversion lignified carbon sequestration fertilization "crazy wires" and how rotational grazing can bring more productivity (therefore profit) from the land. And somewhere in there, Dad became our biggest fan.

After our journey, we returned to our sustenance farming lifestyle and Dad continues to watch our show. After a couple more years Dad's 86 and on his deathbed. He'd lived a good life and was quite thoughtful and reflective towards the end. On one particular day, he made a comment about how I was moving the cows and the sheep. I'll never forget the spirit of what he said… *"You've got it looking good up there. The way you move the animals around has really helped."* Days later he passed. Here was a man who hired one of my friends to remove all my "crazy wires," while I was gone on the farm tour, and now he was commending my results. What happened? I spoke no convincing or argumentative words. In fact, my only defense was a guidebook and a plan in action.

In the end, time told the story, and this story happened to have a beautiful ending.

You're reading this because you want to live a more meaningful life. A life where you call the shots, work outside with your family, make a great living all while leaving the earth a better place (not worse). It's ironic, you probably want to leave your "town job" for a better life working the land.

Times have changed.

Now, Joel shows a younger generation that farming can be done on small parcels of land (leased, even). He's proven it can be extremely lucrative and employ the entire family. Oh, and the BIG kicker!? It doesn't stink. It actually leaves the people, the animals, and the environment better than they were.

Fortunately for you and I, Joel has already taken decades to fine-tune these strategies, and he's so graciously shared the steps inside his invaluable books like *"Salad Bar Beef"*, *"Pastured Poultry Profits"* and others. No doubt, one of the keys to success with Joel's systems is the mobile infrastructure. I'd find myself studying the pictures in his books and doing my best to implement them, and I ended up uncertain if I was doing it right. Now, I couldn't be more excited that Chris Slattery and Joel Salatin have partnered to put together step by step plans for building said infrastructure. The perfect guidebooks are already here. Now, I'm thrilled to say we have the perfect companion.

Justin Rhodes
Permaculturalist, film producer,
author and teacher

photo courtesy of Rachel Salatin

"Could you please send me blueprints for the Eggmobile?" I wish I had a nickel for every request like this. Of course, it's not just Eggmobiles; it's broiler shelters, Gobbledygos, Millennium Feathernets, cattle Shademobiles and all the other mobile infrastructure our family has developed over the years.

But I have a problem: I can't draw a stick man. A true curse is having me on your team playing Pictionary. My bride Teresa says when I draw a state, it looks like every other. I have no artistic bone in my body.

Many years ago an engineer friend made actual blueprints of both the cattle Shademobile and the Eggmobile, but then people said they were too complicated and they couldn't read actual blueprints. We've taken lots of pictures and put them in books and on the internet, but somewhere between verbal descriptions and pictures lies a step-by-step construction visual.

The excuse I always gave for not offering such a guide was that I wanted folks to scrounge up what they had and go from there. I've seen Eggmobiles made out of school buses, tilted over old-time hay loaders and even upside-down yachts. Ingenuity is amazing. But more common is the person who just wants to take a proven model and duplicate it.

The real reason I didn't offer a guidebook was because I'm not a visual person at all. I can't tell the top from the bottom of a blueprint. Topographical maps are like hieroglyphics. My idea of torture would be to put me in a room with a 1,000 piece puzzle and tell me to do it in a year. I'd hang myself.

Remember those standardized aptitude tests where they had a diagram with a missing component and you were to choose A, B, or C that would fit in that spot? I literally scored less than 20 percent on those things. Try as I might, I could not figure out what fit where. But on the section "what is the best title for this paragraph" I scored 99 percentile. I'm 100 percent verbal and seemingly don't have a visual gene in my body.

This is why function over form is my mantra. Many people think our infrastructure is amazing but it isn't pretty. I see the healthy animals and fertile soil and I really don't care if the infrastructure is pretty; the question is does it make healthy animals, plants, and soil? You can pretty up these designs with paint and perfection, but when you don't have any money, poor-boy is the way to go.

My mentor Allan Nation, founder of Stockman Grass Farmer magazine, always said "a profitable farm has a threadbare look." I recite those words every day, and laugh all the way to the bank. These designs have made us a lot of money over the years because they're cheap, functional, scalable, and simple.

Meanwhile, hundreds of folks have asked us for these designs. When Chris Slattery left his promising engineering career to apprentice at Polyface several years ago, we suddenly had a first class engineer on board who could do everything we couldn't. Another common business axiom is "success will always require you to partner with people who are different than you." My, how true. Chris spent nearly a year and a half with us, gaining intimate understanding in both the use and background of all our designs. When we needed duplicates, he helped build them.

Most apprentices simply enjoy them, but now I realize Chris internalized and literally metabolized these designs. He can see stress points and angle braces that nobody else can see. As you'll see, these designs are the end of the long line of trial and error. The basic idea of mobility and scalability seems straightforward, but doing it efficiently and economically can be a different story.

Of course, when he was here Chris heard requests from folks about these designs and it must have been maddening for him to listen to my apologies and "aw shucks" responses. As a guy who reads blueprints for fun, he must have been embarrassed at my lack of professionalism. Ouch.

What a joy, then, when he approached me about taking on this project. For the record, I did not initiate this; it's all Chris' idea. He asked me to do the text (ah, more words, my comfort zone) and give the evolution and main principles of a lifetime of designs. That part has been fun and I've proof read it carefully. But the drawings? I can't tell a board from a brick, so if anything is wrong there, blame Chris, not me.

I'm incredibly grateful to partner in this project with him. He's been a dynamo and creative genius throughout and I hope this design book encourages thousands of farmers and wanna-be farmers to manage their land, their animals, and their bank accounts better.

Joel Salatin
Summer, 2020

photo courtesy of Chris Slattery

On May 28, 2014 I turned in my identification badge and walked through security for the last time. I had just quit what could have been a promising career in the nuclear energy sector. Just a few days before that I had rented out my home to a tenant, so I was now more or less homeless and unemployed. Two days later I was enroute to the small town of Swoope, Virginia with nothing more than a suitcase, muckboots, and a dream to farm.

That summer I worked alongside 10 of the most respectable interns and was blessed to have been selected to stay onboard and lead the next wave of incredible individuals as an apprentice. Those 16 months at Polyface literally changed my life.

While working for the Salatins I became intimately involved with the construction and maintenance of their infrastructure. Being the "engi-nerd" that I am, I would spend off-time measuring and sketching the various designs on graph paper as a resource for when I started my own venture someday. Little did I know that these sketches would be the basis for this entire book.

As I made my rounds through the Polyface office one Saturday afternoon to scope out what bakery treats dear Wendy (Polyface store manager and media maven) had brought in with her that day, she eagerly hugged me and handed me the phone. On the other line was a gentleman asking for construction details for the Millennium Feathernet. After a long conversation with the man, Wendy thanked me profusely and lamented that she received calls like that all of the time; these were questions she could not answer and half-jokingly told me to write a book on it. This got me thinking that this information needed to be formally captured and shared.

After my tenure at Polyface I made it a point to visit as many farms in my travels as I could. As I traveled around I was stunned by how many people imitated—and failed—to replicate the Salatins' infrastructure effectively. I also met another faction of people who had the deep passion to farm, but scantly knew the difference between a flathead and Phillips screw driver. For these people the biggest road block preventing them from achieving their dream was how to build the infrastructure. For everyone else, it was the minute details that prevented them from replicating Polyface's proven designs successfully.

That aside, I cannot dismiss that I have also seen and read about some spectacular, creative infrastructure that works, and works well. I am ecstatic, and I know Joel is too, that in many cases people have taken what he has done, run with it and dare I say even improved upon it. After all,

isn't that how all of civilization advances? I dream of someday showcasing the genius behind the best farming innovations worldwide. It's a wonderful progression, and I hope that this book inspires more of it!

Our target audience for this book, however, are not those innovator-types. While they can certainly glean from the breadth of content we have shared within these pages, our primary goal with this book is to arm the novice with a roadmap and the confidence to build a structure that is proven to work.

Creating a step-by-step format has proven to be very challenging; especially considering how large and complicated many of these projects are. For you masterful builders out there--humor me--as I am sure you will see many shortcuts I didn't take, and even better ways to do some things. The proverbial saying, "all roads lead to Rome" comes to the front of my mind. Ultimately, there are endless ways to build things, but I've done my best to map these steps out as concisely as I knew how.

As I left the corporate world behind me that afternoon in May 2014, I never dreamed I'd be using my formal education or technical skills again. It has been such a blessing to be able to meld my mechanical aptitude and 10+ years of

3D modelling/CAD experience with my passion for raising food in a holistic, healing manner.

I am incredibly grateful to Joel and his family not only for the opportunity to join forces on this book, but also for the opportunity to become part of their extended family over my 16 month stay with them. I also owe much gratitude to my own family who picked up my slack, time and time again over these past two years, especially as deadlines approached. Without their loving support this book would never have been possible.

In conclusion I will say that I assume no credit for any design in this book. Apart from squaring all of Joel's proudly non-90° angles and the occasional standardization of materials to consolidate/minimize waste, these are all still Polyface originals. As an added bonus, I even tried to pick Joel's brain for potential improvements that he would make along the way.

My primary role in this project was to capture the simplistic, yet genius evolution of things that "simply work" at Polyface in a manner that I hope everyone can replicate!

Happy Building!

Chris Slattery
Summer, 2020

photo courtesy of Rachel Salatin

JOEL SALATIN and his family operate Polyface Farm in Virginia's Shenandoah Valley. His parents purchased the farm in 1961 and developed the basic principles of design and production that now show 60 years' refinement. He gravitated toward communication activities in high school and college, graduating with a BA degree in English, and after a brief journalism hiatus returned to the family farm full time Sept. 24, 1982. Editor of *The Stockman Grass Farmer* magazine, he writes and speaks around the world on food and farm issues. With a long track record of innovation and excellence, Polyface holds educational seminars, farm tours, day camps and events to encourage duplication and understanding. This is his 14th book.

photo courtesy of Chris Slattery

CHRIS SLATTERY is a mechanical engineer turned farmer/homesteader. With years of experience in CAD/3D modeling coupled with a lifetime of learning from helping his father—a highly skilled tradesman—Chris brings to the table a unique skill set. The amalgamation of his love for engineering, regenerative agriculture, and hands-on building experience has led him to team up with world famous farmer/author Joel Salatin of Polyface Farm to create a first of its kind comprehensive design manual for all of the farm's most iconic infrastructure contraptions. Chris currently resides on his family's 112 acre farm in South Central Kentucky, where he enjoys improving the land, and raising happy, healthy livestock. His dream is to continue writing and showcasing the genius of DIY farmers/homesteaders from around the globe. If you've got a farm contraption/contrivance/invention you want to share with the world, feel free to contact Chris at: showcaseyourgenius@gmail.com.

POLYFACE DESIGNS

A Comprehensive Construction Guide
for Scalable Farming Infrastructure

Joel Salatin & Chris Slattery

BROILER SHELTER

photo courtesy of Jessa Howdyshell

The famous Polyface broiler shelter began as a rabbit run for my older brother's rabbit operation when we were kids. But rabbits dig and we could never keep them in. Adding insult to injury, rabbits also run and they can squeeze under things. Unlike a chicken that escapes, you can't run down a rabbit and catch it. Just mentioning it brings back childhood memories that make me start twitching.

When we finally gave up with the rabbits, Dad pulled these boxes up into the rafters of the barn to await repurposing for something else. After all, we were farmers, and we never throw anything away. I started a flock of laying hens

when I was 10 years old. A couple of years later, my chicken operation outgrew my coop and I needed more space. Dad suggested we pull those old rabbit runs down and put chickens in them.

That simple suggestion birthed pastured poultry, Polyface style. We expanded those original 8ft x 12ft runs into 10ft x 12ft, which was as big as we could make them without being too heavy to move easily. We now have about 200 of these shelters. Each one handles 75 broilers.

The single biggest error people make is weight. If you're coming to a pastured chicken project from a background in cattle or other livestock, the tendency is to use 2x4s rather than

1x1-1/2 material. I think I've seen almost every design imaginable. Tall ones, A-frames, great big ones. I've seen them built out of PVC (I have one rule about building anything out of PVC--don't do it), steel, and fiberglass. I've seen plenty of PVC shelters hanging in trees.

The goal is light enough to move easily by hand but heavy enough not to blow away in a breeze or get pushed around by a coyote or fox. Anyone contemplating a different design should answer the following questions:

1. Can you easily take it down the public road to another property?

2. Is it short enough to conform to undulations

in the land so you don't have to use sandbags or other techniques to plug up big holes?

3. Can you move it easily by hand, without starting an engine?

4. If it does get destroyed, is it easy to replace?

5. If something on it breaks, is it easy to fix?

6. How many chickens are affected at once if something catastrophic happens?

7. Can you move 4,500 chickens by hand in 60 minutes without starting an engine?

8. Is your capital cost for one season's production less than $1.25 per chicken?

9. Does your covering material last forever?

10. Can you get into pastured broiler production with no tractor or truck?

These are the questions to ask yourself whenever someone starts making fun of this simple broiler shelter. These are all proven, established principles and benchmarks here at Polyface. You're welcome to come and see us operate this model if you're skeptical. To be sure, only a handful of pastured poultry operations in the world meet our benchmarks. The most common reason is a shelter too heavy. The second is a dolly that doesn't work efficiently. The third is simply poor technique, which includes moving, feeding, watering, and overall management.

Amazingly, many people get stuck on "how do you get them out?" They're crawling around underneath trying to catch chickens and all sorts of crazy stuff. We simply use two pieces of quarter inch plywood, like a paddle, and bring the chickens around to the front, then slip the other panel in to close off the capped rear, putting a prop stick behind. This partition keeps all the

birds in the front of the shelter. We remove both doors and then pick up the chickens as fast as 1, 2, 3. Easy peasy.

We've tried bigger shelters but in our undulating terrain we can't keep the bottom flush to the ground. Too many depressions and holes. The smaller footprint is far more forgiving on uneven ground.

Some of our shelters are 30 years old; they can last a long time. We use lightweight pressure treated lumber (not hardwood, but softwood); the bracing is critical. Don't cheat. Every brace is critical; if one breaks, fix it immediately or you'll stress other parts of the shelter. Perhaps the most important but misunderstood part is the aluminum roofing.

Steel roofing is too heavy and too absorptive of heat. Aluminum is reflexive and dissipates sunlight rather than concentrating it. Canvas or fabric is lighter, but it too absorbs heat and doesn't breathe. Ever been in a tent on a sultry hot day? It's downright stifling. The aluminum never rusts, is light, but reflective. These are all critical properties in making the box this small but not too hot in the summer yet cozy enough for the shoulders of the season.

With poultry netting, the bigger the span the more fragile it is. With only a 2ft span and stapled tightly, it's hard to tear up. Most predators are opportunists; if something is difficult, they just move on to another target. Interestingly, to my knowledge we've never had any predator get up on top and climb in from the top. Apparently they are too stupid to realize that those chickens they see in there on the ground could be accessed

some other way besides a straight line. Rather than climbing up on top and coming in that way, they just keep circling and trying to figure out how to get in on ground level.

These shelters have several significant benefits:

1. If wood breaks, you just scab on another piece and fix it.

2. Wood is a renewable resource, unlike plastic or steel.

3. We can throw 8 of these on an 8ft wide 24ft long gooseneck low boy trailer and transport them from property to property, down the public road, fairly easily. We stand them on edge and tie them down so they can't sway.

4. You need no machine to get into this business; nothing. No truck, no tractor; nothing but the dolly and some muscles.

5. Once you get proficient at building them, you'll see that you can slap these together efficiently and cheaply. If you need more, you can scale up without breaking the bank, and you can have infrastructure in place fast.

6. To my knowledge, in 50 years of using them, we've never had a single hawk take a chicken. That's pretty cool.

Now let's deal with a couple of nuances before turning you loose on the diagrams.

1. **Heat.** The middle roofing panel on the vertical back end is removable. We put a piece of poultry netting across the rear and when the summer gets hot, we simply

CHICKENS

unscrew that one panel and create a breezeway. Even on extremely hot days the birds stay comfortable.

2. **Cold.** Cornish Cross chicks at 21 days old can handle 20 degrees F in these shelters with no supplemental heat. The secret is the number. We've found that if we have more than 90 in the group, the middle one or two get suffocated as the chicks huddle tighter together to stay warm. If we have fewer than 50, a couple on the outside will die from cold. The sweet spot seems to be about 75—enough mass to generate heat but not so much that the middle ones gets compressed.

3. **Rain.** The birds naturally have enough sense to get under the covered end when it's raining. But if they get soaked, like in a hurricane, we fluff out a couple of paddies of dry hay in the back of each shelter. The chicks immediately climb up on that dry hay and their body heat dries them out. This is only a problem when the birds are very small; it's not a problem when they're big. In saturated conditions, little ones get wet but not muddy because they aren't heavy enough to tromp up the grass sward. The bigger birds, being heavier, can muck up the ground quicker. But they can handle a lot more wetness than a little bird.

4. **Predators.** The first line of defense is a healthy diversified ecosystem. We fence out our woods and riparian areas to stimulate moles, voles, chipmunks, rabbits, and squirrels. Secondly, we graze or mow

photo courtesy of Chris Slattery

around the shelters to reduce long grass that predators enjoy for cover. Skunks and possums don't like to walk across long expanses of open ground because a great horned owl can pick them right up. If we have a rogue unhappy with our wildlife-friendly habitat, we go to more extreme measures, including trapping and shooting.

5. **Wind.** Only once in 50 years have we had significant wind damage, and that was a derecho, which is such a weather anomaly it only occurs once a century. Once in awhile a door blows off. There again, if you know a heavy wind is coming, just set some water buckets on the door; or a rock. If you start making hinges and other things, it'll complicate load out and add weight. Because these shelters are squatty, they

simply do not catch wind like something higher.

6. **Running over chickens.** Always move the shelters toward morning sun. Never move them against the capped end. Chickens are drawn to light so they naturally want to go toward what they can see. Move in a stop-and-go motion. Pull forward a foot, then stop for a moment; let the birds catch up, then move another foot. This is like learning dance steps; with practice, you'll get the hang of it. But if you get frustrated and blame "the stupid chickens," you'll never develop mastery.

7. **Hills.** Don't pull them up steep hills. Take them to the top and pull them downhill, or pull them on the level. Don't make it hard on yourself; go with the flow.

Tools

❑ Hand Saw

❑ Socket Driver Set
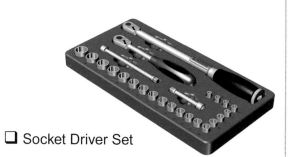

❑ Table Saw

❑ Stapler

❑ ¹/₂" Staples
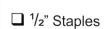

❑ Power Drill

❑ Driver Bits (for screws)

❑ ¹/₈" & ¹/₄" Drill Bits

❑ Marking Pencil

❑ Tape Measure

❑ Speed Square
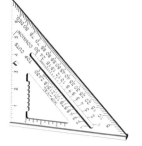

❑ 8" Tin Snips

❑ Hammer

Hardware

DECK SCREWS

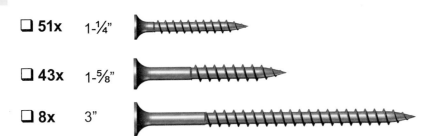

- ❏ **51x** 1-¼"
- ❏ **43x** 1-⅝"
- ❏ **8x** 3"

ROOFING SCREWS

- ❏ **1x** 1-½" to 2" Roofing Screw
- ❏ **125x** 1" Roofing Screws*

Use 1" roofing screws wherever possible. There will be areas where longer screws are needed to properly fasten, but it is good practice to minimize the number of screws that penetrate through the wood and create hazards for livestock and handlers.

BOLTS

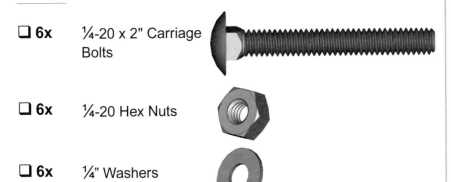

- ❏ **6x** ¼-20 x 2" Carriage Bolts
- ❏ **6x** ¼-20 Hex Nuts
- ❏ **6x** ¼" Washers

Materials

- ❏ **≈30ft** Roll of 24" Chicken Wire
- ❏ **≈6ft** Roll of 72" Chicken Wire
- ❏ **≈8"** Lightweight Chain
- ❏ **1x** 5 Gallon Bucket
- ❏ **≈6ft** Scrap Garden Hose (or plastic pipe)
- ❏ **≈4ft** ¼" Water Tubing
- ❏ **≈40ft** 12-½ Ga Medium Tensile Wire
- ❏ **1x** S-Hook
- ❏ **1x** Plasson® Broiler Drinker (see pages 24-25 for details)

Wood Cutlist

Wood scraps are highlighted in red.
Do not discard until project is completed.

QTY SIZE

☐ **2x** Pressure Treated 2x4s 12 feet long (cut in half lengthwise)

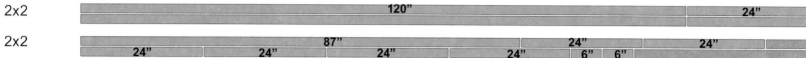

2x2

2x2

☐ **2x** Pressure Treated 1x6s 12 feet long

1x6

1x6

☐ **5x** Pressure Treated 1x6s 12 feet long (cut in half lengthwise)

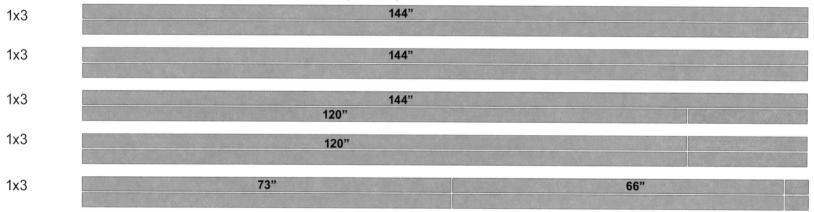

1x3

1x3

1x3

1x3

1x3

☐ **3x** Pressure Treated 1x6s 12 feet long (cut in thirds lengthwise)

1x2

1x2

1x2

Roofing Cutlist

QTY SIZE

❑ **2** Two Corrugated
Aluminum Panels
48" x 14 feet long

*It may prove difficult to
procure corrugated aluminum
panels in your area. We
are fortunate to have a
supplier locally; Virginia
Frame Builders & Supply in
Fishersville. This aluminum
can also be ordered online.
Companies like Corrugated
Metals, Inc. carry aluminum
and may ship to your region.
Aluminum roofing will be the
most expensive component on
this project, but it is also the
most critical. As mentioned
before, the material properties
of aluminum are significantly
lighter than steel and reflect
heat instead of absorbing
it. These pens remain
remarkably cool, even in the
heat of the summer.*

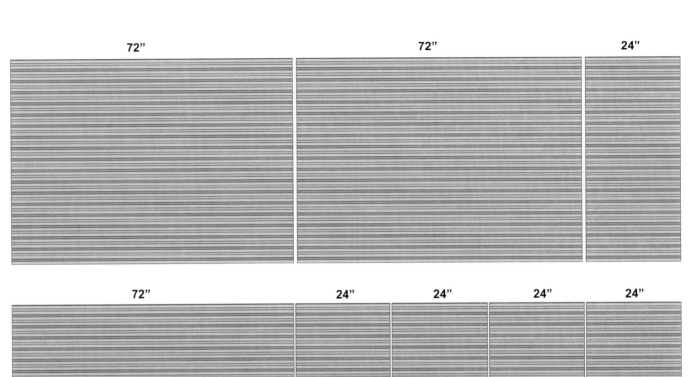

❑ **1** 24" x 6ft Corrugated
Steel Roofing
(or other available
roofing scraps)

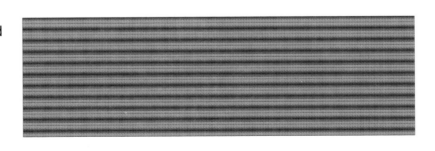

Build The Frame

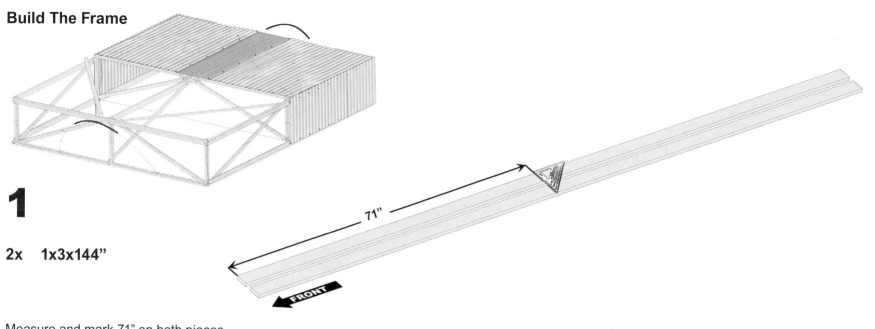

1

2x   1x3x144"

Measure and mark 71" on both pieces.

2

3x   2x2x24"
6x   1-⅝" Deck Screws

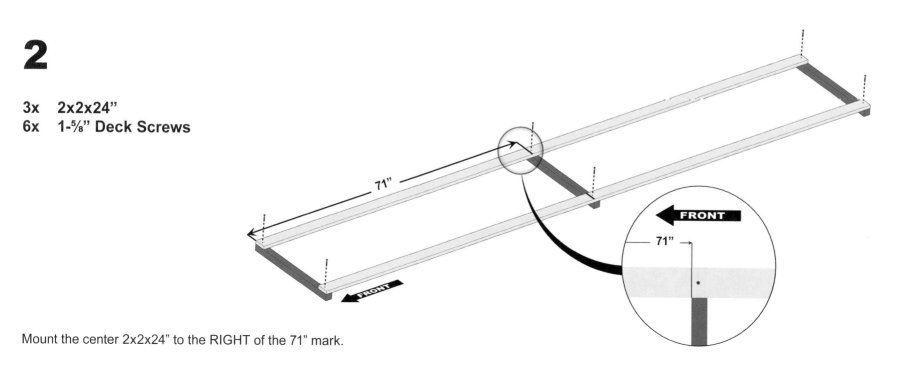

Mount the center 2x2x24" to the RIGHT of the 71" mark.

3

4x   1-¼" Deck Screws
2x   1x2x72"

Flip the frame over and attach diagonal braces, making sure everything is square.

4

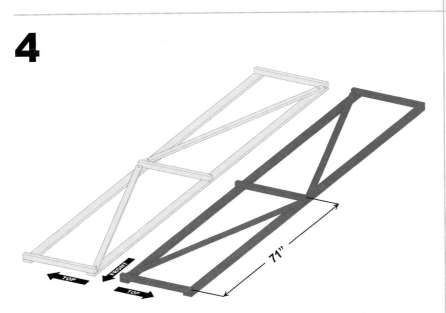

Modify steps 1-3 as needed to create a mirrored side as shown. Pay special attention to the 71" dimension to ensure it is oriented correctly.

5

1x   1x3x120"
1x   2x2x120"

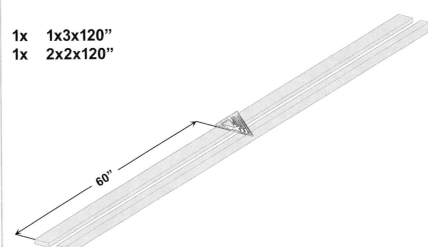

Measure and mark 60" on both pieces.

6

1x   2x2x24"
1x   3" Deck Screw
1x   1-⅝" Deck Screw

This time center the 2x2x24 piece on the marks.

7

Repeat steps 5-6 to create a second side.

8

4x   3" Deck Screws

71"

71"

It is important that the orientation of the sides are correct. Verify the 71" dimension from Step 2 to ensure they are facing the correct direction.

9

4x   1-⅝" Deck Screws

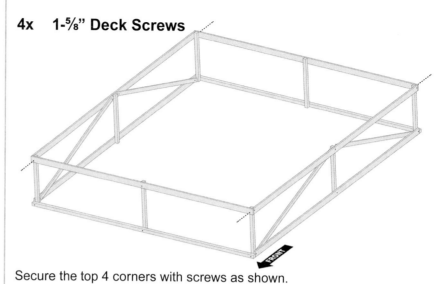

Secure the top 4 corners with screws as shown.

10

1x   1x3x144"
2x   1-⅝" Deck Screws

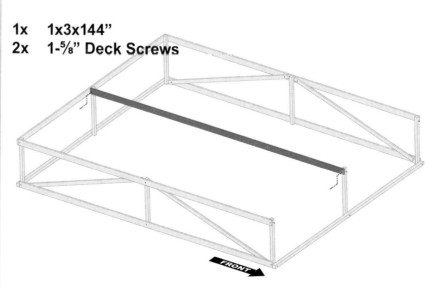

Secure the central ceiling support in the orientation shown.

11

4x   **1x2x60"**
4x   **1-¼" Deck Screws**
4x   **1-⅝" Deck Screws**

1-¼" Screws go in the tops, while the 1-⅝" screws are for fastening the bottoms.

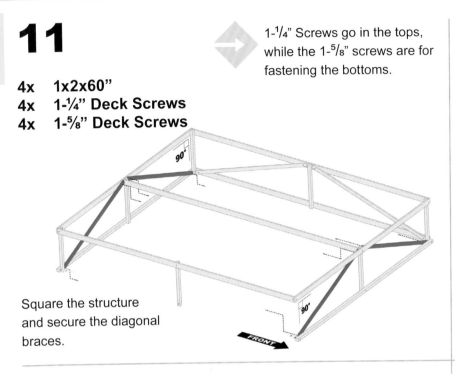

Square the structure and secure the diagonal braces.

13

2x   **2x2x6"**
4x   **1-⅝" Deck Screws**

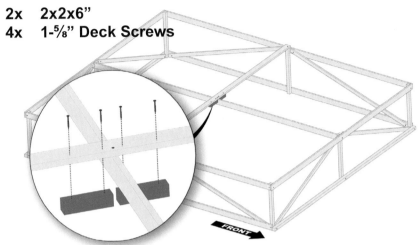

Attach the 2x2x6" blocks on either side of the center as shown.

12

1x   **1x3x120"**
3x   **1-⅝" Deck Screws**

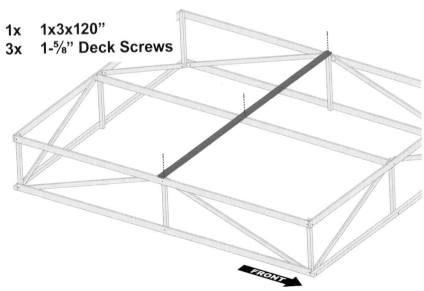

14

2x   **1x2x60"**
4x   **1-⅝" Deck Screws**

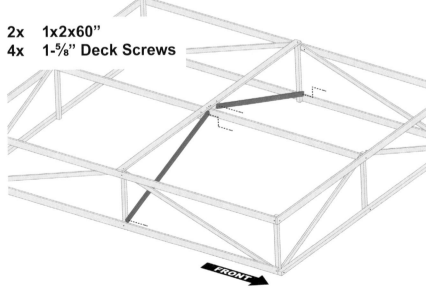

Attach the two center diagonal braces.

15

≈12-14ft   **12-$^1/_2$ Gauge Medium Tensile Wire**

1x   **2x2x21-$^3/_4$ to be used as a prop stick**

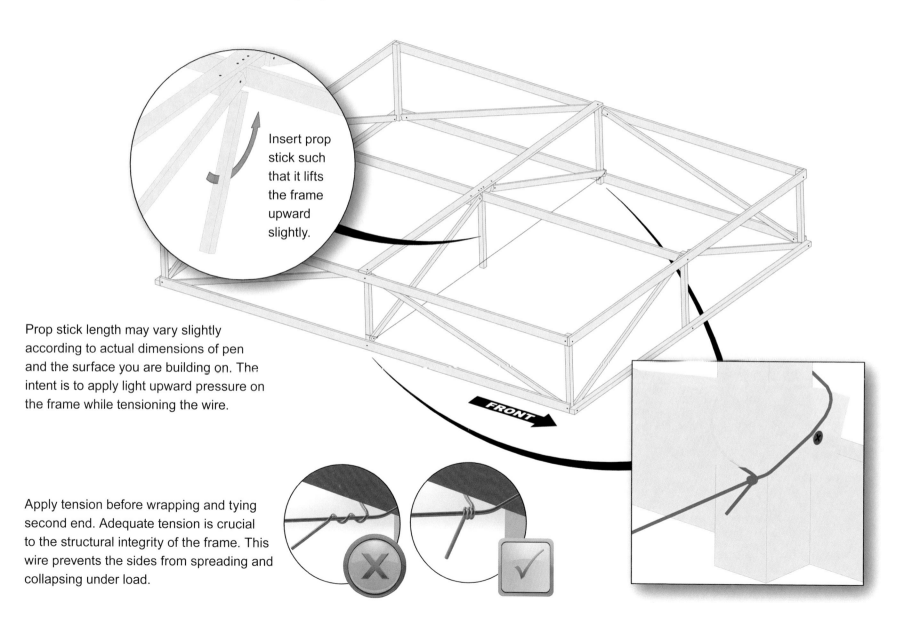

Insert prop stick such that it lifts the frame upward slightly.

FRONT

Prop stick length may vary slightly according to actual dimensions of pen and the surface you are building on. The intent is to apply light upward pressure on the frame while tensioning the wire.

Apply tension before wrapping and tying second end. Adequate tension is crucial to the structural integrity of the frame. This wire prevents the sides from spreading and collapsing under load.

16

2x   1x2x72"
2x   1-¼" Deck Screws

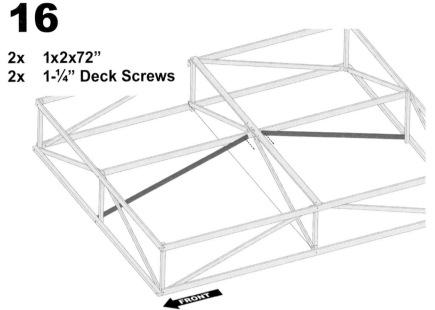

Add additional bracing as shown and screw the tops in place first.

17

2x   1-⅝" Deck Screws

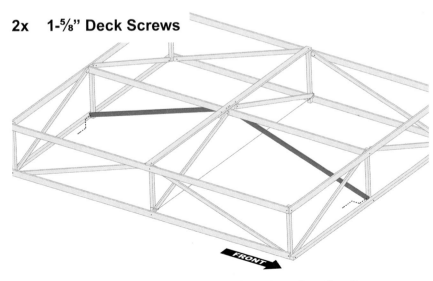

Then secure the bottoms of the braces from the other direction.

18

3x   1x2x84"
6x   1-⅝" Deck Screws

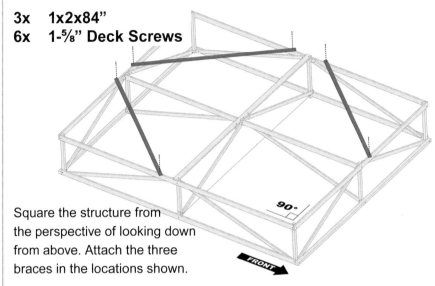

Square the structure from the perspective of looking down from above. Attach the three braces in the locations shown.

19

1x   2x2x87"
2x   3" Deck Screws

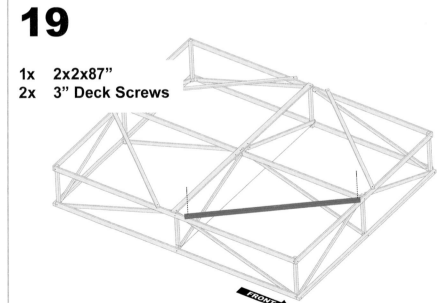

Lastly, secure the 2x2 brace in the remaining quadrant. This brace is thicker to support the weight of the drinker and water bucket.

20

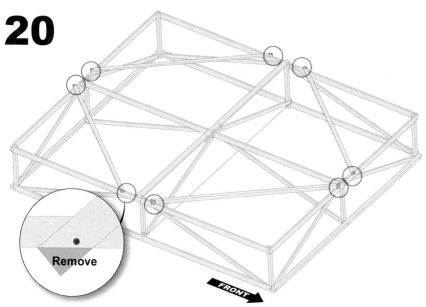

Remove

With the four top braces in place and secure, cut the excess over hanging material off, marked here in green.

21

≈22ft   Roll of 24" Chicken Wire

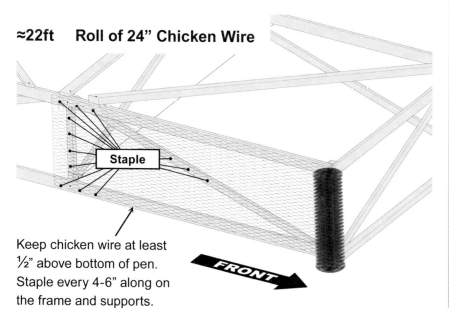

Staple

FRONT

Keep chicken wire at least ½" above bottom of pen. Staple every 4-6" along on the frame and supports.

22

Fastening chicken wire is easier with two people: one person unrolls while applying tension and the second person staples.

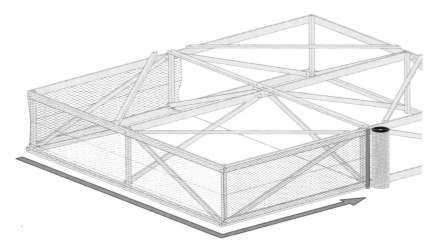

Continue wrapping and stapling wire. Cut at the blue line using tin snips.

23

≈7ft Roll of 24" Chicken Wire

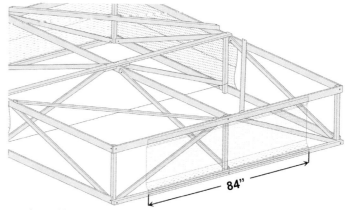

84"

Fasten chicken wire with staples on the back side of pen as shown. Go back over the staples and tap them in with a hammer if they didn't penetrate all the way into the wood.

CHICKENS

24

≈12ft **12-¹⁄₂ Gauge Medium Tensile Wire**
3ft **Scrap Garden Hose or Pipe**

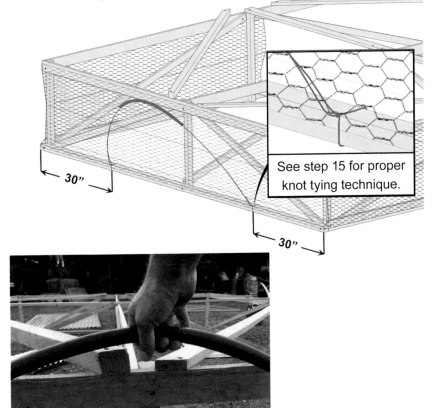

See step 15 for proper knot tying technique.

30"

30"

Joel's Tip

Our benchmark is 60 seconds per shelter, translating to moving 4,500 birds in 60 minutes without starting an engine or needing a partner. If you're not hitting that benchmark, you may be lazy, have too heavy a shelter, be moving it the wrong direction, or have an improperly adjusted handle.

25

photo courtesy of Chris Slattery

The handle height should barely make the shelter come off the ground when you stand straight up. The whole idea is to reduce back strain. Like shearing a sheep, moving these shelters is all about technique, not strength. Part of that technique is handle height so you're not bent over pulling the shelter. You want to be able to lean back and have the shelter follow you easily. A good starting point is to set the handle height to about 4 to 6" above the top of the frame when pulled vertically upwards. This may need to be adjusted up or down depending on your height.

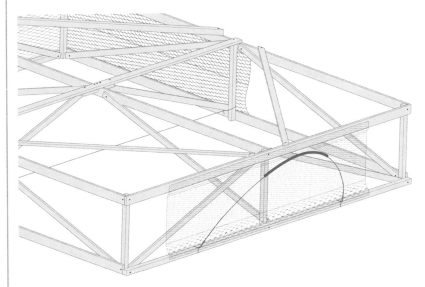

Repeat step 24 to add a handle on the back side of the pen.

26

1x   48x72" Aluminum
≈28x   1" Roofing Screws

Use 1" roofing screws wherever possible. There
will be areas where longer screws are needed,
but avoid screws that penetrate the wood and
create hazards.

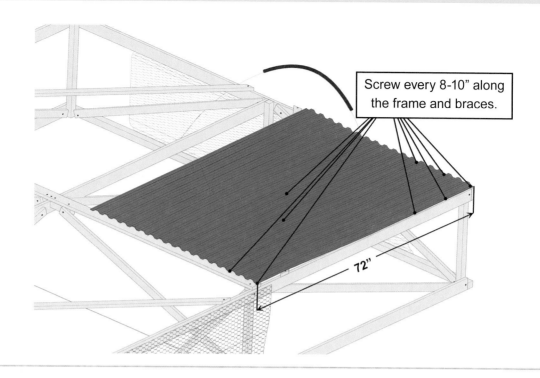

Screw every 8-10" along
the frame and braces.

72"

27

1x   48x72" Aluminum
≈28x   1" Roofing Screws

Screw roof every 8-10" along the frame and
braces on the other side as shown.

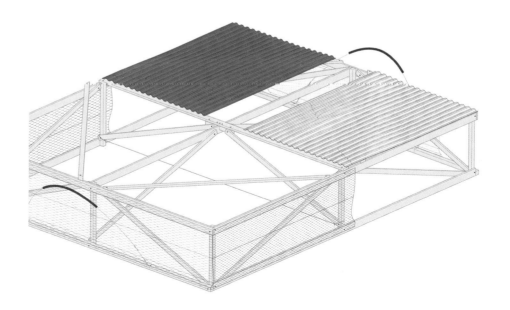

CHICKENS

28

1x   24x72" Steel
≈18x 1" Roofing Screws

We use steel in the center of the roof to minimize the aluminum needed. If you can find it, run two 3x10ft long aluminum pieces width wise and you won't need steel in the middle. Ultimately, it's all about using what is available to you.

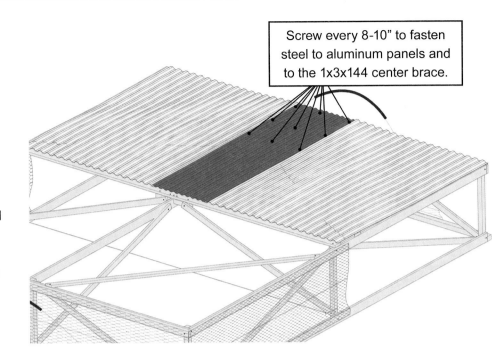

Screw every 8-10" to fasten steel to aluminum panels and to the 1x3x144 center brace.

29

1x   48x24" Aluminum
≈14x 1" Roofing Screws

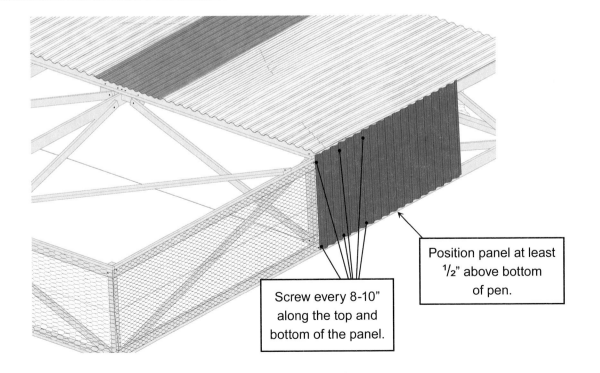

Position panel at least ¹/₂" above bottom of pen.

Screw every 8-10" along the top and bottom of the panel.

30

1x   **48x24" Aluminum**
≈14x   **1" Roofing Screws**

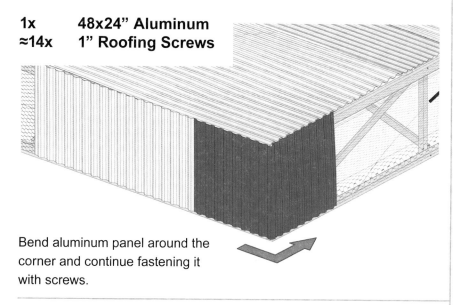

Bend aluminum panel around the corner and continue fastening it with screws.

31

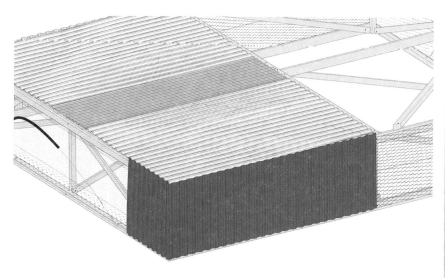

Repeat steps 29-30 on the other side of the pen.

32

≈8"   **Lightweight Chain**
1x   **1-½" Roofing Screw**

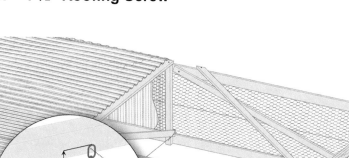

Fasten chain approximately 17" inwards along the 2x2 brace. This distance and the height of the chain may vary with the style of drinker you use. We use Plasson® Broiler drinkers exclusively. See the end of this chapter for details on the setup of these drinkers.

33

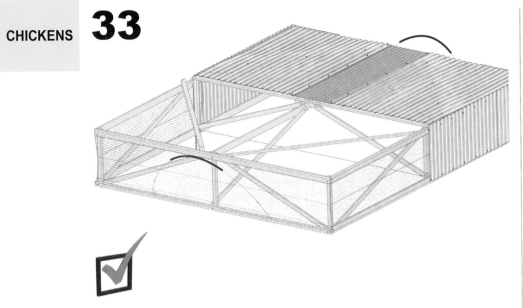

The frame is complete!

Build The Metal Lid

34

2x   1x6x54"
2x   1x6x73"
4x   1-¼" Deck Screws

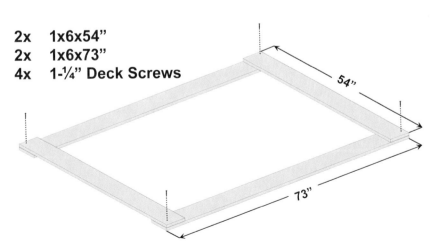

Fasten only one screw in each corner.

35

12x   1-¼" Deck Screws

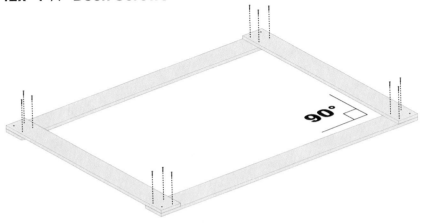

Square the frame and then fasten the remaining screws in each corner.

36

≈2x   Scrap 1x2 or 1x3 pieces
≈12x   1-¼" Deck Screws

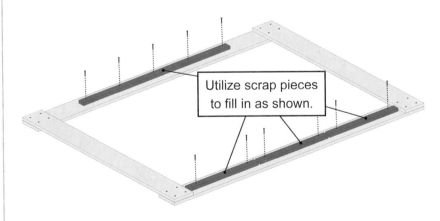

Utilize scrap pieces to fill in as shown.

37

≈2x   **Scrap 1x2 or 1x3 pieces**
≈10x   **1-¼" Deck Screws**

Utilize scrap pieces to fill in as shown.

39

The metal lid is complete!

38

1x   **48x72" Aluminum**
≈24x   **1" Roofing Screws**

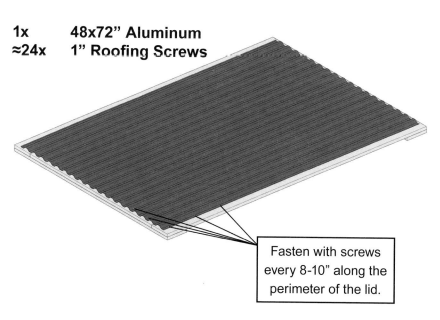

Fasten with screws every 8-10" along the perimeter of the lid.

40

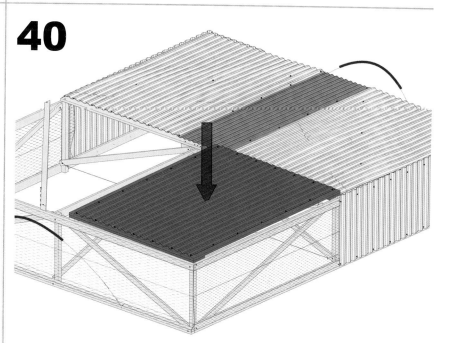

The metal lid simply sits on top of the structure as shown.

Build The Wire Lid

41

4x ¼-20 x 2" Carriage Bolts
4x ¼" Washers
4x ¼-20 Hex Nuts
2x 1x3x66"
2x 1x3x73"

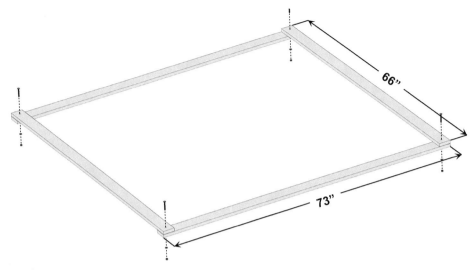

Drill ¼" holes in the four corners. Fasten carriage bolts
and hardware as shown.

42

1x 1x2x94"
2x ¼-20 x 2" Carriage Bolts
2x ¼" Washers
2x ¼-20 Hex Nuts

Square frame and drill ¼" holes on both ends as shown.
Fasten with hardware.

43

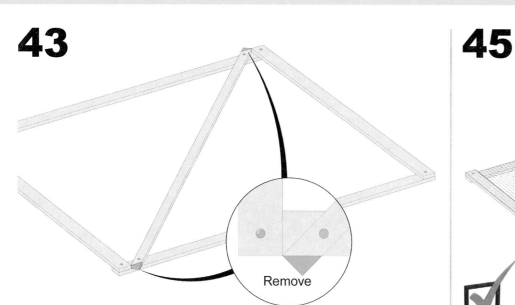

Cut off overhanging excess shown in green.

Remove

44

≈66" Roll of 72" Chicken Wire

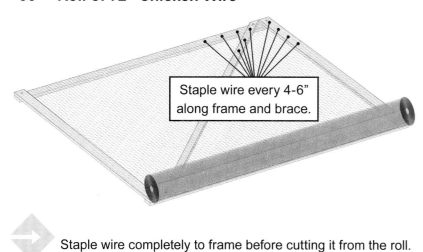

Staple wire every 4-6" along frame and brace.

 Staple wire completely to frame before cutting it from the roll.

45

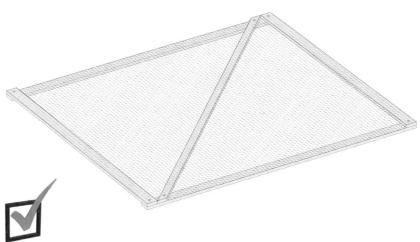

The wire lid is complete!

46

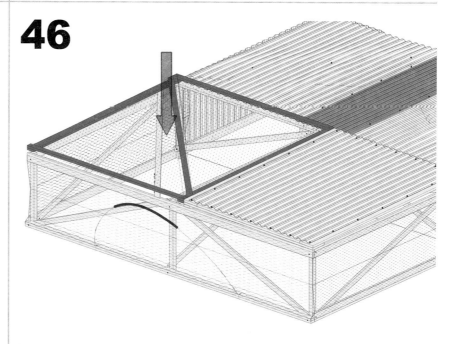

The wire lid simply sits on top of the structure as shown.

Water System

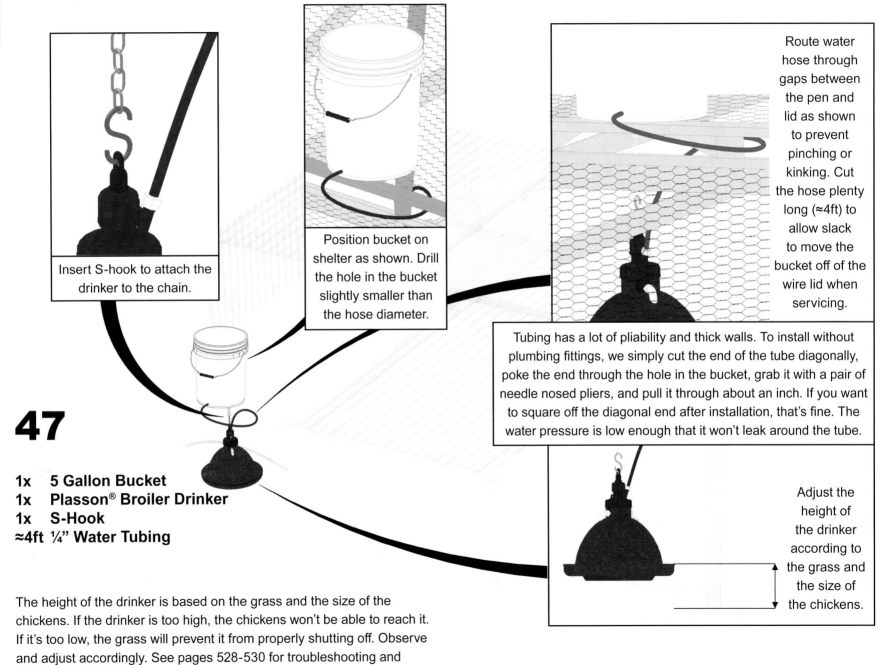

Insert S-hook to attach the drinker to the chain.

Position bucket on shelter as shown. Drill the hole in the bucket slightly smaller than the hose diameter.

Route water hose through gaps between the pen and lid as shown to prevent pinching or kinking. Cut the hose plenty long (≈4ft) to allow slack to move the bucket off of the wire lid when servicing.

Tubing has a lot of pliability and thick walls. To install without plumbing fittings, we simply cut the end of the tube diagonally, poke the end through the hole in the bucket, grab it with a pair of needle nosed pliers, and pull it through about an inch. If you want to square off the diagonal end after installation, that's fine. The water pressure is low enough that it won't leak around the tube.

Adjust the height of the drinker according to the grass and the size of the chickens.

47

1x 5 Gallon Bucket
1x Plasson® Broiler Drinker
1x S-Hook
≈4ft ¼" Water Tubing

The height of the drinker is based on the grass and the size of the chickens. If the drinker is too high, the chickens won't be able to reach it. If it's too low, the grass will prevent it from properly shutting off. Observe and adjust accordingly. See pages 528-530 for troubleshooting and adjustment tips on the bell water drinker.

Why Plasson® Broiler Drinkers?

In the brooder house, we use nipple waterers but that's a protected system with no ability of gunk or contamination to enter. The open buckets allow flies and contamination to enter, which means we need to easily and quickly be able to determine if the watering system is working. The lip of the bell waterer either has water in it or it's empty which fulfills the check requirement quickly. Fortunately, we don't have to fill the water ballast in these waterers. In normal factory houses, the heavy ballast keeps the birds from bumping them and spilling water on the floor. But in the field, these spills are not a problem so we can fill the ballast half full and it works fine.

Some folks wonder why carry water at all. Why not just plumb the system into the water line out at the edge of the field and stretch these tubes across the field from shelter to shelter? Do you know how easy it is trip over things in the field? I've never wanted things strewn over the field between shelters. Furthermore, if something breaks, none of the chickens have water. With the bucket, if something happens on one it doesn't affect all the shelters so it's less risk. Furthermore, the tubing can get extremely hot on a summer day; a clear tubing alternative grows debris.

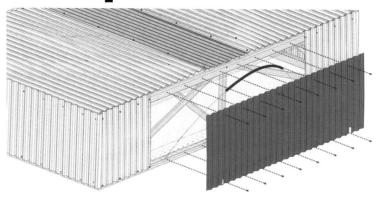

48 Optional

The removable back panel is key to maintaining optimum circulation in the warmer months. During the cooler temperatures of spring and fall we put on the backs. This creates a cozy area for the chickens to escape the wind. This feature extends our growing season and increases animal comfort in all seasons.

49

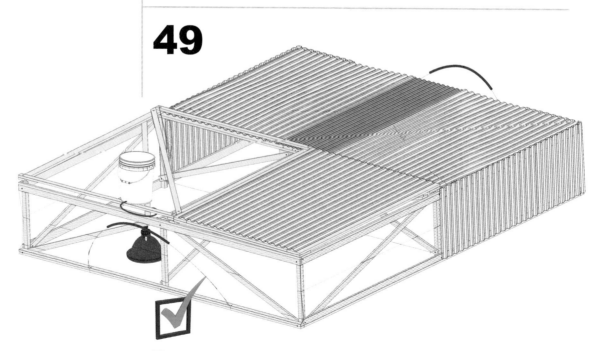

The mobile shelter is complete!

CHICKENS

photo courtesy of Jason Pope

SHELTER DOLLY

When we first started moving broiler shelters, it was a two-person job. One person on the front and one on the back. When Teresa and I were first married, she helped me a lot outside with stacking hay and yes, moving broiler shelters.

One day she strained her back and I had to come up with an alternative to lifting the trailing edge of the shelter. I made a simple dolly that was a glorified pry bar on wheels that she could stick under, push down, and I'd pull the shelter forward. That worked for awhile but it still complicated chores for both of us to have to be in the same place at the same time. Daniel finally

got old enough that he could kind of hang on that dolly and provide enough downward weight that it would bounce along the ground when I pulled the other end.

When my brother the engineer was home one time for Christmas I told him I needed a one-person design. I needed to make moving these shelters a one-person deal. He went to work on designs and eventually came up with what we use today. It's so simple it's hard to believe it took so long to conceive of it.

I'm always surprised by how reluctant many folks are to build this dolly. I've seen folks cut plastic pipe and put that on the shelter bottom boards to help them slide. I've seen people affix wheels, both permanent and movable. If you have anything that keeps the bottom board off the ground, it invites predators. Sticking a wheel onto one corner and then putting a wheel on the other corner, bending over and taking it off when you're done--it's way too time consuming and awkward.

And wheels are expensive. If you're running 200 shelters and we had some sort of wheel contraption on each one that would raise and lower, those 400 wheels would be $5,000 just in wheels. It adds up.

So the one-person dolly was a real game changer. It's a simple tool that works on every shelter, equally well, on either end. Its design lets

you use it like a pry bar to ease the shelter up. With skill, it's a fast, noiseless, efficient procedure.

With this dolly, one person can move a shelter every 60 seconds. One person, without starting an engine, can move 60 shelters an hour; at 75 birds per shelter, that's 4,500 chickens. In one hour without any machine except this dolly. Game changer? You bet.

The trick is to stand sideways to the shelter. Your body becomes the lever and the fulcrum is the dolly axle. Put the prongs under the shelter, then step between the shelter and the dolly, sideways, pushing down on the dolly handle while lifting the shelter handle. It doesn't take much strength, but it does take some skill.

A sweet spot exists on the size of the dolly wheels. The smaller the wheels, the easier they get stuck in depressions, like a cow hoofprint. The bigger the wheel, the less leverage you have. Of course, if the wheels get too big, the diagonal piece that the bottom board slides on is too far away from the right angle of the axle. Like most of these things, a size sweet spot exists.

One day, when pastured poultry displaces Tyson, Cargill, Pilgrim's Pride, Perdue, and Foster Farms, these dollies will be available in every hardware store in the country. Until then, you can make them yourself or have a welder make them for you. The dolly is the most indispensable tool on our farm. Well, maybe next to the chainsaw.

CHICKENS

Tools

☐ Socket Driver Set

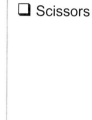

☐ Power Drill

☐ Marker

☐ Scotch Tape

☐ Tape Measure

☐ Speed Square

☐ Scissors

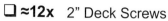

☐ 4ft Level

☐ Welder

☐ Grinding Wheel

☐ Cut-Off Wheel

☐ Angle Grinder

Hardware

☐ **≈12x**   2" Deck Screws

Length of screws is determined by size of wood blocks used.

☐ **2x**   ½-13 x 2" Hex Head Bolts

☐ **2x**   ½-13 Nylock Nuts

Materials

☐ **2x**   7" Lawnmower Wheels

Cutlist

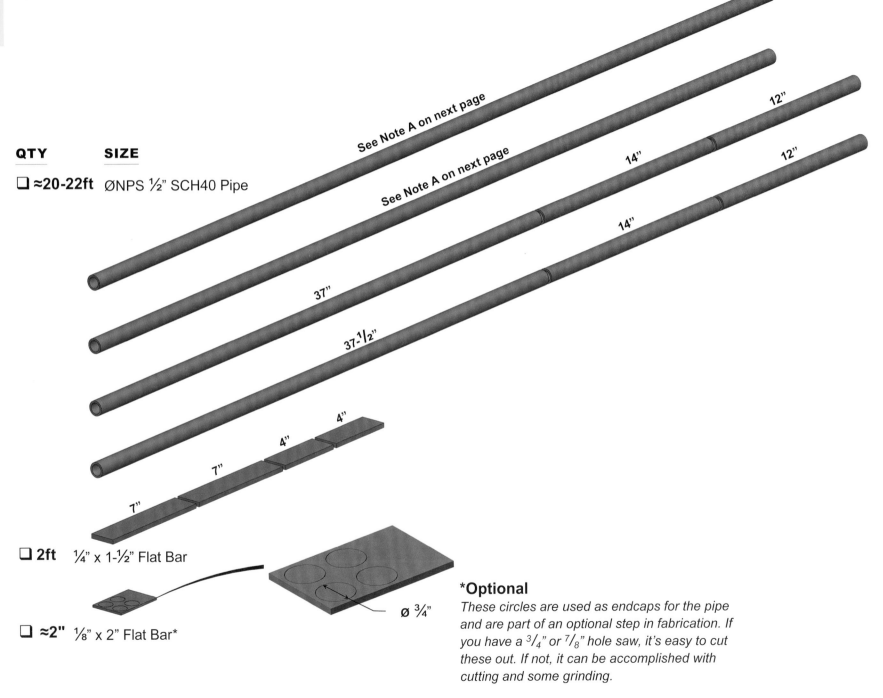

QTY	SIZE
☐ ≈20-22ft	ØNPS ½" SCH40 Pipe

See Note A on next page

See Note A on next page

12"

14"

12"

14"

37"

37-¹/₂"

4"

4"

7"

7"

| ☐ 2ft | ¼" x 1-½" Flat Bar |

| ☐ ≈2" | ⅛" x 2" Flat Bar* |

Ø ¾"

*Optional

These circles are used as endcaps for the pipe and are part of an optional step in fabrication. If you have a ³/₄" or ⁷/₈" hole saw, it's easy to cut these out. If not, it can be accomplished with cutting and some grinding.

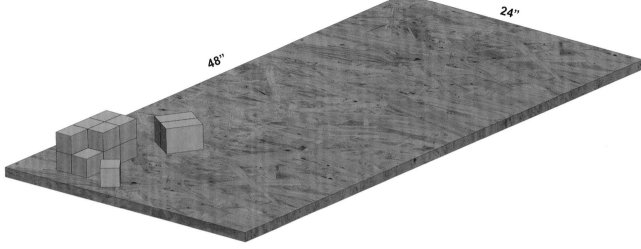

QTY   SIZE

☐ **1x**   24 x 48" Plywood

☐ **≈12x** Miscellaneous Scrap
            Wood Cut Into Blocks
            ≈2-3" long

☐ **1x**   Copy of 3 Sheet Fabrication Template (from pages 555-558)

☐ **4x**   Copy of Pipe Templates (from page 554)
            90° Notch
            45° Notch
            48.7° Notch
            45° Miter

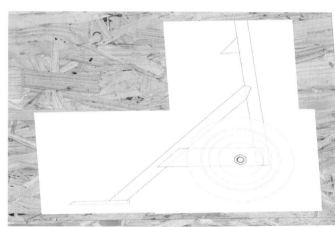

Note:

A) The pipe length is based on the height of the individual. For optimal ergonomics, measure the height of the user's shoulder and make the pipe that length. If this dolly will be used by multiple people of varying heights, then choose a height that is middle of the road. We find 57-60" is a good height for most people.

Shoulder Height

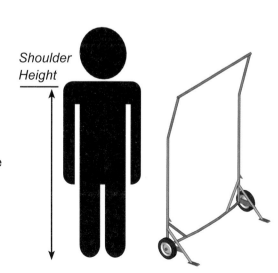

CHICKENS

1

1x   90° Notch Template
2x   ØNPS ½" SCH40 Pipe

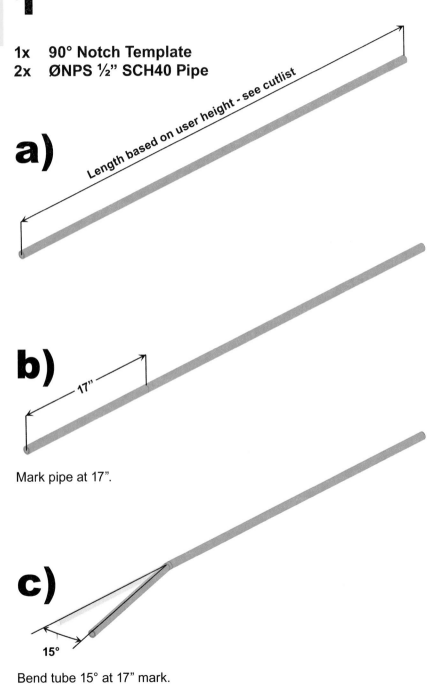

a)

Length based on user height - see cutlist

b)

17"

Mark pipe at 17".

c)

15°

Bend tube 15° at 17" mark.

d)

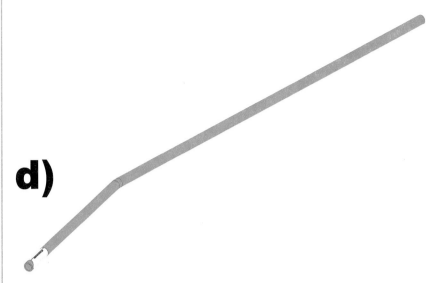

Tape and trace template for 90° notch around tube. With pipe laying flat on work surface, orient the notch pattern so the solid black centerline is facing upwards at the 12 o'clock position.

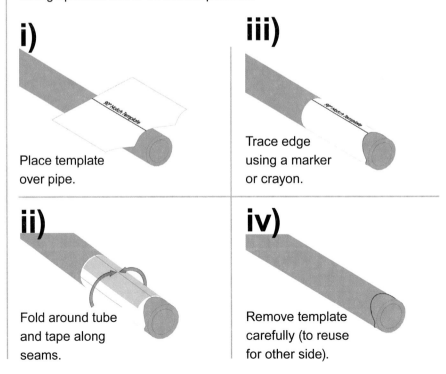

i)
Place template over pipe.

ii)
Fold around tube and tape along seams.

iii)
Trace edge using a marker or crayon.

iv)
Remove template carefully (to reuse for other side).

1e)

Cut and grind notch to size. Test fit using another tube at a 90° angle and continue grinding as necessary.

f)

Drill ⅛" hole anywhere towards the center of the tube. This hole acts as a vent for heat and gases as both ends become welded shut. It will need to be plugged with material later.

g)

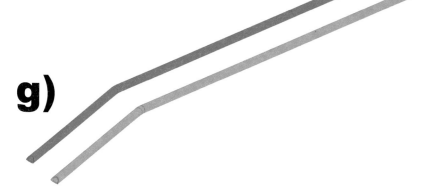

Create a second handle using the same process.

2

2x   90° Notch Template
1x   ØNPS ½" SCH40 Pipe, 37" Long

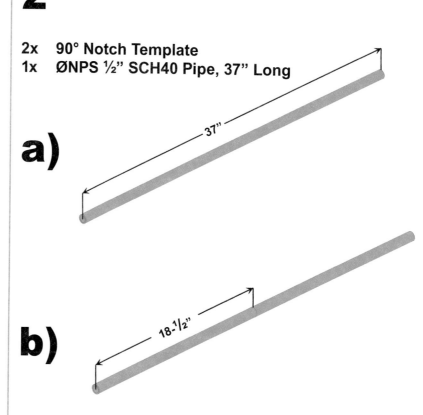

a)

37"

b)

18-½"

Mark center of the pipe.

c)

9"   5-½"   5-½"   9"

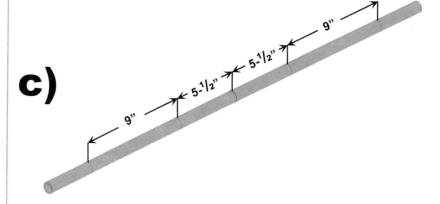

Measure from centerline and mark 4 additional locations.

2d)

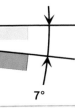

Bend ≈7° at first mark.

7°

e)

Bend ≈7° at second mark.

7°

f)

Repeat process for bends on opposite side.

g)

Both notch patterns are 90°. With pipe laying flat on work surface, orient the notch patterns so that the solid black centerlines are facing upwards at the 12 o'clock position.

h)

Cut and/or grind notches to fit.

i)

Drill ⅛" vent hole anywhere in the center region of the pipe.

3

2x   **45° Notch Template**
2x   **ØNPS ½" SCH40 Pipe, 12" Long**

a)

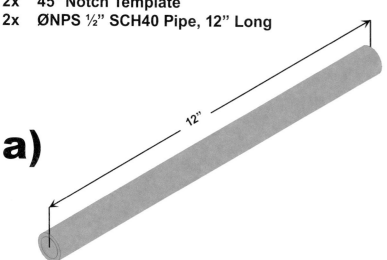

12"

b)

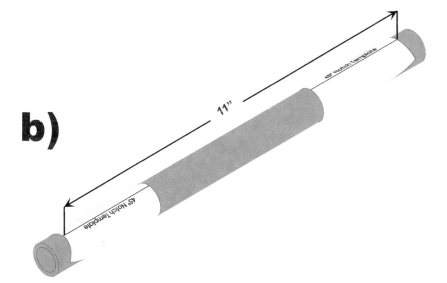

11"

45° Notch Template

45° Notch Template

Both notch patterns are 45°. Orient both patterns so that the solid black centerlines are facing upwards at the 12 o'clock position.

c)

Cut and/or grind notches on both ends (shown in green) to fit.

d)

Drill ⅛" vent hole anywhere in the center region of the pipe.

e)

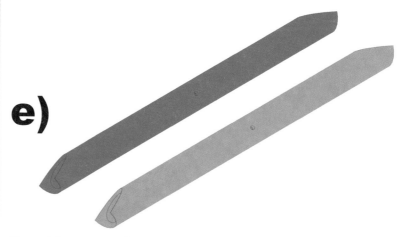

Repeat the process for a total of two pieces.

Broiler Shelter Dolly

4

2x ØNPS ½" SCH40 Pipe, 14" Long
1x 45° Miter Template
1x 48.7° Notch Template

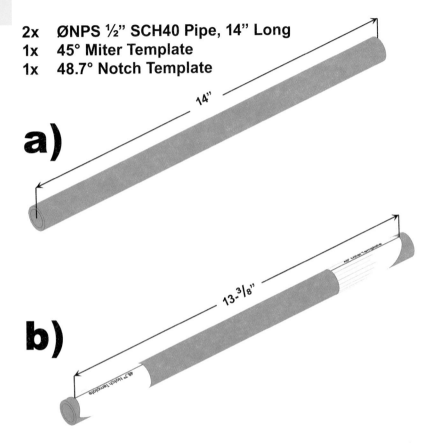

a)

14"

b)

13-³⁄₈"

One pattern is a 48.7° notch and the other is a 45° miter. Orient both patterns so that the solid black centerlines are facing upwards at the 12 o'clock position.

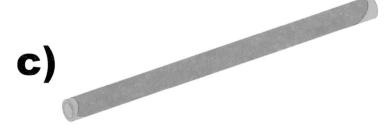

c)

Cut and/or grind notches on both ends (shown in green) to fit.

d)

Drill ¹⁄₈" vent hole anywhere in the center region of the pipe.

e)

Repeat process for a total of two pieces.

5

1x ØNPS ½" SCH40 Pipe, 37-¹⁄₂" Long

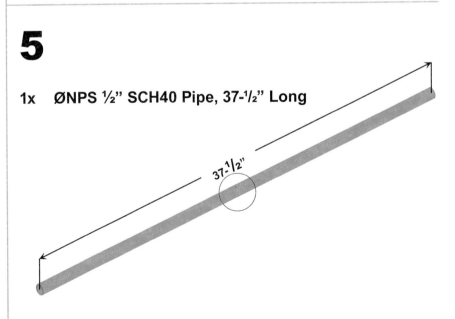

37-¹⁄₂"

Drill ¹⁄₈" vent hole anywhere in the center region of the pipe.

6

2x   ¼" x 1-½" Flat Bar, 4" Long

a)

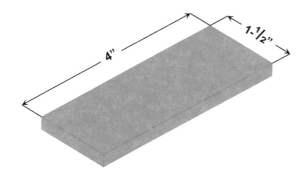

b)

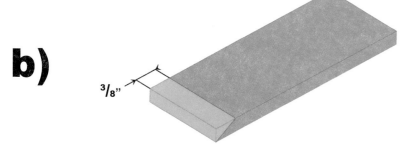

Cut or grind a chamfer. This does not need to be exact.

c)

Repeat process for a total of two pieces.

7

2x   ¼" x 1-½" Flat Bar, 7" Long

a)

b)

Cut off green corner as shown.

c)

Save the scrap triangle
piece for use in a later step.

d)

Repeat process for a total
of two pieces.

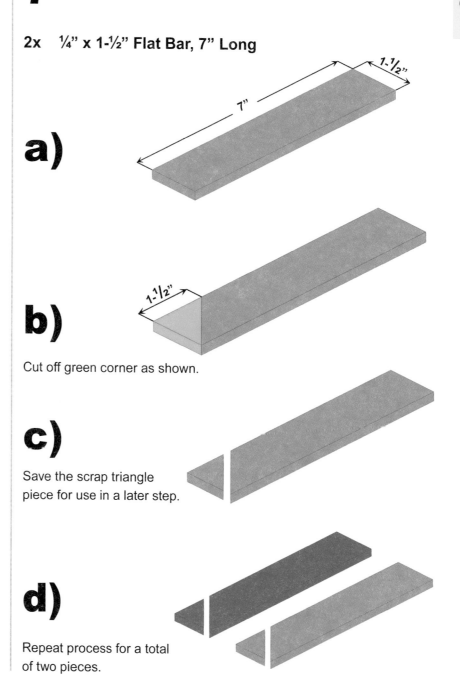

CHICKENS

8

1x   24x48" Plywood Sheet

9

1x   Copy of 3-Sheet
Fabrication Template

Align the three templates and tape them to the board. This composite layout is scaled 1:1 and will serve as a guide for you to lay out and hold pipe members in position until they are tack welded. (Once members are tacked in place, the structure will be removed from the wood and paper to prevent fire during the rest of the weld process.)

10

Extend the right-most line on the template using a straight edge.

11

2x   Scrap Wood Blocks
2x   Wood Screws

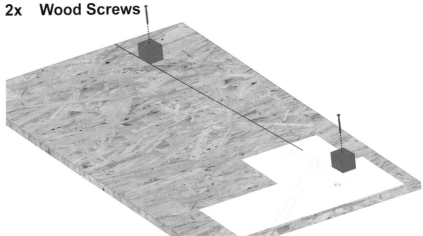

Fasten scrap wood blocks along line.

12

1x   Pipe From Step 1

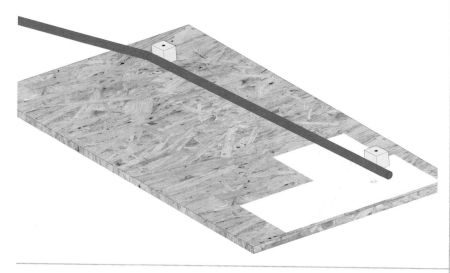

14

1x   Pipe From Step 4

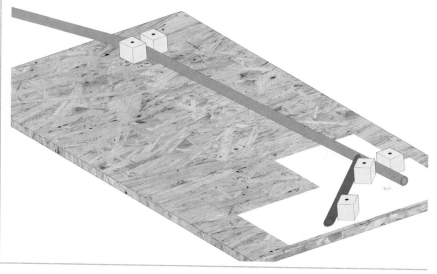

13

3x   Scrap Wood Blocks
3x   Wood Screws

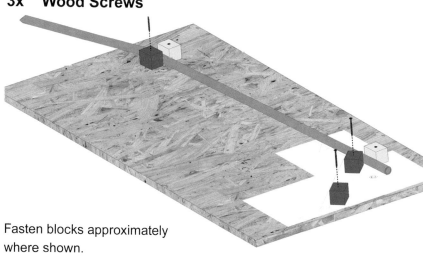

Fasten blocks approximately
where shown.

15

2x   Scrap Wood Blocks
2x   Wood Screws

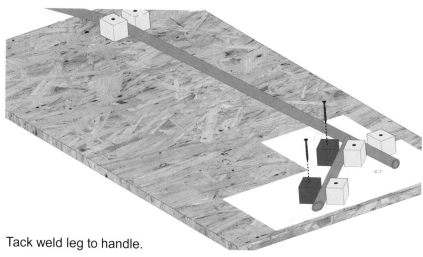

Tack weld leg to handle.

16

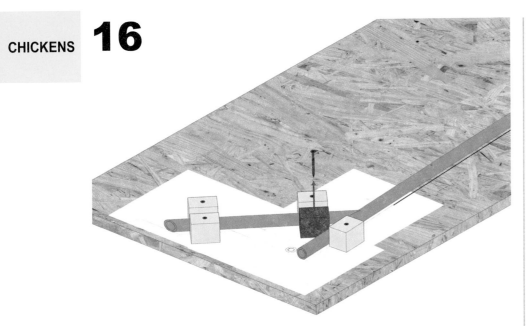

After tack has cooled, remove the block...

17

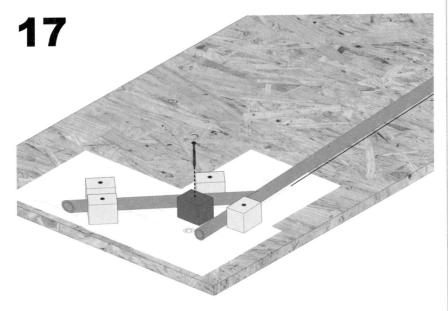

...and reorient as shown.

18

1x   Flat Bar From Step 7

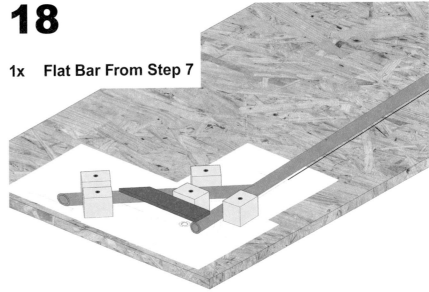

Align flat bar over the template.

19

1x   Scrap Wood Block
1x   Wood Screw

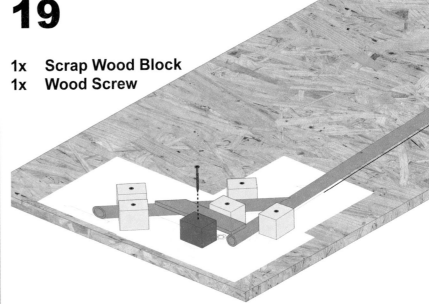

Fasten block on other side of bar to hold in place.

20

1x   Scrap Triangle From Step 7

Tack weld triangle to handle.

21

1x   Flat Bar From Step 6

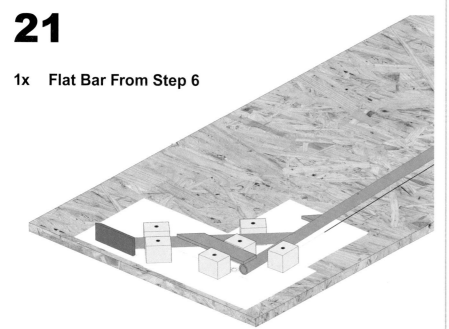

Tack weld foot to leg.

22

Remove from board and finish welding around all joints

23

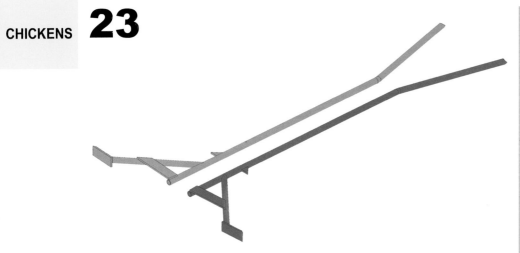

Repeat steps 12-22 to create an opposite hand version as shown.

25

1x   **Pipe From Step 2**

90°

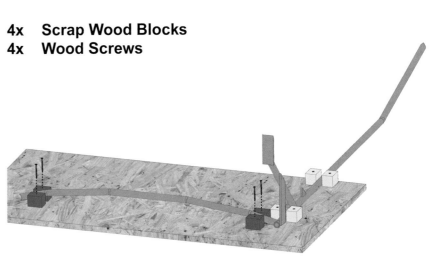

Make sure the pipe is square with handle side. If the plywood has square corners, you can use them as a guide.

24

4x   **Scrap Wood Blocks**
4x   **Wood Screws**

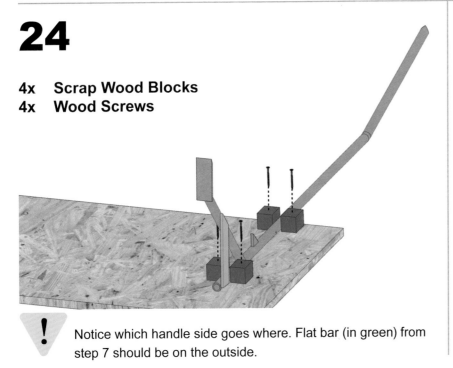

⚠ Notice which handle side goes where. Flat bar (in green) from step 7 should be on the outside.

26

4x   **Scrap Wood Blocks**
4x   **Wood Screws**

Fasten blocks to hold the pipe in place.

27

Make sure second side is square with the rest of the dolly. If the plywood has square corners, it is helpful to use them as a guide.

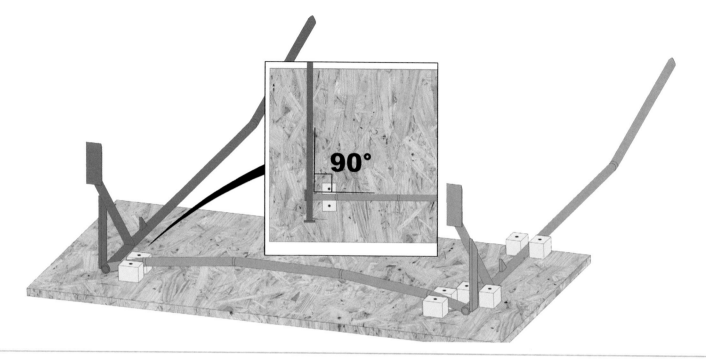

28

4x   Scrap Wood Blocks
4x   Wood Screws

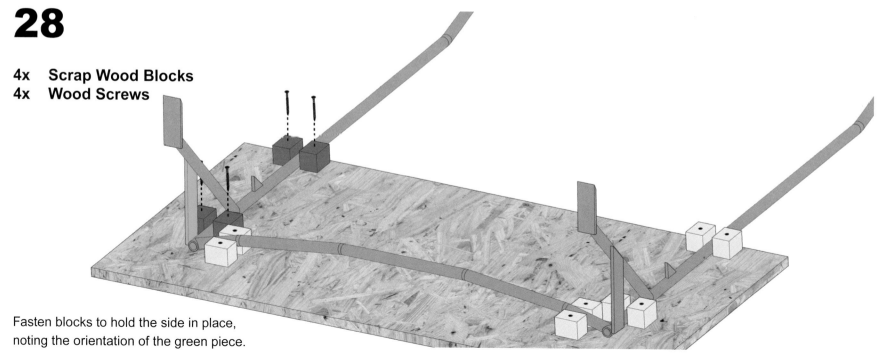

Fasten blocks to hold the side in place, noting the orientation of the green piece.

29

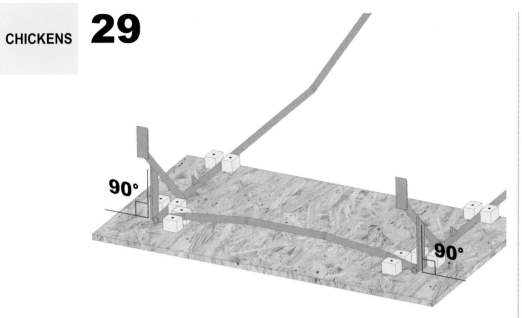

Make sure both handle sides are square and tack weld them in place.

30

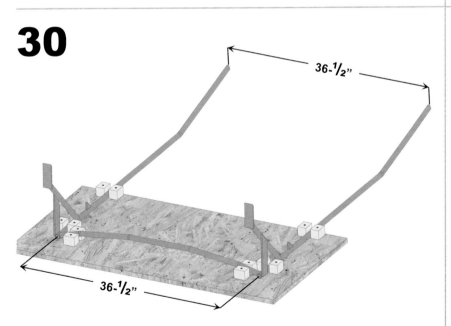

Whatever the bottom measurement ends up being, make sure the top is the same. Adjust and re-square as necessary.

31

2x   Pipe From Step 3

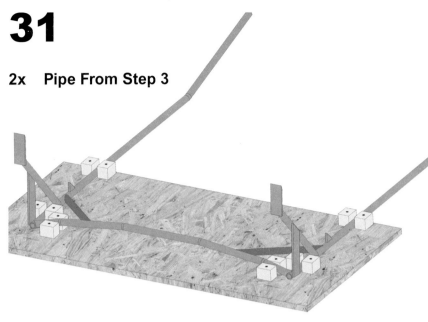

Once the dolly has been squared up, tack the two angle braces in place.

32

1x   Pipe From Step 5

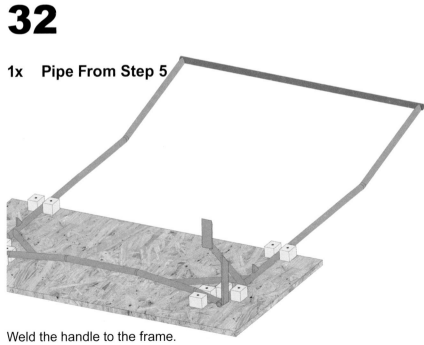

Weld the handle to the frame.

33

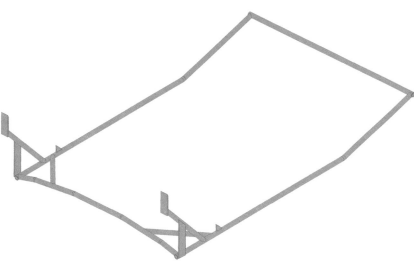

Remove from the plywood and finish welding around all of the joints.

34 Optional

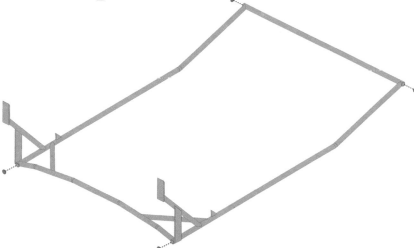

The end caps are optional, but make for a cleaner look and will prevent the insides from rusting out through prolonged exposure to the elements. Weld plug all vent holes after this step.

35

2x   Scrap Wood Blocks

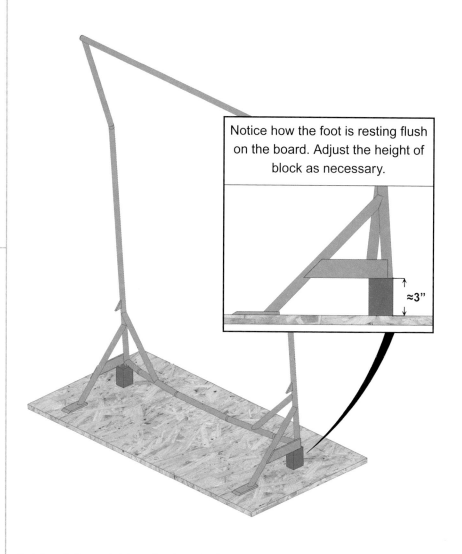

Notice how the foot is resting flush on the board. Adjust the height of block as necessary.

≈3"

Set the dolly upright on the board. Place blocks approximately 3" tall under it as shown. The objective is to get the feet resting flush on the ground so you may need to trim or extend the block to achieve this.

36

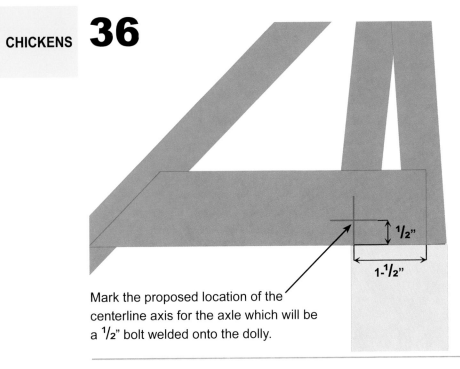

1/2"

1-1/2"

Mark the proposed location of the centerline axis for the axle which will be a 1/2" bolt welded onto the dolly.

38

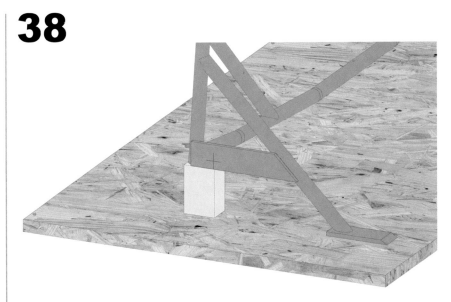

Repeat the process to determine the axle center location on the other side.

37

Set wheel on surface as shown and look through the center axle hole. It should line up with the mark you just made. Adjust the mark up or down if needed, but do NOT alter the location to the left or right.

39

1x   ½ -13 x 2" Hex Head Bolt

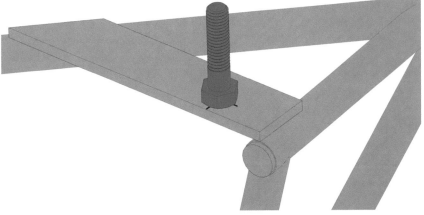

Center the bolt on the marking and clamp it until you've tacked it in place. Then weld around the entire circumference.

40

1x   ½ - 13 Nylock Nut
1x   7" Lawnmower Wheel

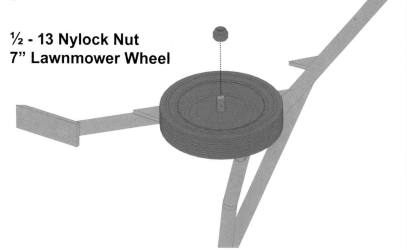

After it has cooled, add the wheel and loosely tighten the nut. There should be no excess wobble but it should not be binding tight either.

41

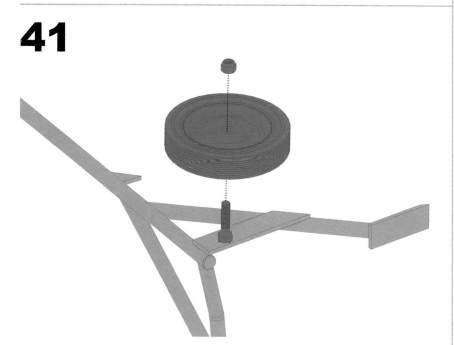

Repeat steps 39-40 to mount the wheel on the other side.

42

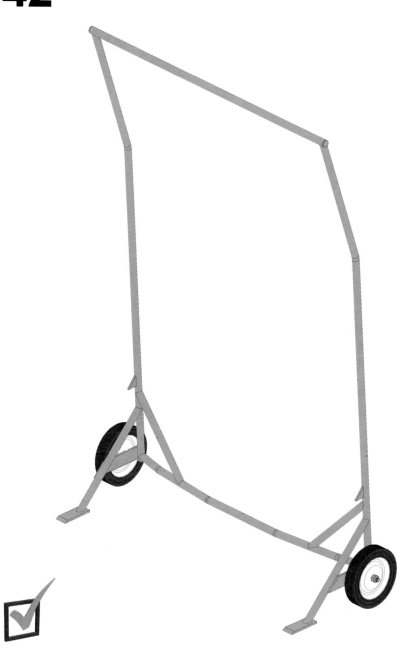

The broiler shelter dolly is complete!

CHICKENS

photo courtesy of Jean Shutt

EGGMOBILE

My first Eggmobile was 6ft X 8ft on bicycle wheels. It had 40 chickens and I pushed it around the field. The birds did not free range because I had it up by the house and didn't want them tearing up the garden and flower beds. That was long before electric poultry netting was invented.

I made six 12ft x 4ft panels out of $^3/_8$ inch light steel rod, and put regular poultry netting on them. Then I fastened three of them together with simple wire loops so I had two sets of triples. By folding them in on each other, I could carry the three gates at once. I had a door in each corner of this little squatty Eggmobile and I'd set up these six gates in a hexagon, like the leaf of a clover, to give the birds a yard. Every couple of days I'd move the yard around to another corner of the Eggmobile.

After going all the way around in 8 days, I'd push the Eggmobile up the field and reset the panels. The hexagonal configuration held them upright. One day I happened to move them onto a spot where the cows had grazed a couple of days before and what to my wonderment did the chickens do but attack those cow paddies like they were ice cream. Lights went off in my head. The chickens tore up every cow paddy in the area, eating out the fly larvae and spreading the cow pies out so that they covered four times the area of the original pie.

It was an epiphany and I set to work immediately building a 3-pt. hitch contraption to set the Eggmobile on so I could move it longer distances. Yes, I could push it around, but it was almost too heavy to push and it was certainly almost too much weight for the bicycle wheels. After spending the rest of that season moving it behind the cows and enjoying all that biological sanitation and fertilizer spreading, I decided the next step had to be a bona-fide trailer upgrade.

Although the 3-pt. hitch setup worked, it was awkward. Anyone who has ever hooked up something on a 3 pt. hitch knows it can be a bit dicey. So I built the first Eggmobile on a mobile home axle, 12ft x 20ft and put 100 chickens in it. They greatly reduced the flies on the cows, ate grasshoppers and crickets, spread out the cow pies, and laid eggs to boot. That was a true holon.

The eggs were spectacularly high quality, which stimulated sales, so the next year we doubled the chickens to 200. When we went to 300, we had trouble with them going in the single access door, so we cut another door in and went on up to 400. That was all the Eggmobile could handle. The next upgrade was a second one hooked to the first one, an Eggmobile train.

We have numerous Eggmobiles now, some 8ft x 16ft and some 12ft x 20ft. The smaller ones handle 250 birds each. I think we could hook three of those together, but we haven't done that yet. Why make a train rather than one humongous Eggmobile? The answer is simple: they become hard to maneuver and very heavy on the land. In spring, when the soil is wet, trying to pull a big heavy Eggmobile across the pasture can be impossible. With two, you can always unhook one and the axle weight isn't so much. They're also much more maneuverable in tight spots because of the pivot point.

We do not recommend an Eggmobile for acreages under 50. The reason is that the totally free range chickens will roam out 200 yards from home. That's a circle 400 yards in diameter. On small acreages, you can never move the Eggmobile into a totally new, unfamiliar area. If the birds stay around a familiar area too long, they find a new home base that is NOT the Eggmobile.

photo courtesy of Kate Simon

That favorite home base is usually the lawnmower handle on the back porch or the tractor steering wheel in the shop. And you know what a chicken leaves wherever she roosts.

On small acreages, we recommend either a modified broiler shelter (with nest boxes in it) or a small Millennium Feathernet. One day a week you can open it up and let the chickens go as far as they want. By having only one day of exposure to the outbuildings and back porch, they won't ever get familiar enough with those areas to settle down. This is chicken psychology 101.

If you have trouble with the chickens laying in the fence row or anywhere besides the Eggmobile, it's because they aren't being moved often enough far enough. They're getting too familiar with their surroundings.

The single biggest make or break component of the Eggmobile is training them to go in. We've learned that only laying hens work with it; if a chicken isn't laying, her nesting instinct is not high enough to go back to it. When we put a new flock in the Eggmobile, we lock them in for a day. They have to lay, eat, and drink in there—they are quite unhappy, but they need to learn this is their new home.

The next morning at daybreak we let them out. They timidly exit the Eggmobile and in a couple of hours begin thinking about feed and where to lay their eggs. A chicken might not have the biggest brain, but it's enough to remember where she spent yesterday and where she laid her

CHICKENS

egg. So she goes up the chicken ladder and lays her egg in the nest box. The point is to try to get her to go in and out a couple of times that first day.

As the sun starts to set that first day, we go out with a long bamboo pole or something similar and sit under the Eggmobile to keep the birds from settling down underneath. They instinctively know they are vulnerable if they settle down for the night in the field so they're looking for security. The first night more than half will go in and the rest you'll have to catch and put in. The second night, 90 percent will go in. The third night all but a couple will go in. If one doesn't go in the fourth night, put her in but expect her to be the first casualty of the season.

Once they're trained, they're good for the season. If you neglect one of these three nights, it will take an additional week to train them. Take the training seriously and you'll be rewarded for the whole season. Can you put netting or aprons around to keep them from going underneath? We've tried that. It's way more work than just sitting under there and shooing them away.

Can you surround the Eggmobile with the electrified poultry netting like the Millennium Feathernet? Yes, but if you're going to use the netting, you may as well go with a Millennium Feathernet because it has no rubber tires, no axle, much lower center of gravity, and is easier to build. Eggmobiles are for extensive pasture sanitation; for pastured egg production, the Millennium Feathernet is a better option. But

both photos courtesy of Janis Stone

if you have cows and you want to increase their hygiene and leverage their manure, nothing works like an Eggmobile.

Note that the Eggmobile carries a substantial amount of feed. To reduce chores and running across the field, we have these bulk feed boxes so we only have to bring feed to the Eggmobile a couple of times a month. For water line, we use ³/₈ inch air hose because garden hose reels can hold so much more air hose than garden hose. That lets us park farther away from a water valve.

Going to a flat roof rather than a slanted roof simplified the construction substantially. The Eggmobile is always on a bit of a slant, so as long as the sheet of roofing extends from side to side in one sheet, it doesn't matter which direction the water runs off the roof. A flat roof is much easier to build than a slanted one.

Originally we made the nest boxes

accessible from the outside, but again, that complicated construction. It's just as easy to gather the eggs from inside. Better an airy inside than a complicated outside access.

If you have too much trouble with predators grabbing sleeping chickens through the slatted floor, you can staple poultry netting on the floor. It actually gives extra traction when you walk and definitely acts as a double layer of protection.

Automatic doors are now available that work fairly well. They're expensive, but if your Eggmobiles are far away, they're worth the expense. And remember, the eggs are only one part of the income from this marvelous contraption. I've decided it would be worth having one even if the chickens never laid an egg. Few things are as beautiful as seeing those chickens spread out on a new area, attacking cow pies and eating grasshoppers in the early morning. It's poetry.

Tools

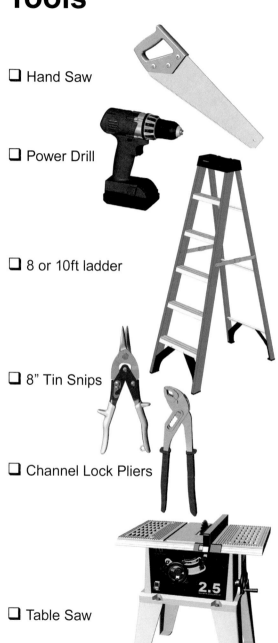

- ❑ Hand Saw
- ❑ Power Drill
- ❑ 8 or 10ft ladder
- ❑ 8" Tin Snips
- ❑ Channel Lock Pliers
- ❑ Table Saw

- ❑ Welder
- ❑ Circular Saw
- ❑ Grinding Wheel
- ❑ Cut-Off Wheel
- ❑ Angle Grinder
- ❑ Socket Driver Set

- ❑ Stapler
- ❑ ¹/₂" Staples
- ❑ String Line
- ❑ Marking Pencil
- ❑ Tape Measure
- ❑ Speed Square
- ❑ Hammer
- ❑ ¹/₂" HSS Drill Bit
- ❑ Driver Bits (for screws)
- ❑ ¹/₂" Auger Bit
- ❑ 4ft Level

Hardware

DECK SCREWS

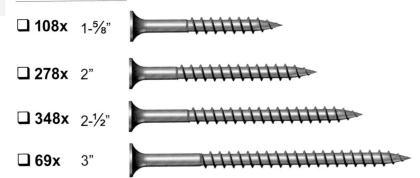

- ❏ **108x** 1-⅝"
- ❏ **278x** 2"
- ❏ **348x** 2-½"
- ❏ **69x** 3"

NAILS

- ❏ **478x** 20D Galvanized Nails*
- ❏ **4x** 8D Framing Nails
- ❏ **84x** 8D Ribbed Nails*

It is recommended that these nails be either spiraled or ribbed for superior holding power.

OTHER

- ❏ **8x** ⅜ x 1-¾" Hex Head Lag Screws

- ❏ **335x** 1-½" Roofing Screws

- ❏ **27x** ½" Washers

- ❏ **16x** ⅜" Washers

- ❏ **4x** ⅜"-13 Nylock Nuts

- ❏ **24x** ½"-13 Hex Nuts

- ❏ **1x** Male Hose End (see note G on page 52)

- ❏ **1x** Female Hose End (see note G on page 52)

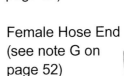

- ❏ **1x** Hose Reel & Mounting Hardware (see note D on page 52)

BOLTS

☐ **4x** ½-13 x 2"
Carriage Bolts

☐ **20x** ½-13 x 3"
Carriage Bolts

☐ **1x** ½ x 4" Hex Head
Lag Screws

☐ **2x** ⅜ x 2" Lag Screw
Eyebolts

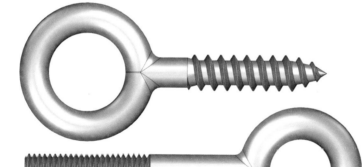

☐ **4x** ⅜-16 x 2-½"
Eyebolts

☐ **10x** Strap Hinges

4"

3-¹/₂"

Materials

☐ **4x** Nesting Boxes
(see note E on page 52)

☐ **50ft** Roll of 24"
Chicken Wire

☐ **3x** 12" Bungee Cord

CHICKENS

☐ **1x**   Adjustable Clevis Hitch
Part 18078

☐ **1x**   Bolt-On A-Frame
Channel-Up
Part 15349-52
(see note B)

☐ **1x**   6 Gallon Waterer (see note C)

☐ **1x**   Trailer Jack (see note A)

☐ **3x**   Bulk Feeder
& Lids
(see note F)

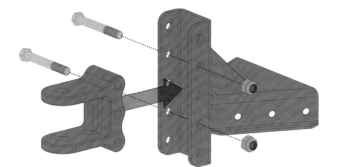

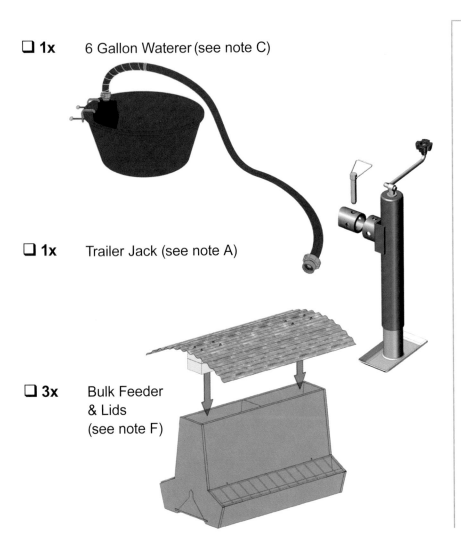

A savvy person could come up with a "poor man's" hitch that would do the job. However, in our effort to simplify this build for a novice and provide a final product that will be safe and durable, we have opted to showcase an off the shelf option that can handle the abuse. A bonus of this option is the flexibility to change the hitch type (ball, clevis, pintle, etc) according to one's needs. It also allows adjustability in height which is beneficial, especially when hooking egg mobiles together.

Notes:

A.)   The trailer jack we prefer is one with a tube mount configuration. The small steel tube can be easily welded to a chassis and has proven to be a very durable design. In 2020, this jack could be found at Tractor Supply stores for less than $60. (Part # 11630699)

B.)   In 2020, these specific trailer parts can be found online from Croft® Trailer Supply. You will need to purchase the appropriate hardware to mount the hitch. Refer to installation manual for details.

C.)   Refer to page 531-533 for details on procurement and setup of poultry drinkers.

D.)   Do not skimp on a hose reel. Buy something that is sturdy and well built. NORTHERN® Tool + Equipment has some good options. For more holding capacity try filling your reel with $^3/_8$" rubber air hose instead of a standard garden hose. Don't worry about the reduced flow, because the demand is not very high. Ask your local hardware store for the appropriate plumbing fittings to convert from one to the other.

E.)   See page 461 for more on nesting box selection and setup.

F.)   See pages 456-458 for more details on bulk feeders and the creation of lids for them.

G.)   You will need replacement hose ends because you will be reversing the fittings on your hose and reel (i.e. replace the male end with a female and vice versa).

Wood Cutlist

Wood scraps are highlighted in red.
Do not discard until project is completed.

QTY	SIZE
☐ **16x**	2x6x10ft
☐ **4x**	2x4x8ft
☐ **8x**	2x4x10ft

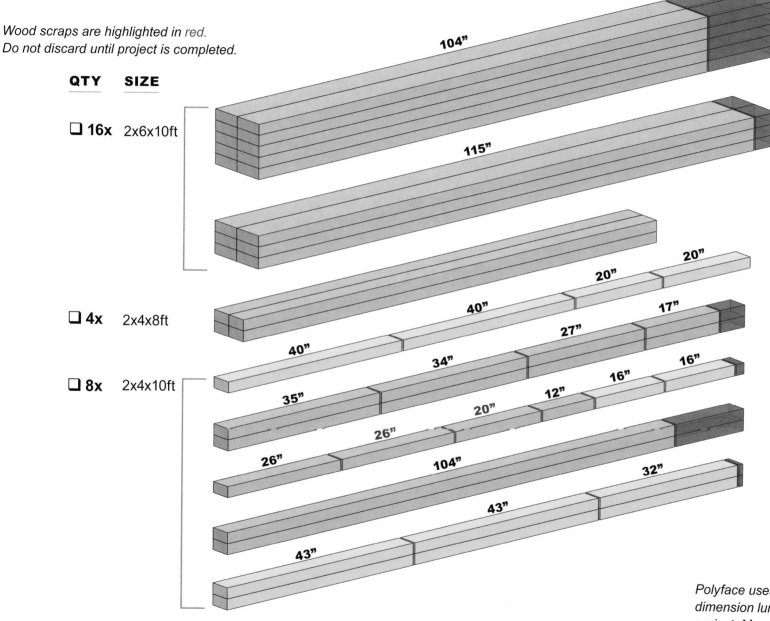

104"
115"
20" 20"
40" 27" 17"
40" 34" 16" 16"
35" 16"
26" 20" 12"
26" 104" 32"
43"
43"

 Items highlighted in yellow are for building the grain bin. If you are not building a grain bin you will not need these pieces.

⚠ *Measurements in red are approximates. Cut each piece to fit.*

Polyface uses nominal dimension lumber for this project. Meaning, a 2x4 is actually 2" x 4" as opposed to a dimensional 2x4 which is 1-1/2" x 3-1/2".

Wood Cutlist

Wood scraps are highlighted in red.
Do not discard until project is completed.

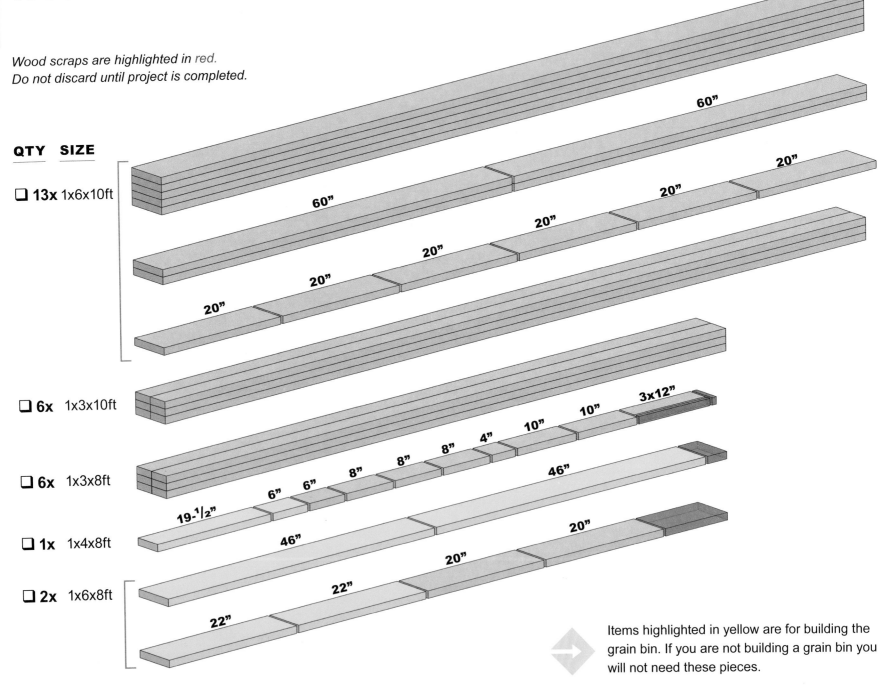

QTY	SIZE
☐ 13x	1x6x10ft
☐ 6x	1x3x10ft
☐ 6x	1x3x8ft
☐ 1x	1x4x8ft
☐ 2x	1x6x8ft

Items highlighted in yellow are for building the grain bin. If you are not building a grain bin you will not need these pieces.

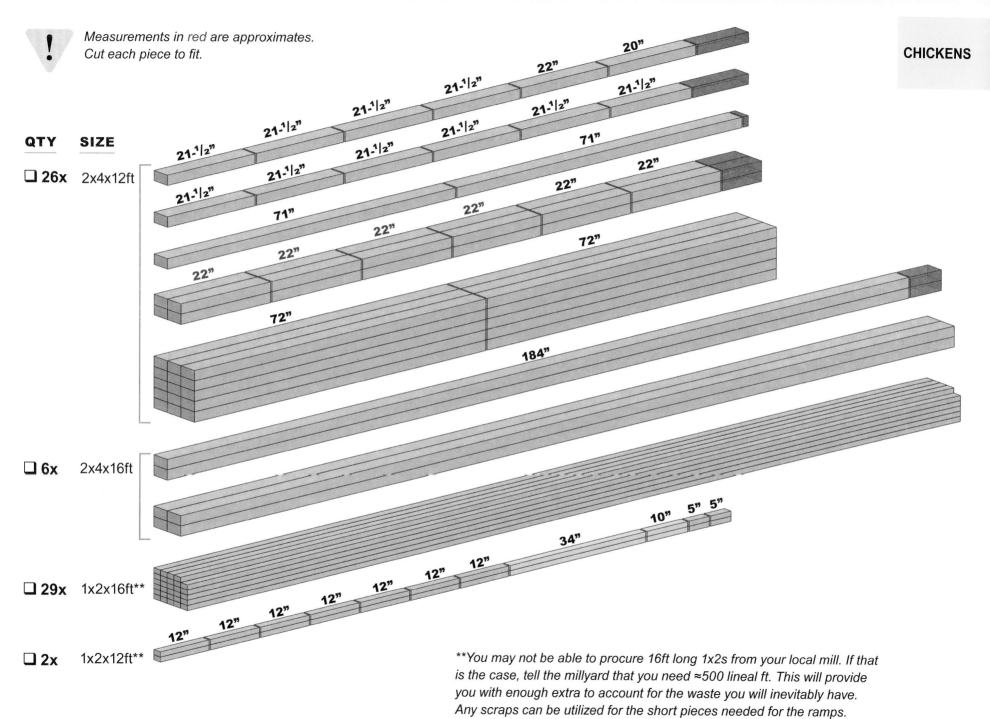

Measurements in *red* are approximates.
Cut each piece to fit.

QTY	SIZE
☐ **26x**	2x4x12ft
☐ **6x**	2x4x16ft
☐ **29x**	1x2x16ft**
☐ **2x**	1x2x12ft**

20"
22"
21-¹/₂"
21-¹/₂"
21-¹/₂"
21-¹/₂"
21-¹/₂"
71"
21-¹/₂"
21-¹/₂"
22"
21-¹/₂"
22"
71"
22"
22"
72"
22"
22"
72"
72"
184"

10" 5" 5"
34"
12"
12"
12"
12"
12"
12"
12"
12"
12"

**You may not be able to procure 16ft long 1x2s from your local mill. If that
is the case, tell the millyard that you need ≈500 lineal ft. This will provide
you with enough extra to account for the waste you will inevitably have.
Any scraps can be utilized for the short pieces needed for the ramps.

55

Wood Cutlist

QTY	SIZE
☐ **1x**	4x8x½" Plywood Sheet
☐ **1x**	4x4x½" Plywood Sheet

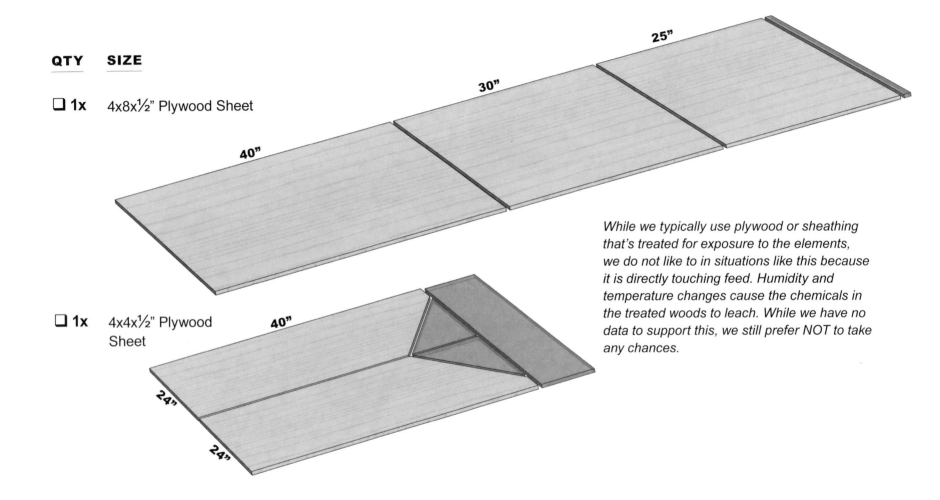

While we typically use plywood or sheathing that's treated for exposure to the elements, we do not like to in situations like this because it is directly touching feed. Humidity and temperature changes cause the chemicals in the treated woods to leach. While we have no data to support this, we still prefer NOT to take any chances.

 Items highlighted in yellow are for building the grain bin. If you are not building a grain bin you will not need these pieces.

Metal Cutlist

⚠️ *Measurements in red are approximates.*
Cut each piece to fit.

75"

36"

75"

75"

QTY | **SIZE**

☐ ≈36" C5x6.7 C-Channel

75"

☐ ≈13ft C5x6.7 C-Channel

75"

20ft

☐ ≈13ft C5x6.7 C-Channel

20ft

☐ 20ft C5x6.7 C-Channel

20 Brackets 3" Long

Drill hole in the center.

☐ 20ft C5x6.7 C-Channel

48"

ø½"

☐ ≈9ft 4x4x⁵⁄₁₆" Angle

☐ 1ft ¾x2" Bar

12"

This piece of steel is used as a drawbar, or hitch, on the back of the Eggmobile. If you never plan to hook two Eggmobiles together, then you will not need this.

12"

ø1"

¾"

2"

The syntax used to name C-Channel may seem confusing at first. It is based on a unified standard upheld by the American Institute of Steel Construction (AISC).

C5x6.7

The "C" designates that the profile is "C-Channel".
The "5" designates a height of 5".
The "6.7" designates that this profile weighs 6.7 lbs per linear foot. (This makes it easy to estimate the weight of your chassis.)
The end user can rest assured that all steel procured to this standard will have the same profile dimensions and properties. This is an American standard, so our international readers will need to procure the closest equivalents with what they have available in their region.

Items highlighted in yellow are for building the grain bin. If you are not building a grain bin you will not need these pieces.

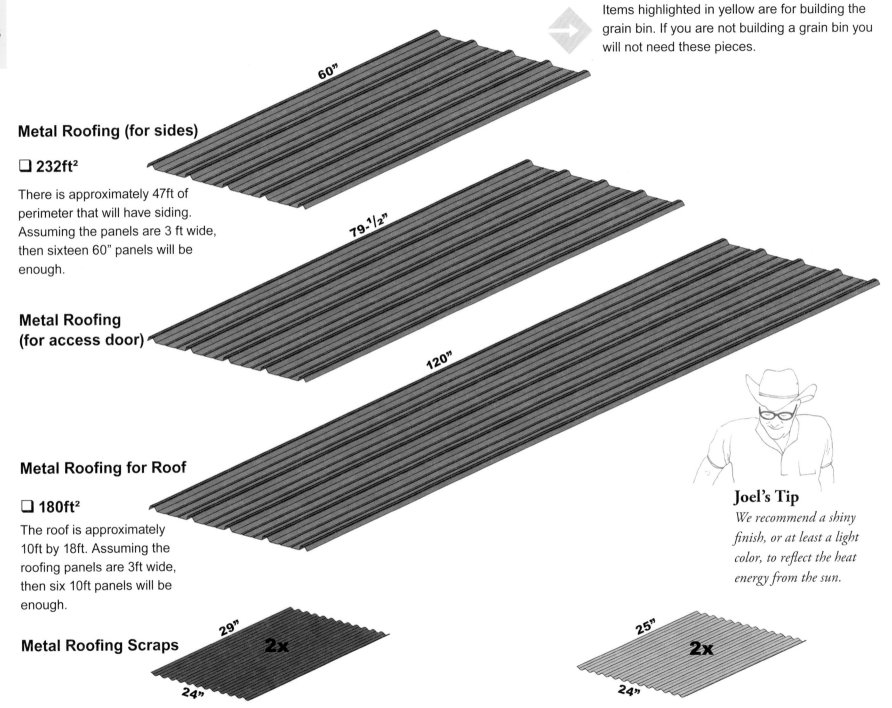

Metal Roofing (for sides)

❏ **232ft²**

There is approximately 47ft of perimeter that will have siding. Assuming the panels are 3 ft wide, then sixteen 60" panels will be enough.

60"

Metal Roofing (for access door)

79-¹/₂"

Metal Roofing for Roof

❏ **180ft²**

The roof is approximately 10ft by 18ft. Assuming the roofing panels are 3ft wide, then six 10ft panels will be enough.

120"

Metal Roofing Scraps

29"

2x

24"

25"

2x

24"

Joel's Tip

We recommend a shiny finish, or at least a light color, to reflect the heat energy from the sun.

Chassis

Although this book illustrates only one example of an Eggmobile chassis, we use different sizes and styles for our dozen working models. Additionally, thousands of variations can be found on the internet, providing examples of innovation and genius. Our biggest fear in presenting this material is that we risk stifling your imagination by prescribing a specific way to build a chassis. If you have a chassis or can procure one, use it. For those with the creativity and the know-how to strike out on your own, glean what you can while sticking to the basic design principles. For the rest of you, this is an excellent and proven design.

Why C-Channel?

We often use large custom milled oak beams to build the chassis, but not everyone has access to affordable large-dimension customized timbers. Our on-farm band sawmill offers that luxury. You can certainly price both options, but generally you'll find steel the cheaper and lighter option. C-channel is less expensive than square or rectangular tube, and typically lighter than its comparable counterparts. It's also easy to weld and does not trap moisture which can lead to accelerated rust. However, if you do have your own mill and/or plentiful hardwood lumber, have at it—big timbers work great.

Some basic things to remember:

We do not recommend hay wagon running gears for two primary reasons. First, the four wheel points require flex when going over uneven ground. Pull a hay wagon through

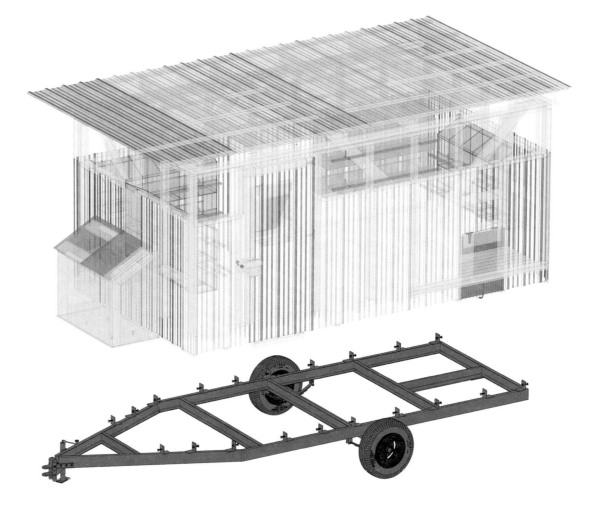

a ditch and you can see the flat bed flex. If you mount a box on that flat plane, it either needs to flex (impossible) or be strong enough to hold a wheel in the air as the egg mobile goes over a low spot. That strain usually rips the walls apart. The trailer axle with mounting point allows the whole box to tilt as it traverses uneven ground. The second reason is that hay wagons are hard to back up. The second pivot point on the tongue makes them much less maneuverable if you have to wiggle through a tight spot. The

trailer is much easier to line up or back up if you have to adjust your position.

Two more minor reasons not to use a hay wagon chassis involve wheel base and height. Hay wagons have a narrower wheelbase than the mobile home axles we typically use. Because the hay wagon chassis is higher than the low mobile home axle, the whole structure has a higher center of gravity. This combination leads to tip-overs. We've seen it happen—it's not a good thing to wake up to in the morning.

This Eggmobile weighs in at a little under 4,000lbs empty. Add 250 chickens and 16 bushels of feed and it will be 6-7000lbs. If you live in a wet climate and/or have soft soil conditions you may want to consider adding a second axle. We cannot specify what to use because there are too many variables, and we typically like to use what's available to us. We can, however, provide some information to help you make your own selection, and things to look at when making your selection.

Choosing an axle

Modern trailer axles fall into two categories: a.) the torsion axle and b.) the straight axle (aka spring axle). We typically use a straight axle configuration for all of our projects. Although you may consider several things when selecting an axle, we'll focus on two features.

1.) Axle weight rating: weight ratings are broken in to classes: 2k, 3.5k, 5.2k, 6k, 7k, and all the way up to 15k. Each class of axle also has standard wheel hubs that fit corresponding tire/wheel rim combinations. Most of these (with the exception of mobile home axles) are standardized and easy to swap and find parts for. Just be aware that your rim-tire combos match the hubs on the axle you choose. What class axle you use ultimately depends on how heavy your final project will be and what kinds of payloads it will be expected to carry. Also note that another way to increase payload capacity is to add a second axle (in tandem). The upside is that you have weight distributed to 4 tires instead of 2 which means lower

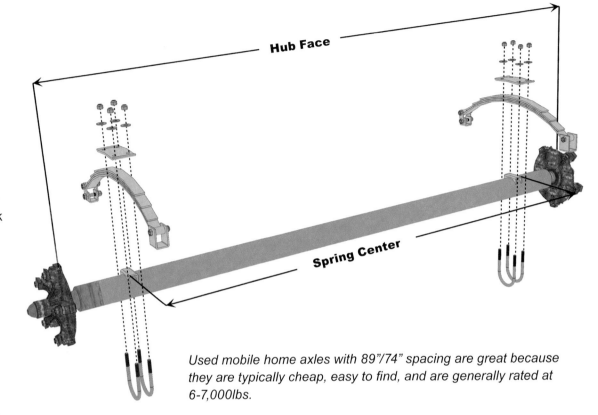

Used mobile home axles with 89"/74" spacing are great because they are typically cheap, easy to find, and are generally rated at 6-7,000lbs.

ground pressure (aka ruts that are not as deep). The downside is the added cost of a second axle and the maintenance of 4 tires instead of 2. It's also important to understand that by not using a leaf-spring suspension system, the load carrying capacity of the axle is reduced by at least 50 percent. Remember that the tongue can carry more than 1,000 pounds.

2.) Common Axle Sizes: Once you have chosen a weight class for your project, you will then need to choose what size axle to buy. Sizes are broken down by two key dimensions: distance between hub faces, and distance between

spring centers (see the included illustration above).

The axle we chose for this project happened to be a used mobile home axle with 89"/74" spacing. Mobile home axles are great because they are typically cheap, easy to find, and are generally rated at 6-7,000lbs. The downside is that they're designed for one-time use—to transport a home to its final destination. Hence, they are made cheaply. However, we have never had any major issues.

As mentioned above, you can put a box on almost anything. But here are some key considerations to make any design more functional.

Ventilation

Since this is generally summer quarters, you want lots of light and air movement. Of course, the slatted floor offers ventilation, but that air needs to exit. Since warm air rises, the obvious best area is high, under the roof. Make sure you cover the opening with poultry netting to keep out predators and wild birds.

Slatted floor

One of the key benefits to this whole structure is never having to shovel manure or put in bedding. The width of the slats and their spacing is critical. If the spacing is too wide, predators can get in. If it's too narrow, the manure won't go through. If the slats are too wide, the manure collects on top. If they're too narrow, they break when you step on them. We use 2" slats with a 1-3/4" spacing and that keeps us from ever having to shovel or scrape anything.

Nest boxes

We install one box per ten hens. They do not use them evenly, and will rotate preferences based on sunlight orientation. Hens like privacy and low light. Morning light streaming into one bank of nest boxes will make them less desirable for that day. Always position the lowest nest boxes above chicken eye level, approximately 18". That discourages mischievous loitering. Make sure the perch boards in front of the boxes hinge so you can

close them up at gathering (around 4 p.m.). The single biggest inefficiency in egg production is dirty eggs; if the chickens sleep in the boxes, you'll have dirty eggs because they poop where they sleep.

Square footage

Because the Eggmobiles serve as primarily bedroom and work room only, the square footage allowance can be low. They spend nearly all their time outside and in inclement weather, they use both the ground underneath the Eggmobile and the inside floor. We've found that .6 square feet per bird is enough for comfort. Don't expect animal welfare certification to agree to this notion, but certifiers don't have any understanding of Eggmobiles.

Predators

If you think you like chicken, wait until you talk to a raccoon. Everything out there that goes bump in the night likes chicken more than you do. But predators are opportunists. If you religiously close it up at night, predators get frustrated at not being able to get a meal. The key to protection is to consistently close the doors at night. That security provides a halo effect that discourages attacks.

Overall cost

Chickens are not fussy guests. Function over form works. Anyone who visits Polyface understands quickly that this is a shoestring bootstrap outfit. I did some consulting for a guy once who built an Eggmobile that cost $30,000. You could run it like an Airbnb. It had expanded

metal flooring strong enough to hold a tractor trailer. It was so heavy he couldn't move it. Be gentle. Use salvage material when possible. Keep it simple and light.

Water and feed

The key to efficiency is bulk storage and not having to haul any more than necessary. Water is too heavy. Lots of people install water storage and it's just too heavy. Install some water lines around your property and use air hose mounted on a spool to access your valves. What you should try to inventory on the Eggmobile is as much feed as possible. You can't send it through your fields with a pipe.

In general, a hen will eat about 4 ounces of feed per day. A 250-bird Eggmobile, then, needs about 60 pounds per day. If you can hold 400 pounds inside and 1,000 pounds in a bin on the outside, that's enough to carry you for 3 weeks. With that schedule, you can often plan a rendezvous near your main farm feed storage when empty to reduce or eliminate runs to the field for a re-fill. A 5-gallon bucket of feed holds about 35 pounds.

Our bulk feeders inside hold 100 pounds apiece. A 250-bird outfit will last nearly a week on that amount. The whole idea is to group your work to reduce chore time. Chores are things that have to be done every day about the same time. If you can spend 15 minutes once a week filling feed bins, that's a lot better than doing it every single morning. The non-fill mornings are a simple move.

CHICKENS

1

1x   C-Channel 20ft

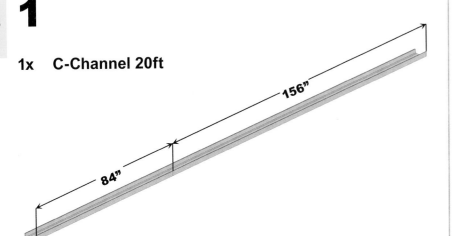

156"

84"

Mark all sides of the C-channel. This will be the location of the notch and bend in the next steps.

2

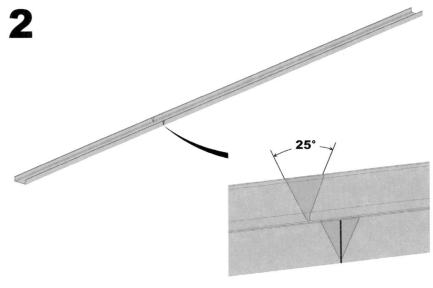

25°

Cut the notches in the shape of a "V" at the angle prescribed.

3

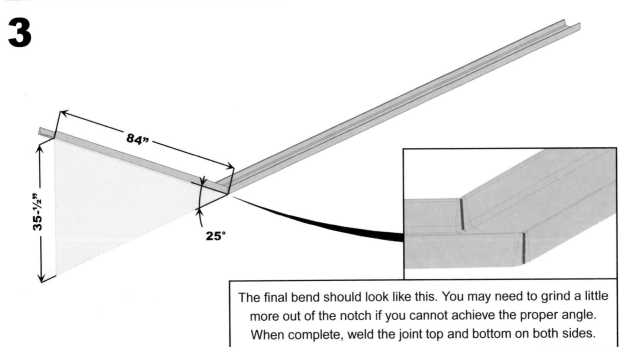

84"

35-½"

25°

The final bend should look like this. You may need to grind a little more out of the notch if you cannot achieve the proper angle. When complete, weld the joint top and bottom on both sides.

With the notch cut, the next step is to create the bend. One way to determine whether or not your angle is correct is to do some simple trigonometry using inputs that we already know. Since we know that we want a 25° angle and that the length of the hypotenuse of our triangle is 84", we can thereby calculate the opposite side of the triangle (vertical measurement) using the following equation:

Sin(25°)*Hypotenuse=Opposite

0.42262*84"=35.5"

4

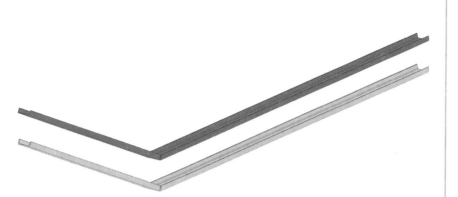

Repeat steps 1-3 to create an identical second side.

5

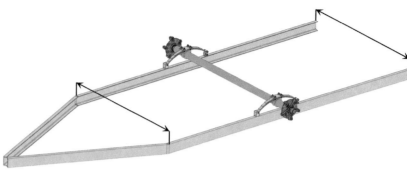

Set the axle on the frame rails and ensure that the rails are parallel by measuring the spacing across the length of the rails.

6

1x   C-Channel 75"

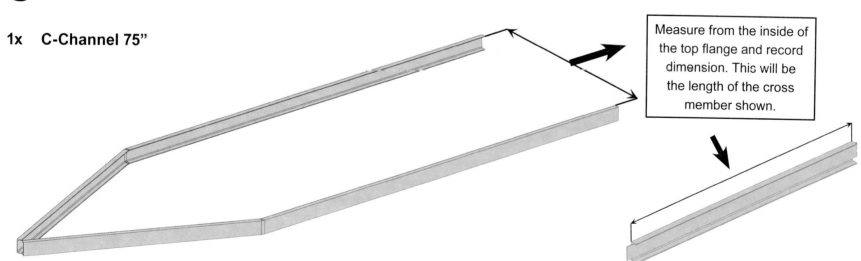

Measure from the inside of the top flange and record dimension. This will be the length of the cross member shown.

Create A Notch

7

a)

1-½"

b)

¾"

c)

d)

Cut notches into the ends of the C-Channel so that it will fit inside of another member perpendicular to it.

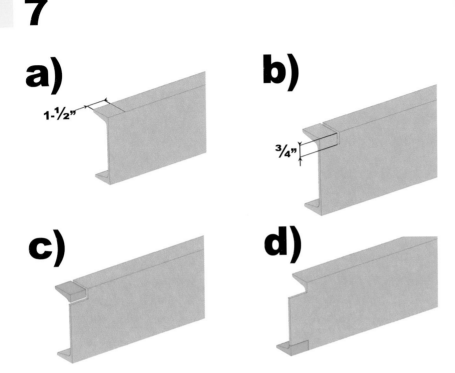

e)

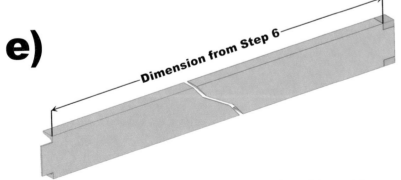

Dimension from Step 6

Using the dimension from step 6, mark the location of the notches on the other side and cut them out. Test fit and trim as necessary.

8

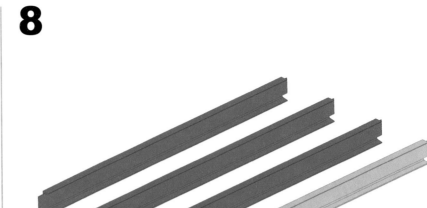

Repeat steps 6-8 to create a total of four cross members.

9

1x   C-Channel From Step 8

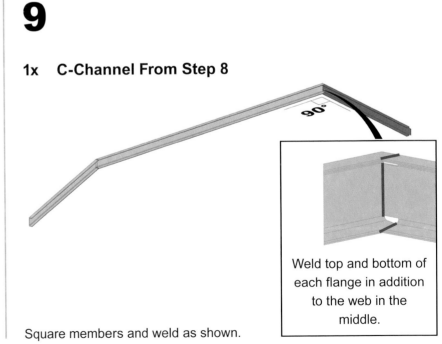

90°

Weld top and bottom of each flange in addition to the web in the middle.

Square members and weld as shown.

10

3x   C-Channel From Step 8

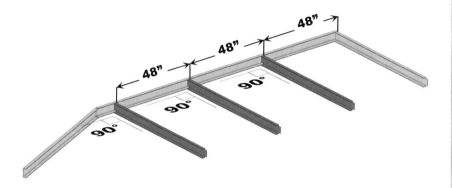

Space remaining members 48" apart. Square them with frame and weld in place.

11

1x   4x4x⁵⁄₁₆" Angle

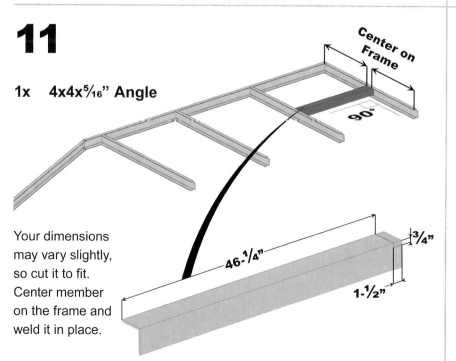

Your dimensions may vary slightly, so cut it to fit. Center member on the frame and weld it in place.

12

1x   C-Channel From Step 4

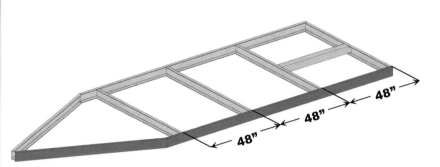

Add the other side of the frame rail. Align everything and weld in place.

13

1x   Bolt-On A-Frame Channel-Up & Mounting Hardware

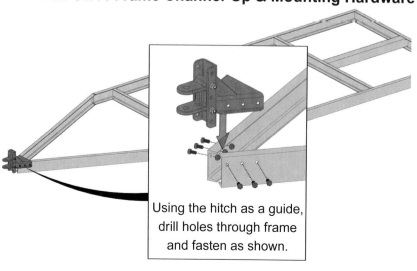

Using the hitch as a guide, drill holes through frame and fasten as shown.

Create A Mitered Notch

14

1x   C-Channel 36"

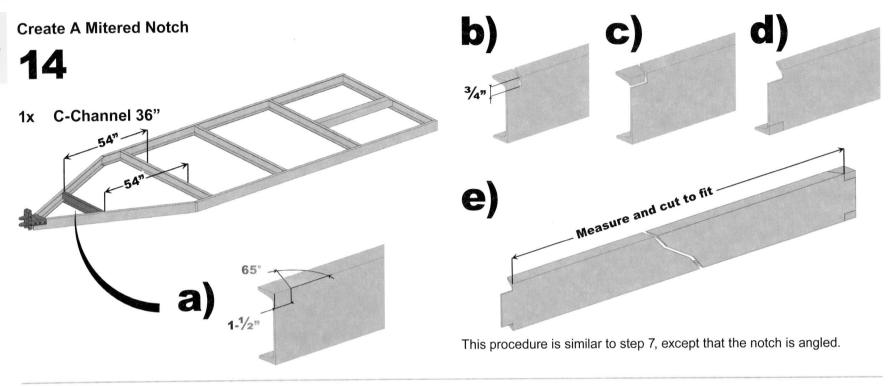

b)

³⁄₄"

c)

d)

a)

65°

1-¹⁄₂"

e)   *Measure and cut to fit*

This procedure is similar to step 7, except that the notch is angled.

14 Optional

1x   ³⁄₄x2x12" Drawbar

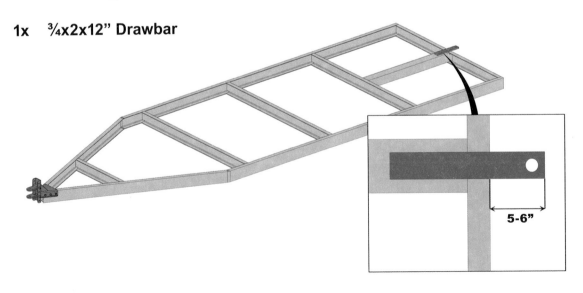

5-6"

This step is optional in case you ever plan to hitch two Eggmobiles together. If you do, then weld a drawbar or some similar thickness metal to the back as shown.

15

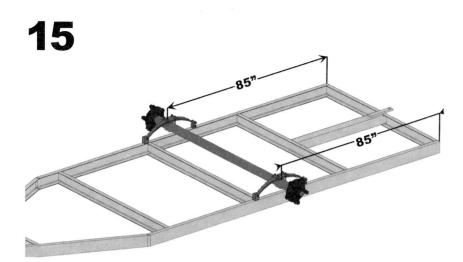

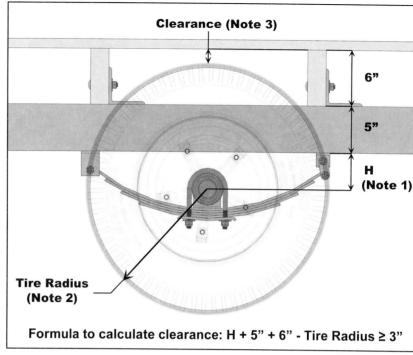

Clearance (Note 3)

6"

5"

H
(Note 1)

Tire Radius
(Note 2)

Formula to calculate clearance: H + 5" + 6" - Tire Radius ≥ 3"

Take time to position the axle square on the frame. An axle that is crooked will cause the trailer to not track straight. This is not a big deal around the farm, but if you plan to transport these Eggmobiles from farm to farm, or long distances, this could become an issue.

!

Before welding, verify that you will have adequate clearance using the following method. If you do not have adequate clearance, consider welding spacers to the chassis frame. In the images shown, we have 2" square tubing which fits nicely on the bottom of the C-Channel.

Note 1: This measurement will vary based on axle class, spring size, and manufacturer. Take this measurement directly from the axle you plan to use.

Note 2: Tire radius also varies. Measure the radius of your tires for this calculation.

Note 3: Ideal clearance is 3-4".

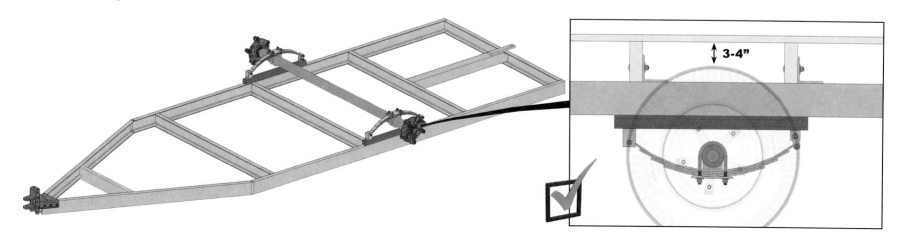

3-4"

16

1x   Weld-on Mount For Trailer Jack

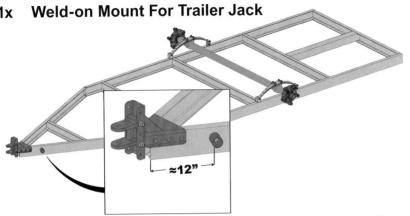

Choose a location relatively close to the hitch to mount your jack. If you plan to mount a grain bin to the front of this then it will be especially important not to interfere with it.

17

Flip the chassis over using a tractor or loader and install the tires/rims and jack.

18

1x   2x6x104"

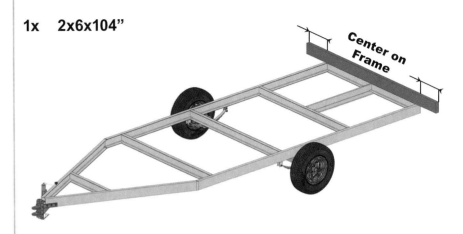

Center the first joist on the frame. It should also be flush against the back end of the chassis.

19

2x   4x4x⁵⁄₁₆" Angle Brackets

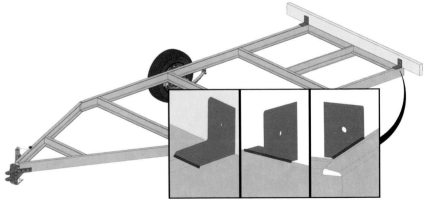

Weld brackets in the three locations shown.

20

2x   ½-13 x 3" Carriage Bolts
2x   ½" Washers
2x   ½-13 Hex Nuts

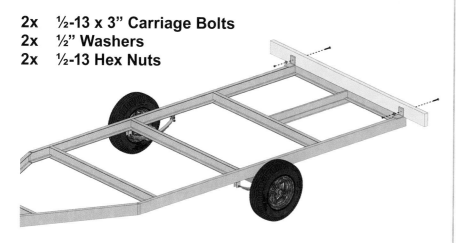

Drill holes through wood using the brackets as guides and fasten with bolt hardware.

22

2x   4x4x$^5/_{16}$" Angle Brackets
2x   ½-13 x 3" Carriage Bolts
2x   ½" Washers
2x   ½-13 Hex Nuts

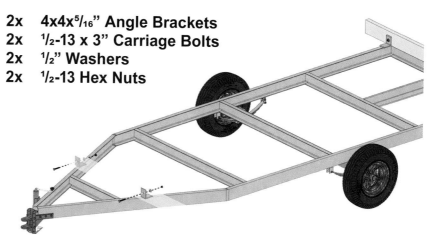

Orient the brackets as shown and weld in place. Then drill and secure with bolt hardware.

21

1x   2x6x104"

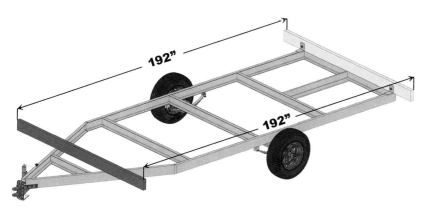

Measure 192" from both corners and center 2x6 on chassis.

23

2x   String Lines
4x   8D Framing Nails

Drive a nail into the edge of each corner. Tie a string to each end and pull taunt. These serve as guides to align the rest of the floor joists.

Eggmobile

24

4x   2x6x104"
8x   4x4x⁵⁄₁₆" Angle Brackets
8x   ½ - 13 x 3" Carriage Bolts
8x   ½" Washers
8x   ½-13 Hex Nuts

Carefully note which side of the 2x6 each bracket is mounted on– it matters for future steps. Also, the spacing of the floor joists may seem odd, but there is a reason. Use the string lines to align each joist and keep the ends straight.

25

2x   2x6x104"
4x   4x4x⁵⁄₁₆" Angle Brackets
4x   ½-13 x3" Carriage Bolts
4x   ½" Washers
4x   ½-13 Hex Nuts

Remember to pay attention to what side of the 2x6 the brackets are mounted!

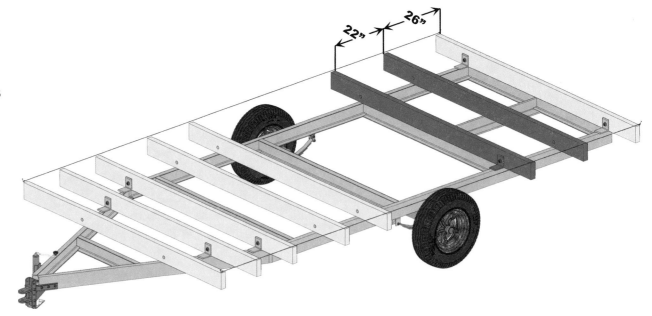

26

2x **2x6x104"**
4x **4x4x⁵⁄₁₆" Angle Brackets**
4x **½-13 x 3" Carriage Bolts**
4x **½" Washers**
4x **½-13 Hex Nuts**

Position final two joists equidistant from the center of the axle at approximately 11" to provide ample clearance for the tire.

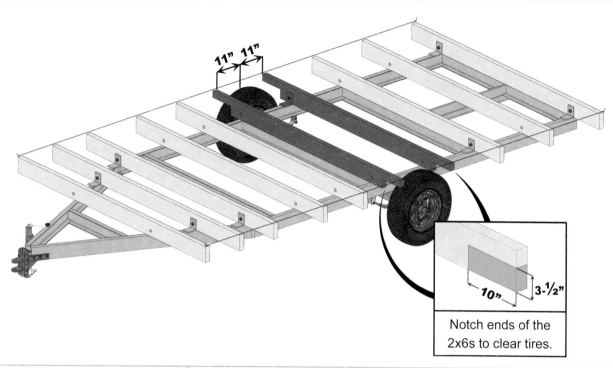

Notch ends of the 2x6s to clear tires.

27

4x **2x4x22"**
16x **20D Galvanized Nails**

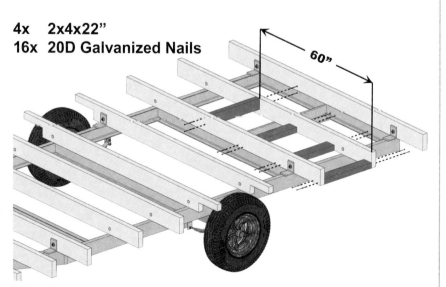

These boards are the basis for a compartment to stow away a ramp.

28

4x **2x4x22"**
16x **20D Galvanized Nails**

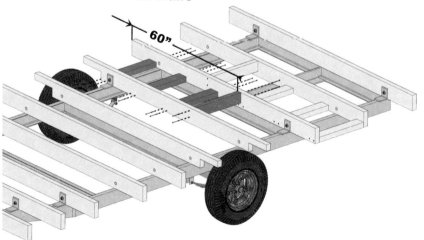

CHICKENS

29

1x   2x4x22"
1x   2x4x26"
8x   20D Galvanized Nails

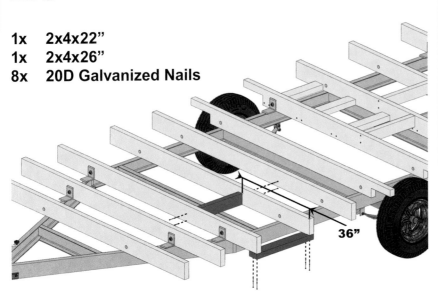

36"

30

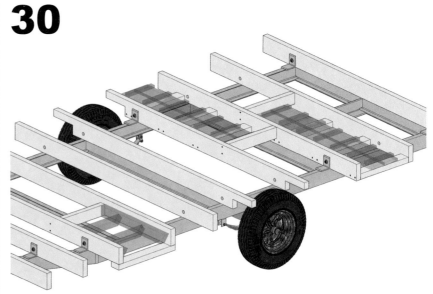

These compartments will stow the ramps and step when they are not in use.

31

1x   1x2x192"
10x 2-½" Deck Screws

Install the slatted floor. Center the first piece lengthwise and work outward in both directions. If the 1x2s are not 16ft long, you will need to cut and piece them and make sure that your joints start and end on the 2x6 joists.

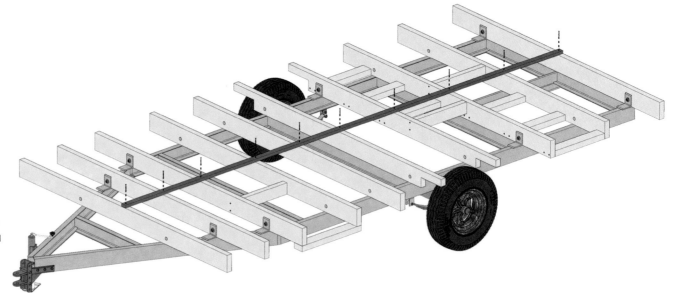

32

14x   1x2x192"
140x   2-½" Deck Screws

 Screwing these in is more time consuming, but it makes replacing a broken floor slat much easier in the future.

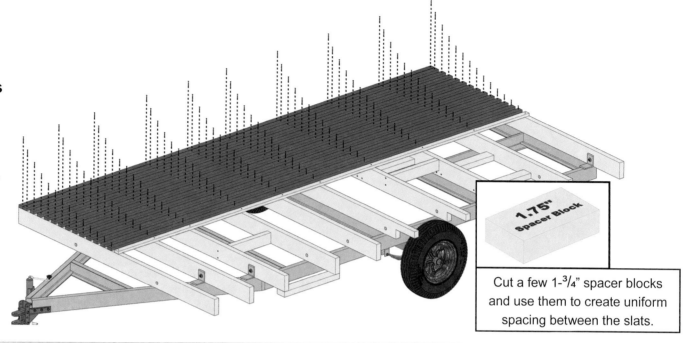

1.75" Spacer Block

Cut a few 1-³/₄" spacer blocks and use them to create uniform spacing between the slats.

33

14x   1x2x192"
140x   2-½" Deck Screws

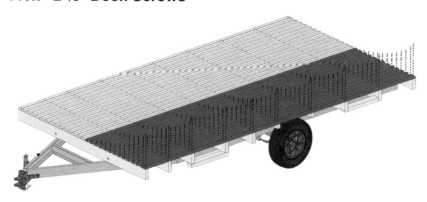

Repeat process for remainder of floor slats.

34

✓

The chassis base is completed!

CHICKENS

35

5x   2x4x72"

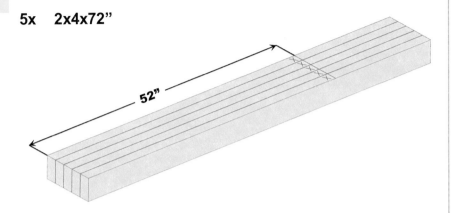

It's time to frame the walls. Line precut wall studs together in order to measure and mark the locations of future horizontal supports.

36

2x   2x4x96"

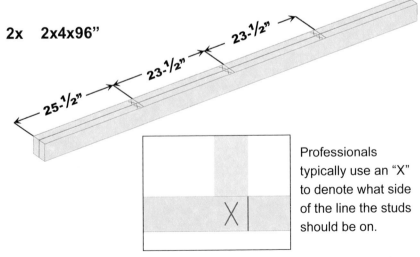

Professionals typically use an "X" to denote what side of the line the studs should be on.

Measure and mark the locations for each stud on the top and bottom plates as shown.

37

5x   **Studs From Step 35**
20x   **20D Galvanized Nails**

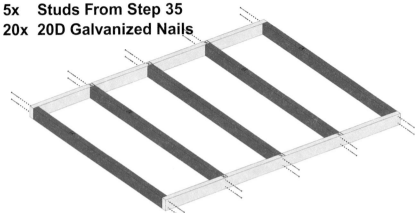

Align the studs on the marks. Remember that the board lines up with the side of the mark that has an "X".

38

Measure and cut each horizontal support to fit.

1x   **2x4x21-½"**
4x   **20D Galvanized Nails**

Install the first horizontal support using the marks as a guide.

39

1x   2x4x21-¹/₂"
4x   20D Galvanized Nails

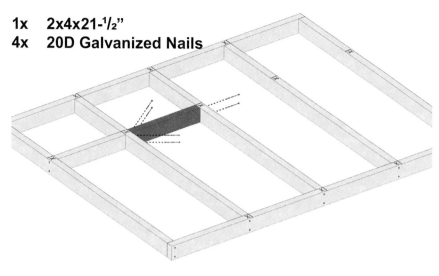

Install the second support. Notice how it is nailed.

40

2x   2x4x21-¹/₂"
8x   20D Galvanized Nails

Repeat the process for the remaining supports.

41

1x   24" Chicken Wire

Staple

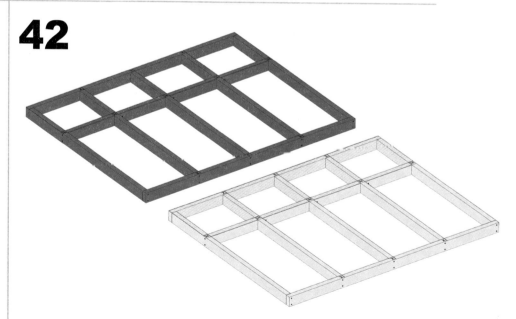

Staple every 4-6" along the perimeter and on the interior supports. Cut chicken wire flush with edge of the wall.

42

Repeat steps 35-41 to create a second wall and staple the chicken wire along the top.

43

8x   20D Galvanized Nails
2x   2x4x21-½"

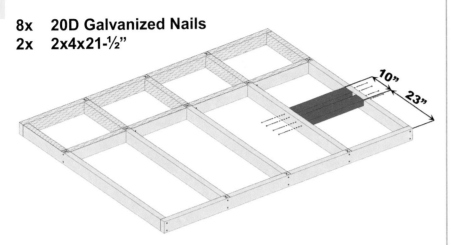

Take one of the two walls and add these supports for mounting the hose reel in a future step. This wall will be the front of the Eggmobile.

44

8x   2x4x72"

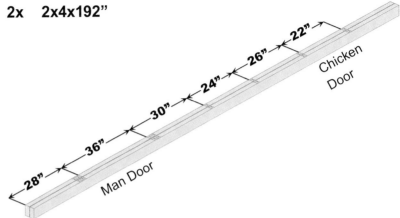

Measure and mark location of future horizontal supports.

45

2x   2x4x192"

Measure and mark the location of the vertical studs along the top and bottom plates. Pay attention to what side of the "X" each mark is on.

46

8x   Studs From Step 44
32x   20D Galvanized Nails

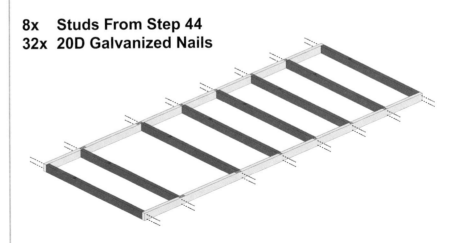

Position studs on the marks. Notice that only the outermost studs in the doorway have been placed.

47

1x 2x4x22"
1x 2x4x26"
8x 20D Galvanized Nails

Install the horizontal supports. Measure and cut each horizontal support to fit.

49

3x 2x4x22"
12x 20D Galvanized Nails

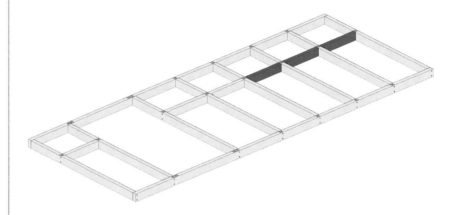

48

1x 2x4x22"
4x 20D Galvanized Nails

50

2x 2x4x22"
8x 20D Galvanized Nails

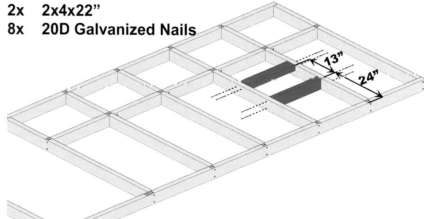

The lower support is a header for the chicken doorway - door hinges will mount to it. The upper support if there to secure an eyebolt in a future step.

CHICKENS

51

2x   2x4x72"
≈16x   20D Galvanized Nails

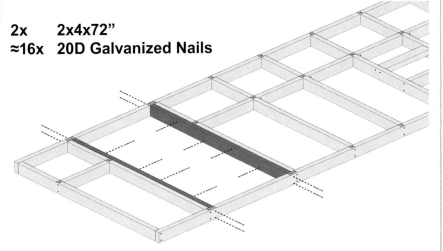

"Sister" the final two studs on the inside of the doorway. These add additional support to the opening.

53

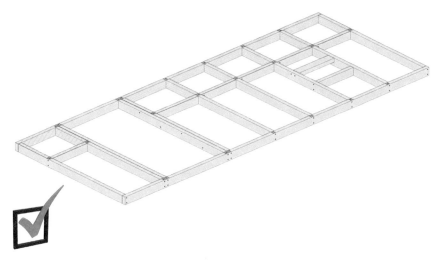

The right wall is complete.

52

≈11ft   24" Chicken Wire

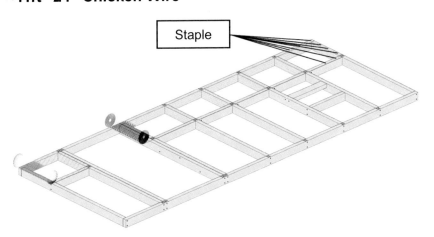

Staple every 4-6" along the perimeter and the interior supports.

54

9x   2x4x72"

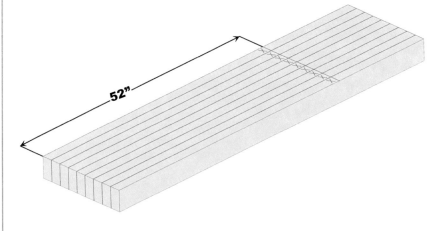

52"

Measure and mark the location of future horizontal supports.

55

2x   2x4x192"

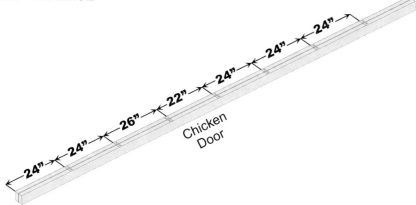

Measure and mark the location of vertical studs along the top and bottom plates.

57

7x   2x4x22"
1x   2x4x20"
32x   20D Galvanized Nails

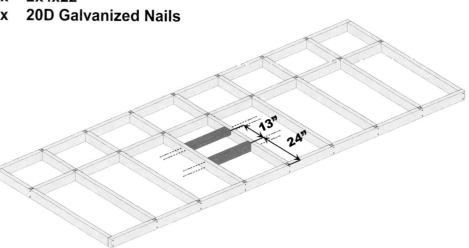

Measure and cut each horizontal support to fit.

56

9x   Studs From Step 54
36x   20D Galvanized Nails

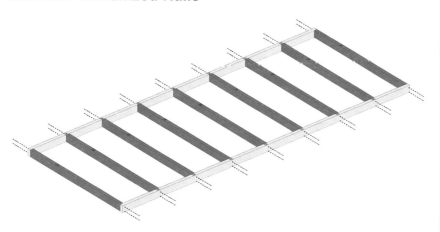

58

2x   2x4x22"
8x   20D Galvanized Nails

59

≈16ft   24" Chicken Wire

Staple

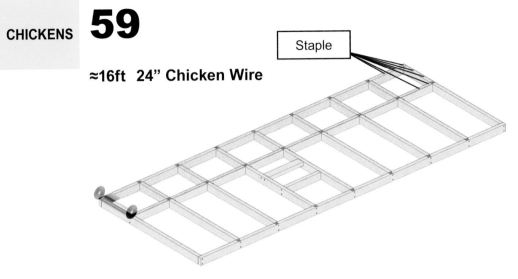

Staple every 4-6" along the perimeter and the interior supports.

60

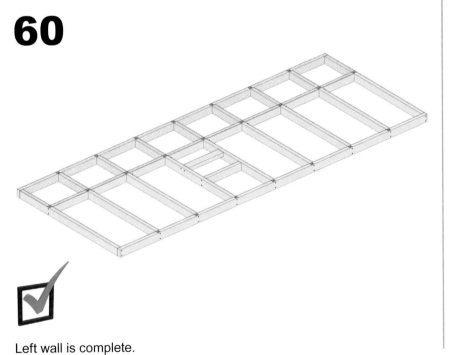

Left wall is complete.

61

Chicken wire always faces out.

1x   **Left Wall**
10x   **20D Galvanized Nails**

62

1x   **Back Wall**
5x   **20D Galvanized Nails**

63

5x   20D Galvanized Nails

Nails can be used but you may find long deck screws to be easier when working between studs.

64

1x   Right Wall
5x   20D Galvanized Nails

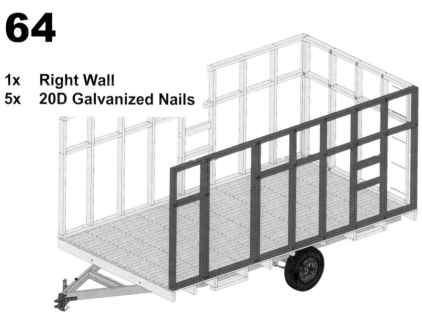

Do not fasten the bottom plate to the floor yet. You need to position the front wall first.

65

1x   Front Wall
10x 20D Galvanized Nails

Fasten corners together.

66

14x 20D Galvanized Nails

Now it is time to secure the baseplates to the floor.

CHICKENS

67

2x   2x4x104"
28x   20D Galvanized Nails

In these next two steps we add a second top plate around the entire wall perimeter. This plate overlaps in the corners and locks everything together.

68

2x   2x4x184"
16x   20D Galvanized Nails

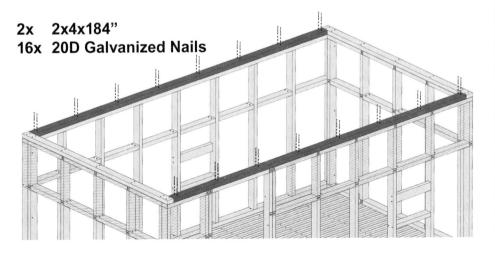

69

5x   1x6x120"

a)

120"

b)

43°

43°

c)

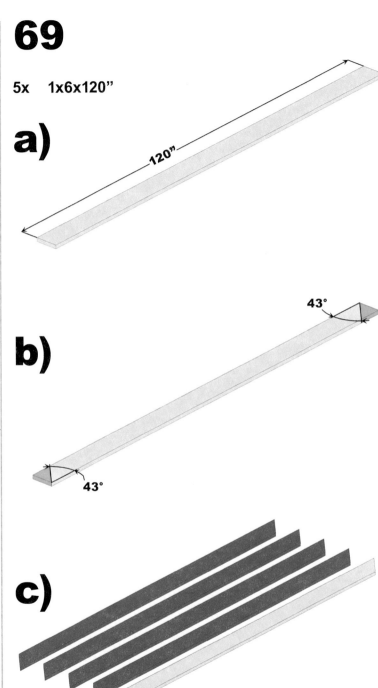

70

1x   Brace From Step 69
6x   2-½" Deck Screws

In these next few steps we are bracing the walls to provide rigidity to the structure. Square each wall and fasten the brace in place.

72

2x   Brace From Step 69
12x 2-½" Deck Screws

71

1x   Brace From Step 69
6x   2-½" Deck Screws

73

1x   Brace From Step 69
6x   2-½" Deck Screws

74

1x   2x4x12"
4x   20D Galvanized Nails

22"

This additional blocking will be used to attach the door latch mechanism.

76

2x   String Lines
4x   8D Framing Nails

Like the floor joists in step 23, set string lines to help you align rafters with each other.

75

2x   2x4x115"
8x   20D Galvanized Nails

We are now on to the roof structure. Center the two end rafters and nail (or screw) them to the top plate.

77

4x   2x4x115"
16x   20D Galvanized Nails

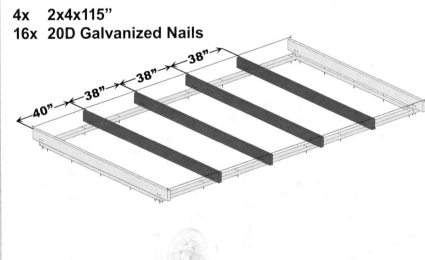

40"   38"   38"   38"

Fasten the remaining rafters as shown.

78

1x   1x3x96"
1x   1x3x120"
14x 8D Ribbed Nails

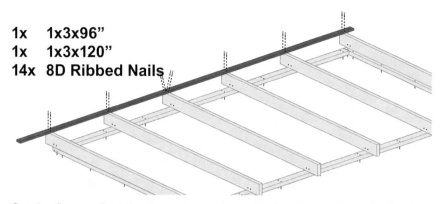

Set the first purlin. We chose to overhang the front more than the back to provide additional rain protection to the grain bin. Note: If you are building a second egg mobile to pull behind another, you will want to switch the overhang to the back. See the picture at the end of this section for details.

79

5x   1x3x96"
5x   1x3x120"
70x 8D Ribbed Nails

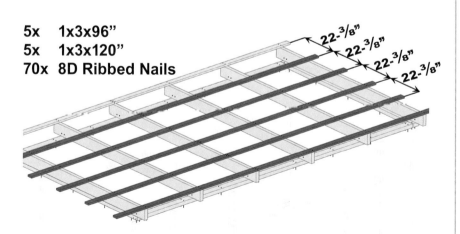

22-³/₈"
22-³/₈"
22-³/₈"
22-³/₈"

Fasten the remaining rafters as shown making sure that they are parallel all the way across.

80

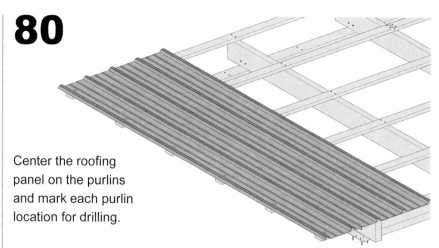

Center the roofing panel on the purlins and mark each purlin location for drilling.

Steps 80 and 81 are optional but help fasten the steel on the roof. Assuming your purlins are evenly spaced then all of your screw holes will line up. You will also find it easier to start the screws with a pilot hole pre-drilled. If you do not pre-drill you'll find it helpful to give the back end of your power drill a good whack to start the screw into the metal.

81

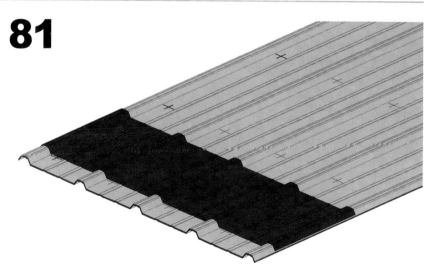

An easy way to transpose your marks across the panel is to use a scrap piece of roofing (assuming it has a square cut). Then stack all of the panels and drill through all of the panels at each drill mark with the appropriately sized drill bit.

82

1x 120" Roofing Panel
24x 1-½" Roofing Screws

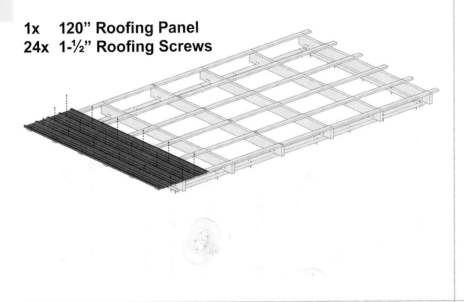

84

≈3x 60" Roofing Panels
≈27x 1-½" Roofing Screws

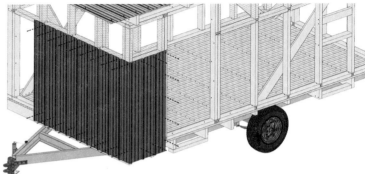

It's time to side the Eggmobile. Start on the left side from the access door. Make sure that the metal panels are oriented so that the next panel always overlaps the previously installed one.

83

≈5x 120" Roofing Panels
120x 1-½" Roofing Screws

Make sure your panels overlap properly and fasten the rest of them.

85

≈4x 60" Roofing Panels
≈34x 1-½" Roofing Screws

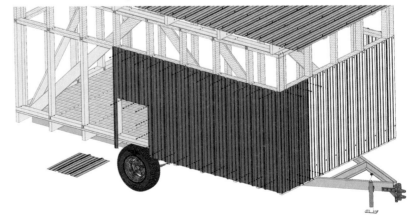

Bend metal around the corners and cut out for the chicken doors.

86

≈4x   60" Roofing Panels
≈32x 1-½" Roofing Screws

Trim panel as needed for the step stow-away opening.

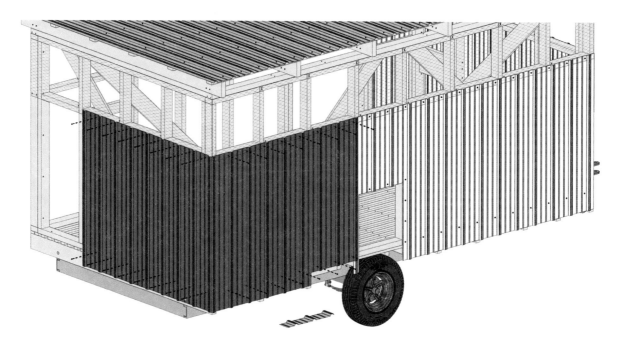

87

≈5x   60" Roofing Panels
≈44x 1-½" Roofing Screws

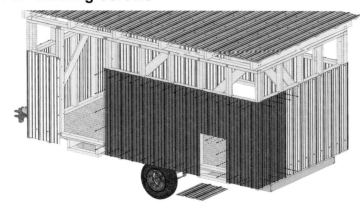

Cut your final piece to fit as you end on the right side of the access door.

88

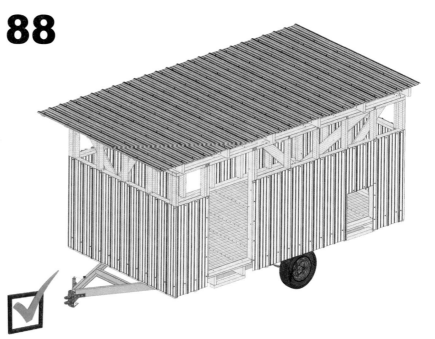

The roof and the walls are complete!

Build An Access Door

89

1x 2x4x35"
1x 2x4x71"
4x 3" Deck Screws

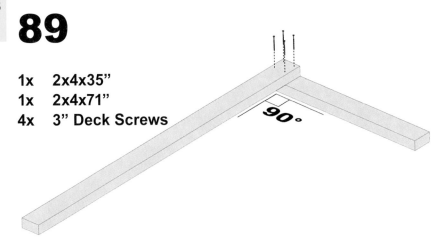

In the following steps we build the door frame. We chose screws because they are easier to work with while trying to keep things aligned and square.

90

1x 2x4x35"
4x 3" Deck Screws

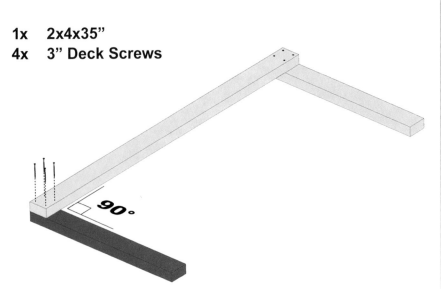

91

1x 2x4x71"
8x 3" Deck Screws

92

2x 2x4x27"
10x 3" Deck Screws

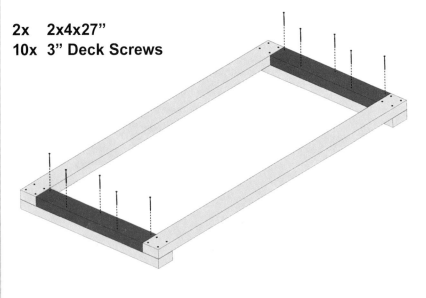

93

3x 4" Strap Hinges
12x 2" Deck Screws

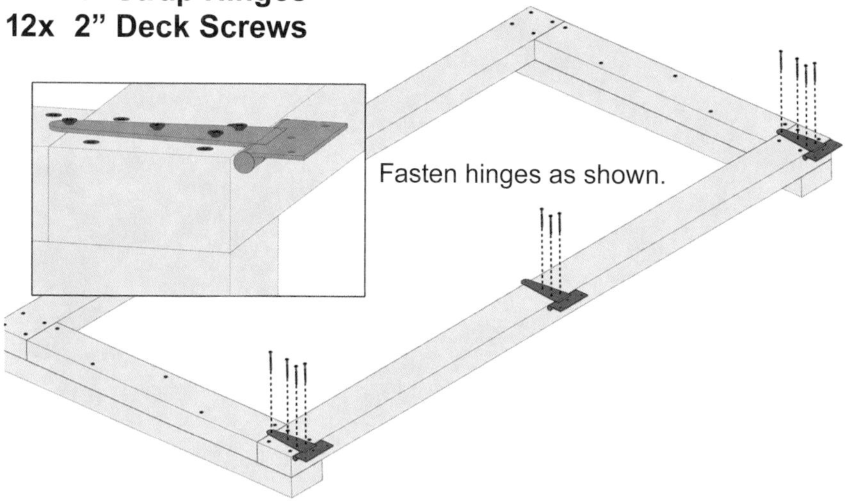

Fasten hinges as shown.

95

6x 3" Deck Screws

94

1x 2x4x72"

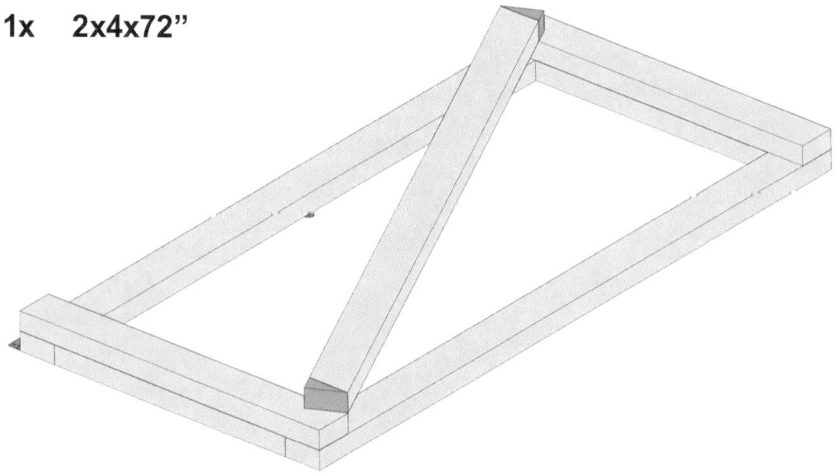

Mark and trim brace to fit. Cut excess off of ends marked in green.

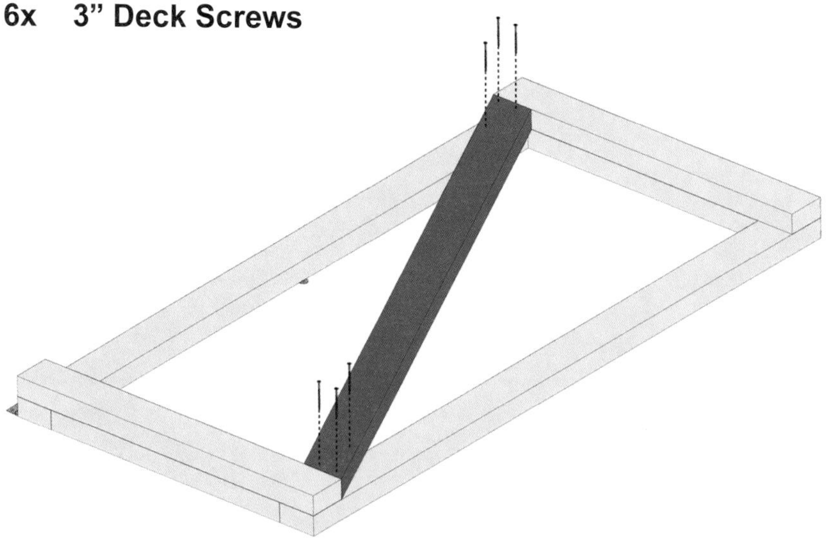

96

1x 2x4x20"
4x 3" Deck Screws

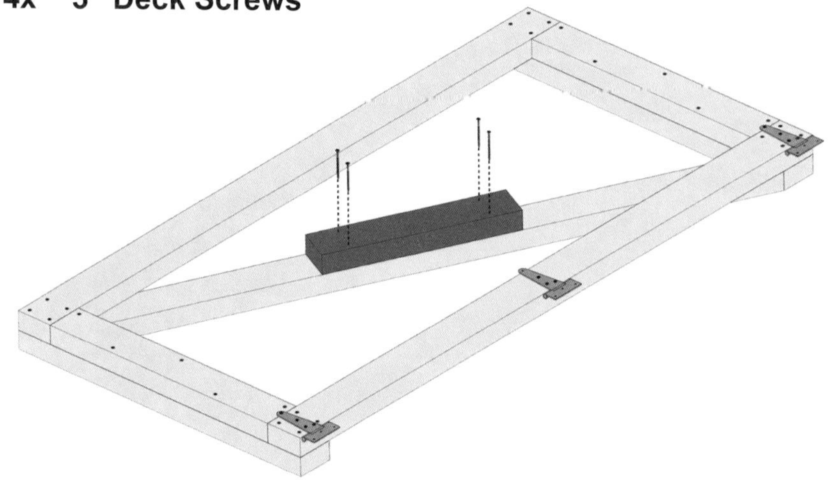

97

1x 79-½" Roofing Panel
17x 1-½" Roofing Screws

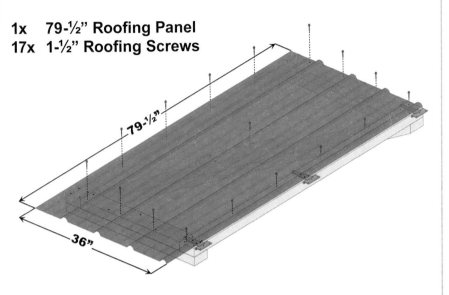

99

1x 1x4x4"
1x 1x4x8"
4x 3" Deck Screws

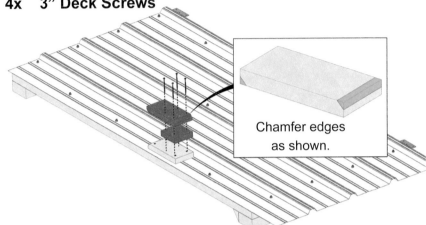

Chamfer edges as shown.

98

4x 1x4x8"
4x 2-½" Deck Screws

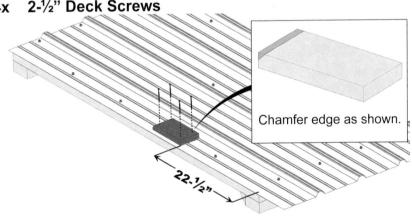

Chamfer edge as shown.

We are constructing the "catch" for the door latch to hold the door shut.

100

9x 2" Deck Screws

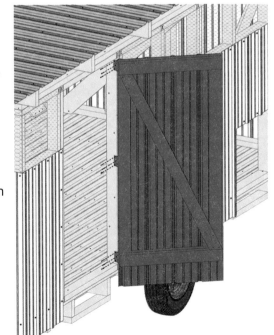

Be sure that gaps are even on the top and bottom. Fasten a few screws and check alignment before fastening the remainder.

101

- 1x 1x4x6"
- 1x 1x3x12"
- 4x 2-½" Deck Screws
- 1x 3" Deck Screws
- 1x ½ x 4" Hex Head Lag
- 3x ½" Washers

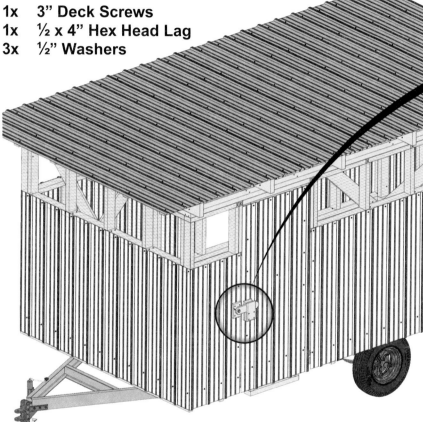

Door Latch Installation

a)

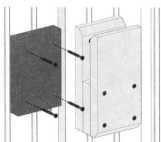

Fasten 1x4x6" block and drill pilot hole for lag screw.

b)

Chamfer edges and drill Ø½" hole as shown.

c)

Do not over tighten. You may need more/less washers to achieve proper clearances.

d)

Add a 3" screw as a stop to hold the latch in the open position.

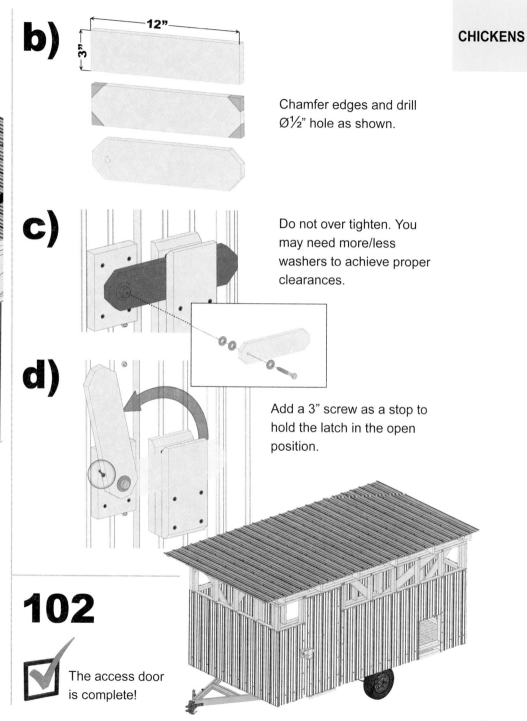

102

✓ The access door is complete!

CHICKENS

Build A Chicken Door

103

2x　1x6x20"
4x　1-⅝" Deck Screws

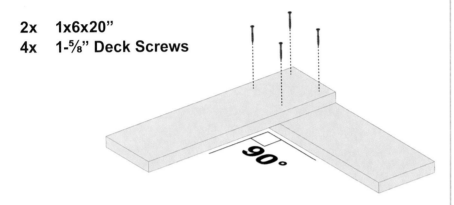

Build the chicken doors.

104

1x　1x6x20"
4x　1-⅝" Deck Screws

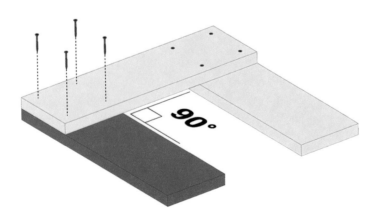

105

1x　1x6x20"
8x　1-⅝" Deck Screws

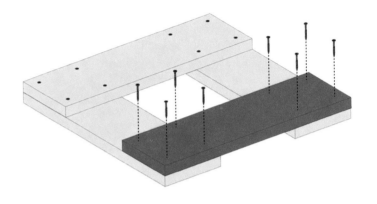

106

2x　4" Strap Hinges
8x　1-⅝" Deck Screws

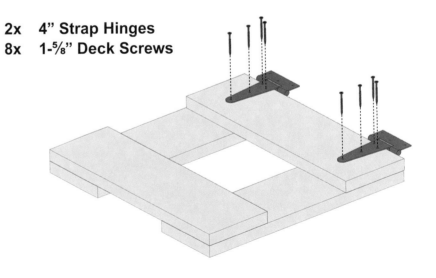

Space hinges as shown.

107

1x   29" Roofing Panel
8x   1-½" Roofing Screws

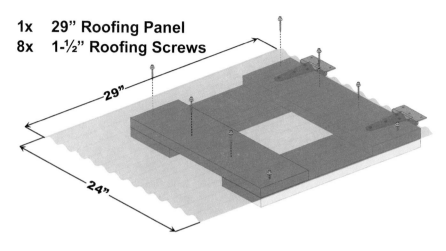

Metal should be flush with hinge side of the frame and centered in the other direction.

109

1x   Bungee Cord

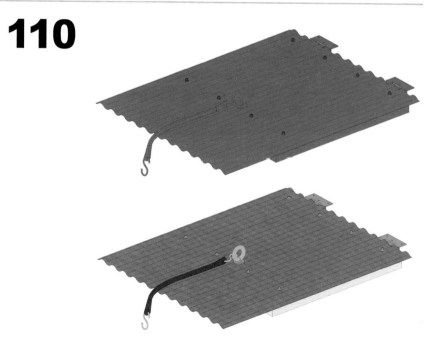

Bend the S-hook closed around the eyebolt.

Insert bungee cord through eyelet and crimp the s-hook closed. Channel-lock pliers work great for this task.

108

1x   ⅜-16 x 2-½" Eyebolt
2x   ⅜" Washers
1x   ⅜-16 Nylock Nut

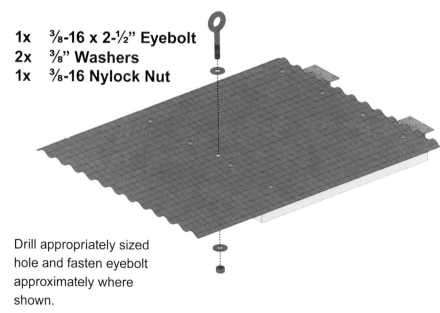

Drill appropriately sized hole and fasten eyebolt approximately where shown.

110

Repeat steps 103-109 to create a second chicken door.

111

6x **2" Deck Screws**
1x **⅜-16 x 2-½" Eyebolt**
2x **⅜" Washers**
1x **⅜-16 Nylock Nut**

 If you find that the bungee is too long, you can shorten it easily by tying a knot in the middle of it.

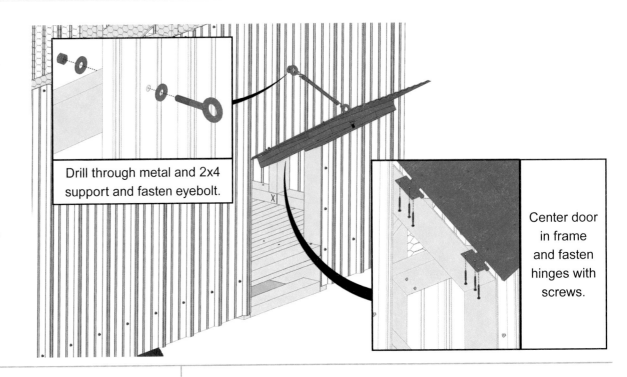

Drill through metal and 2x4 support and fasten eyebolt.

Center door in frame and fasten hinges with screws.

112

1x **⅜ x 2" Lag Screw Eyebolt**

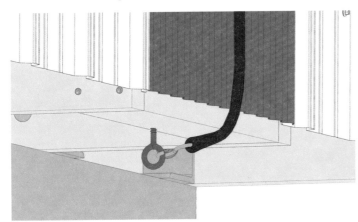

Add an eyebolt lag screw underneath the door. Hook the bungee cord to it to keep the chicken door closed.

113

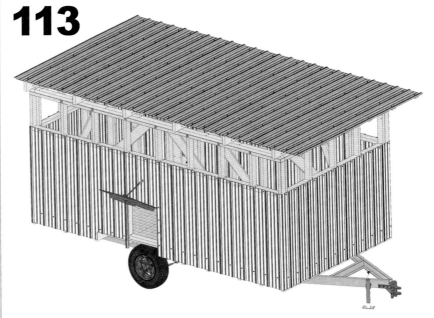

Repeat process on the chicken door on the opposite side.

114

2x 2-½" Deck Screws
1x 1x2x5"

Because the angle braces from steps 70-73 protrude inwards on the walls, it may be necessary to add blocking in certain areas so that the nest boxes have an even surface to mount against. Add blocks as needed to mount the nest box. The location will depend on where you plan to position your nesting boxes.

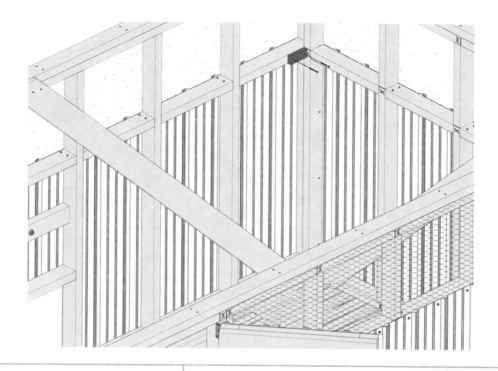

115

1x Nesting Box
2x ⅜ x 1-¾" Hex Head Lag Screws
2x ⅜" Washers

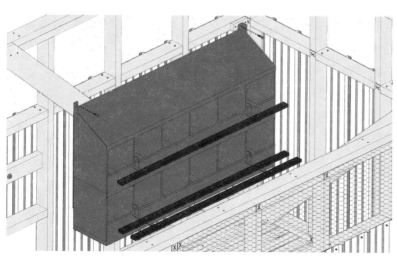

Pre-drill pilot holes for the lag screws.

116

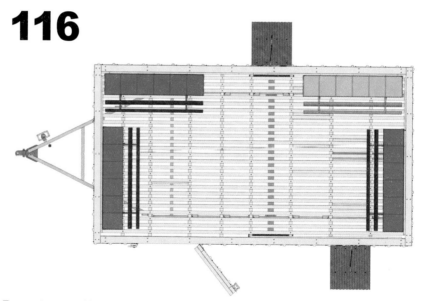

Repeat steps 114 and 115 for 3 other nesting boxes. We arrange ours as shown, but there are numerous options depending on the sizes of the boxes used.

Build A Chicken Ramp

117

2x   1x6x60"

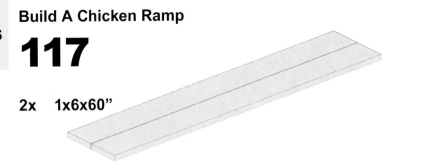

118

7x   1x2x12"
28x   1-⅝" Deck Screws

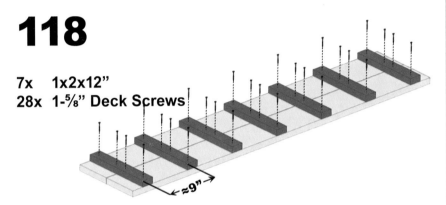

≈9"

Space rungs evenly and be liberal with the application of screws.

119

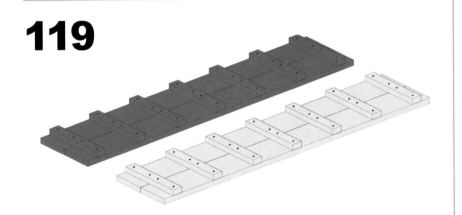

Repeat steps 117 and 118 to create a second ramp.

120

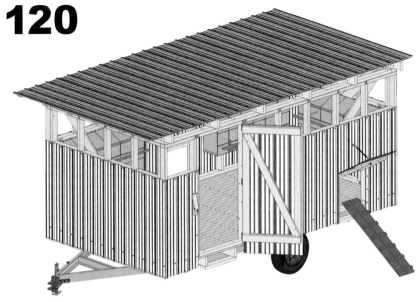

Ramp is in the ready to use position.

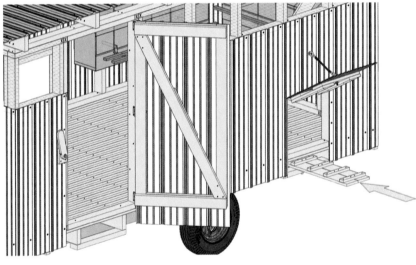

To transport, simply slide ramp into compartment under door. Closing and locking the door will prevent ramp from falling out. Because the axle is in the way on the other side, the compartment could not be directly beneath the door. So, an additional bungee will be needed to prevent the ramp from falling out.

Build A Step Ladder

121

We designed this step for a doorway height of ≈31-½". This measurement may vary on your Eggmobile. Cut your step stringers long and then trim until the top ends nest securely underneath the doorway as shown.

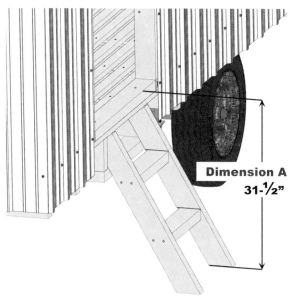

Dimension A
31-½"

122

1x   2x4x34"

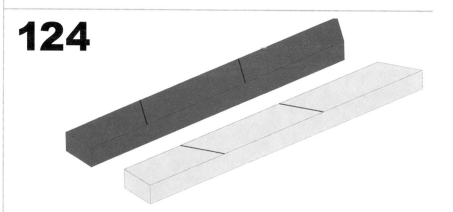

Dimension may vary
33-½"

60°

Miter the foot and remove the area shown in green.

123

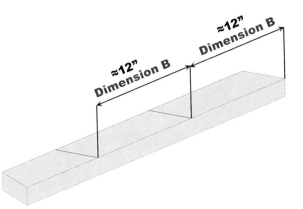

≈12"
Dimension B

≈12"
Dimension B

Mark locations of the step treads. These lines should also be 60° or parallel with the miter you just cut in the previous step.

 If your stringer lengths are different, these 12" step measurements may need to change also. To create evenly spaced step risers, use the following formula:

Dimension A *(from step 121)*   **31.5 ÷ 3 = 10.5**

Dimension B = ⬚ ÷ **Sin(60°)**

Dimension B = **10.5** ÷ **0.866**

Dimension B ≈ **12"**

124

Repeat steps 122 and 123 to create an opposite version of the step stringer.

97

125

2x   2x4x17"
4x   20D Galvanized Nails

Nail step treads to stringers.

126

1x   Stringer From Step 124
4x   20D Galvanized Nails

127

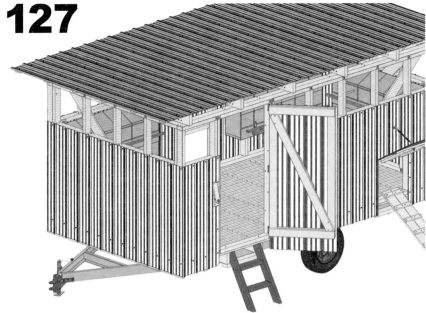

Steps are in the ready to use position.

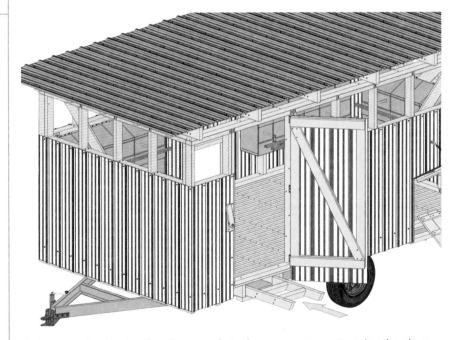

To transport, simply slide the step into the compartment under the door.
Close and lock the door to prevent the step from falling out.

Build A Grain Bin

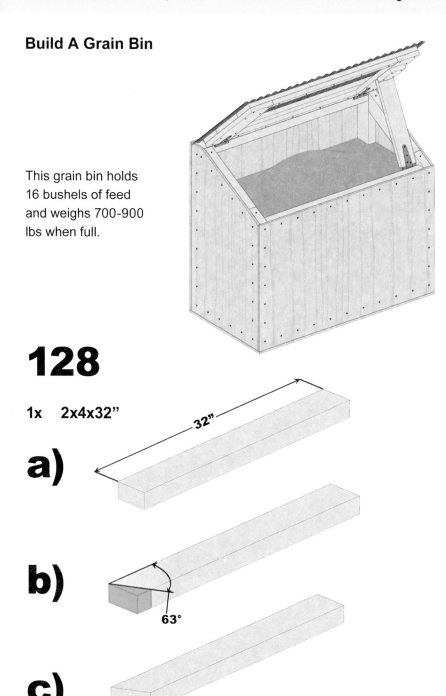

This grain bin holds 16 bushels of feed and weighs 700-900 lbs when full.

128

1x   2x4x32"

a)

32"

b)

63°

c)

129

1x   2x4x16"
1x   2x4x40"
1x   3" Deck Screw

90°

The screw is only there to hold this together temporarily. The strength will come from the plywood siding and a multitude of screws in a later step.

130

1x   2x4 From Step 128
1x   3" Deck Screw

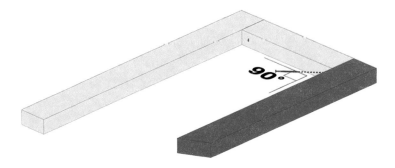

90°

131

1x   2x4x20"

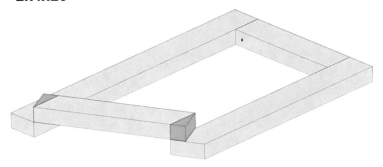

The best way to fit this piece is to place the uncut board on top of where it needs to be and mark where each cut will go. Cut one end and then double check the fit before making the second cut.

133

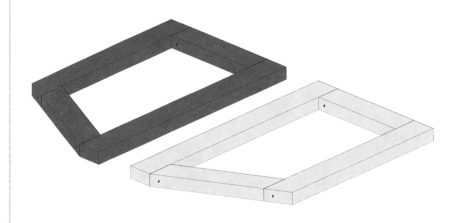

Repeat steps 128-132 to create a second side.

132

2x   3" Deck Screws

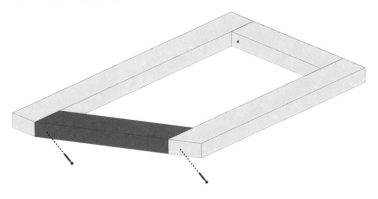

134

1x   2x4x43"
2x   3" Deck Screws

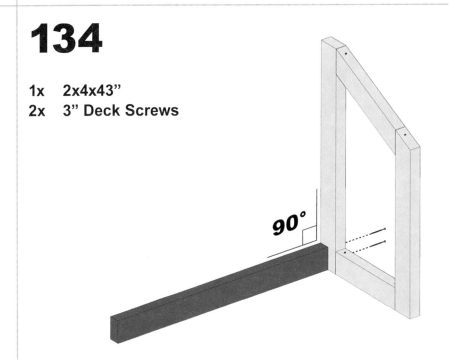

135

2x   3" Deck Screws

90°

137

1x   ½" Plywood 24 x 43"
25x 2" Deck Screws

It is best to cut
this piece of
plywood to fit.
Hold the plywood
against the side
wall and trace it
with a pencil. Cut
accordingly and
fasten every 4-5"
with screws.

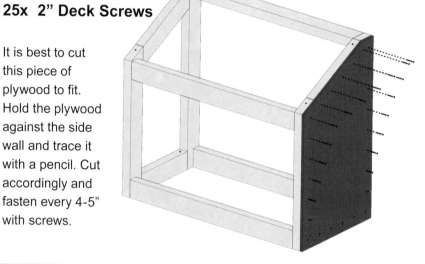

136

3x   2x4x43"
12x 3" Deck Screws

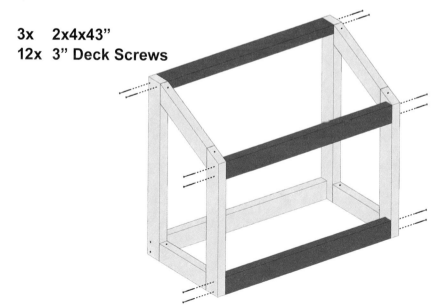

138

1x   ½" Plywood 24 x 43"
25x 2" Deck Screws

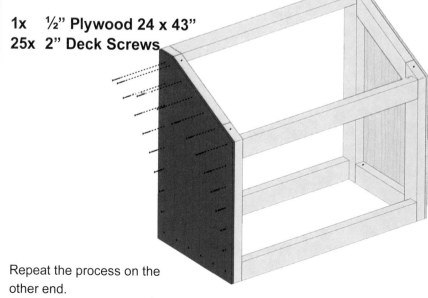

Repeat the process on the
other end.

139

1x **½" Plywood 48 x 30"**
32x 2" Deck Screws

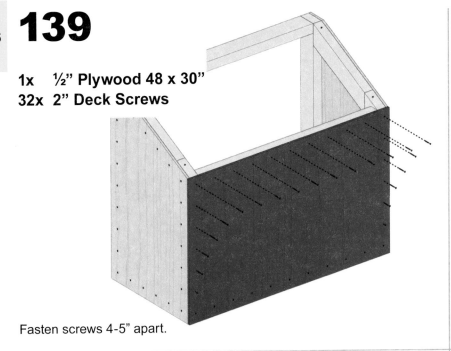

Fasten screws 4-5" apart.

141

1x **½" Plywood 48 x 25"**

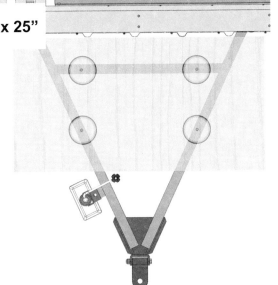

Position the plywood base on the tongue of the trailer and drill holes in 4 locations through the plywood and c-channel flange. Holes should be large enough to accommodate ½" hardware.

140

1x **½" Plywood 48 x 40"**
36x 2" Deck Screws

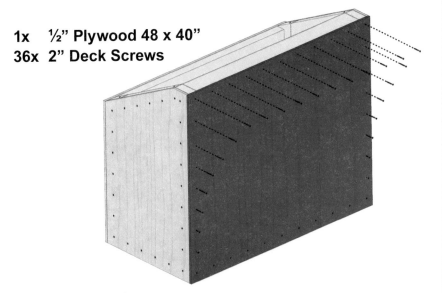

Fasten screws 4-5" apart.

142

44x 2" Deck Screws

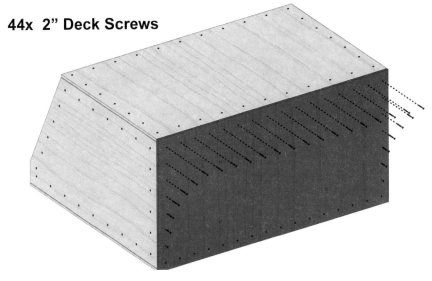

Fasten screws 3" apart.

143

1x 1x4x19-½"
1x 4" Strap Hinge
7x 2" Deck Screws

Install a hinged prop stick for the lid. This is an optional luxury, but definitely a convenience worth the time.

Fasten hinge to the inside of the grain bin as shown.

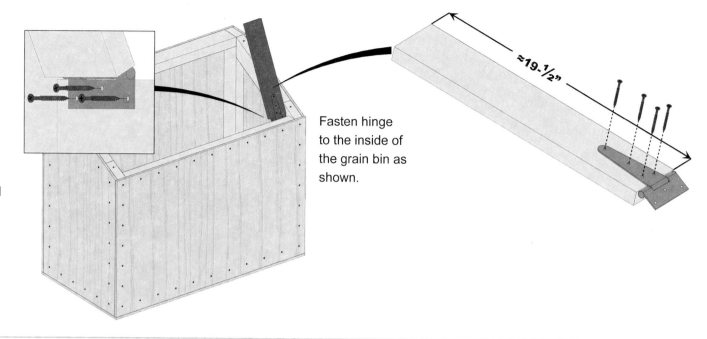

≈19-½"

144

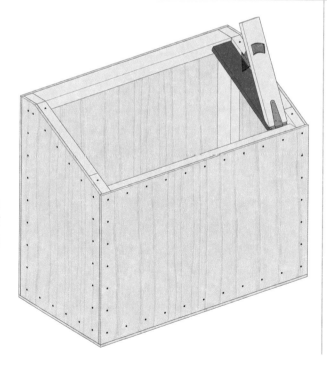

Prop stick should fold in and be flush with the opening of the bin. Trim it to fit as necessary.

145

1x 1x6x22"
1x 1x6x46"
4x 2" Deck Screws

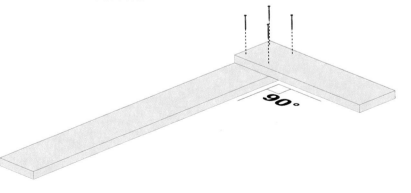

90°

Construct the frame for the lid.

146

1x 1x6x22"
4x 2" Deck Screws

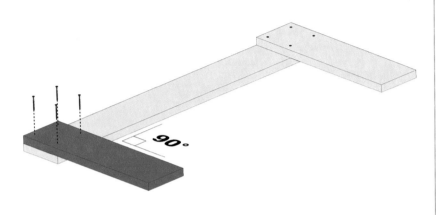

148

1x 1x2x34"
10x 2" Deck Screws

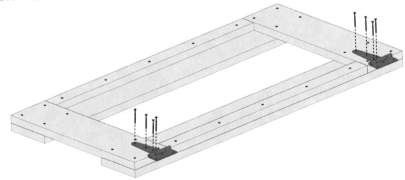

These pieces are fillers only. You may use scrap lumber to fill these spots.

147

1x 1x6x46"
8x 2" Deck Screws

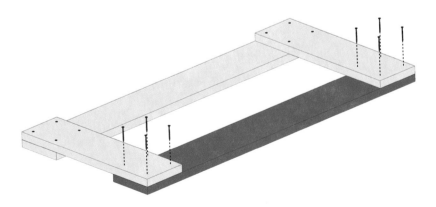

149

2x 4" Strap Hinges
8x 2" Deck Screws

Install the hinges approximately where shown, keeping them 2-3" away from the outside edges.

150

2x 1x2x10"
4x 1-⅝" Deck Screws

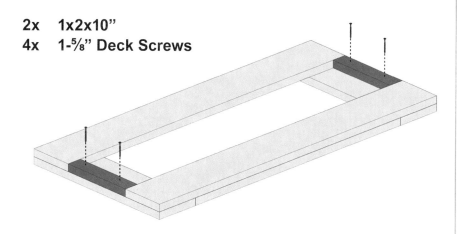

These pieces are fillers only. You may use scrap lumber to fill these spots.

152

2x 25" Roofing Panels
20x 1-½" Roofing Screws

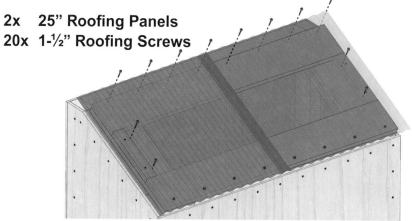

 If you mount the metal too high on the lid it will impede how far the lid will open. Check clearances prior to fastening all of the screws.

151

6x 2" Deck Screws

Attach hinges as shown.

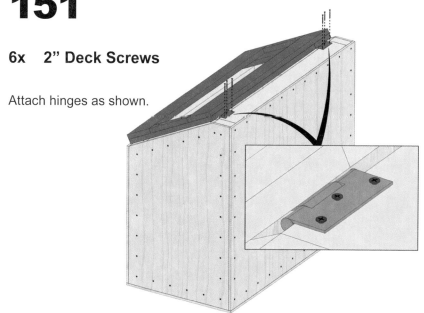

153

2x 1x4x6"
4x 2" Deck Screws

With the lid and prop open to the desired location, fasten the block to the lid as shown. The prop should wedge against the block and hold the lid open.

CHICKENS

154

4x ½-13 x 2" Carriage Bolts
4x ½" Washers
4x ½-13 Hex Nuts

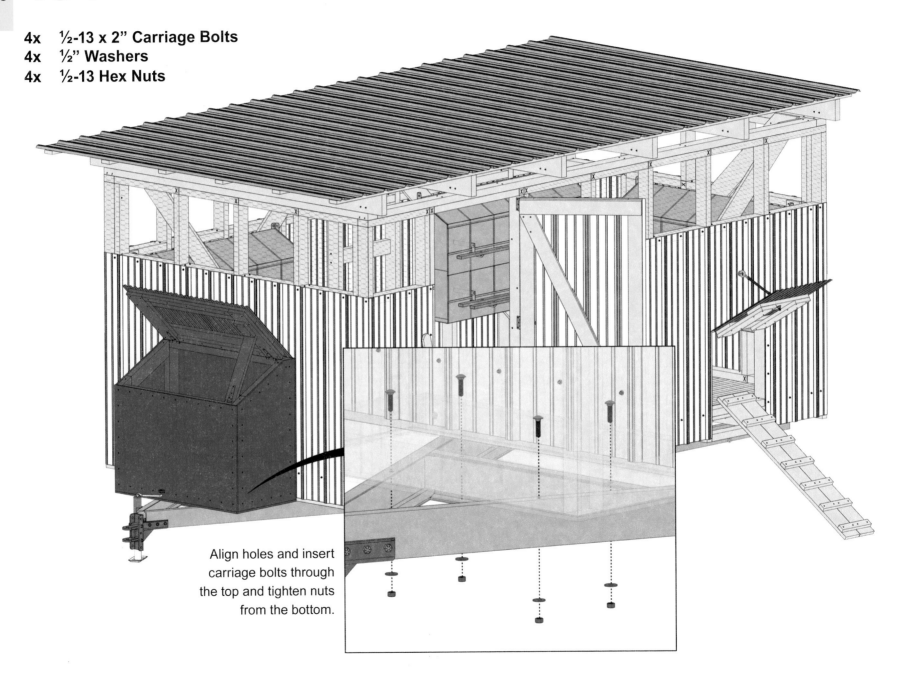

Align holes and insert
carriage bolts through
the top and tighten nuts
from the bottom.

155

2x   1x4x10"
1x   Hose Reel And Mounting Hardware
8x   2-½" Deck Screws

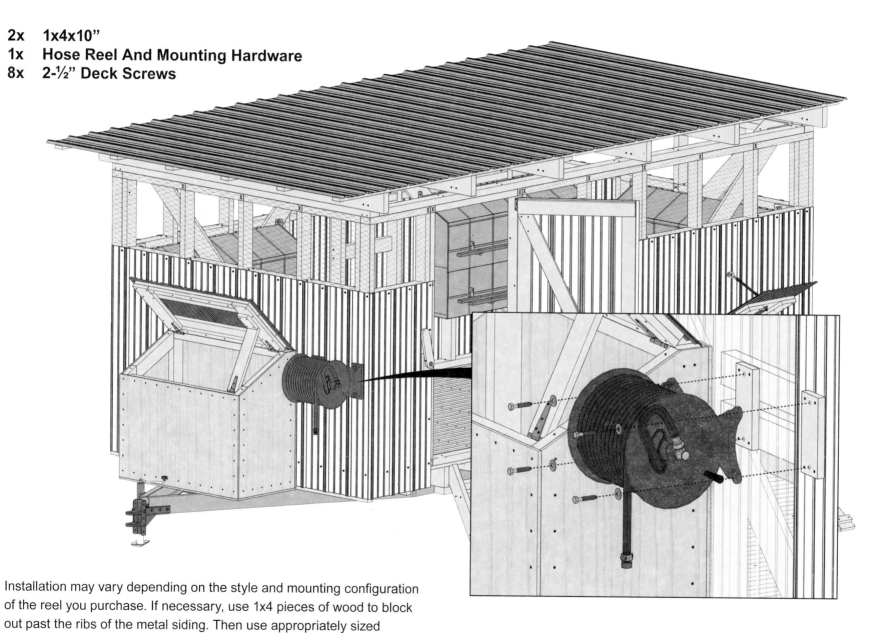

Installation may vary depending on the style and mounting configuration of the reel you purchase. If necessary, use 1x4 pieces of wood to block out past the ribs of the metal siding. Then use appropriately sized fasteners to secure the reel to the wall. We prefer lag screws but the size will depend on the holes in your reel mounting bracket.

156

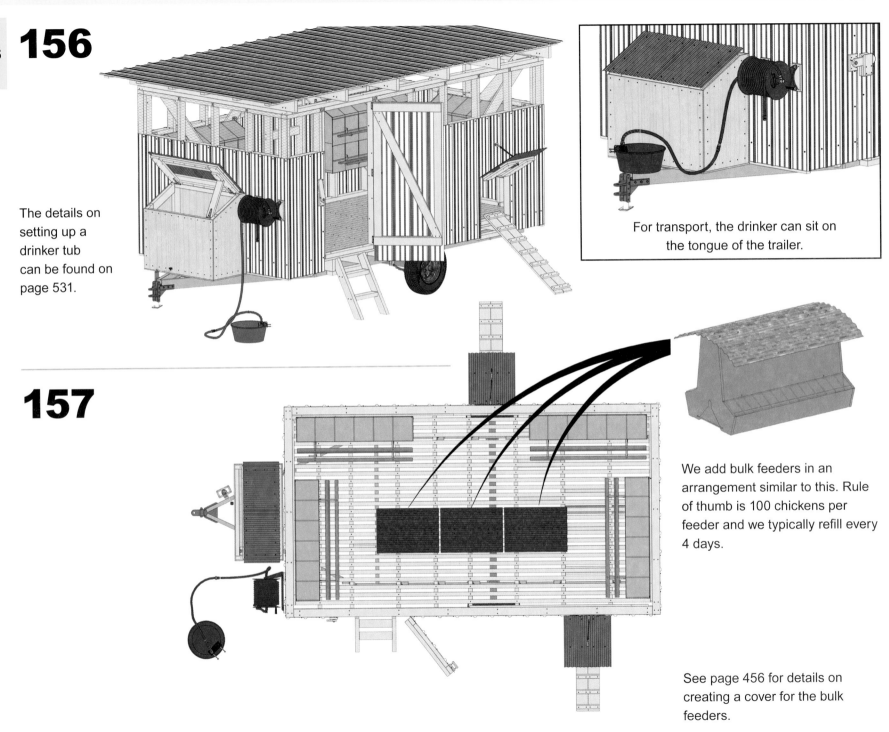

The details on setting up a drinker tub can be found on page 531.

For transport, the drinker can sit on the tongue of the trailer.

157

We add bulk feeders in an arrangement similar to this. Rule of thumb is 100 chickens per feeder and we typically refill every 4 days.

See page 456 for details on creating a cover for the bulk feeders.

158

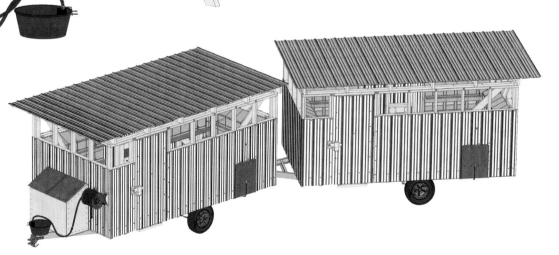

The Eggmobile is complete!

Eggmobile Train

At Polyface we typically hook two Eggmobiles together. When doing so, the rear Eggmobile does not have a hose reel or grain bin. Also note that there is minimal roof overhang off of the front of the second Eggmobile. It is not necessary to protect a grain bin because there is none, and it would create clearance issues between the two units during tight turns.

photo courtesy of Jessa Howdyshell

Named after the Millennium Falcon in Star Wars, the Millennium Feathernet is a mobile laying hen shelter for pastured eggs. It is a two-part system, one being the shelter and the other being the electric poultry netting surrounding it.

It elicits three common questions:

1. Why not just use Eggmobiles?
2. Why surround the birds with the electric netting?
3. Why not use a hoop structure design?

Let's answer these one at a time. The Eggmobile's purpose is pasture sanitation behind the cows. That means we need to move it long distances, sometimes even down the public road, which requires rubber tires. That makes it vulnerable to rolling and even flipping in a hurricane force wind. Because we want the chickens to range as far as possible, we don't want to restrict them in any way.

The netting protects the birds from wandering too far away where they can either get lost or get picked off by predators. The Millennium Feathernet is land intensive rather than land extensive. Because the birds are controlled, it can be utilized around gardens or in populated areas where trespass could be a problem. Predation is definitely a bigger problem with the Eggmobile than with the Millennium Feathernet.

Because the birds are always contained, we don't have to close them up at night, which is a huge savings in labor. Furthermore, we can move the structure and these birds to a new paddock any time of day rather than only first thing in the morning, which the Eggmobile requires. This changes the move from a chore to a simple activity and that's a game changer. A chore is something that has to be done every day at the same time. Chores are the burnout accelerator, so anything we can do to decrease chores is a good thing.

If we let the birds completely free range, they would pre-soil the new area prior to exposure, which would reduce bird sanitation. By controlling access, we can provide completely rested, pathogen-free pasture for the birds when we move them. We move every 3 days rather than daily or every other day like the Eggmobile.

Now to the structure itself. When we first ramped up egg production, we simply retrofitted the broiler shelters with 10 nest boxes tucked inside the back end. We put 50 layers in there and thought we were really slick. Then Michael Plane of Allsun Farm visited us from Australia and told us we were obsolete. He introduced us to the electric netting from Premier and, though he hurt my pride, I realized instantly that he was right.

But what kind of structure was best to pair up with this newfangled electric netting? My mind went first to a hoop structure. They're cheap and light, so we built one and the first day in the field a wind came along and blew it to smithereens. Back to the drawing board. We beefed it up and tried again, but the problem remained that a hoop house has no structural integrity up high. Everything is in the foundation.

For a portable structure, that means braces and cross members down close to the ground. All of this gingerbread down on or close to the ground creates things that people can trip on and things that can catch a chicken and kill her if she doesn't move as fast as you're moving the structure. Not good on either count.

Eventually, Daniel conceived of the current design, which is essentially an X truss supporting an A-frame. The beauty of this design is that all the structural integrity is in the roof, moving the bracing and strength up away from the ground. That means the only things close to the ground are the two pipe skids, and they move parallel to the motion. In other words, if a chicken doesn't move quickly, the whole structure moves over her; nothing is perpendicular to the direction of movement. This design virtually eliminates inadvertently running over chickens.

One of the trickiest balances in portable infrastructure is making it heavy enough and strong enough to handle winds and uneven ground, but not too heavy to move easily with a relatively small tractor (40 horsepower). Although the Millennium Feathernet design is heavier than a hoop structure, it's still light enough to move. We've had it withstand 80 mile per hour winds; portable hoop structures don't, unless you stake them down. Staking adds lots of time and cumbersomeness to the move.

Every time we make a move more cumbersome, it's less efficient and fun to do the move. The more arduous the task, the less liable we are to do it as frequently as it needs to be done. So we let the structure sit too long, then we get a dirt yard, then we get sick chickens and the whole things breaks down. I've seen that on countless farms around the world.

The worst was in Spain on a certified organic egg operation. The mobile houses were so heavy and hard to move that the farmer only

photo courtesy of Chris Slattery

moved them twice a year. These houses had a foot of raw manure underneath and completely denuded ground out 50 yards away. The secret of everything in pastured livestock is frequent movement.

This design is strong. I've had 40 people stand up in the catwalk. I've pulled it around the field in circles, over ditches, through the creek. Nothing phases it. And unlike hoop structures, it's easy to build with wood, which can either be scavenged or purchased from a lumber yard or, in our case, milled on our band saw mill. Our oldest structure is now 15 years old and still going strong; since no wood touches the ground, it's no more weather-vulnerable than wood in a house.

The only negative thing I will say is that it is definitely a fair weather structure. While we could certainly put some sort of cap or wall on one end, it's made to be cool in the summer.

CHICKENS

Since we put our birds in real honest to goodness hoop houses for winter, we aren't trying to make a portable structure that can handle extreme heat in the summer and blizzards in the winter. We get both.

From a design standpoint, sometimes it's easier to have two different structures rather than trying to make one that can handle all weather conditions. If you design for the most outrageous extreme of any weather condition, you'll have so many bells and whistles that it's too complicated to build. Seasonal infrastructure is great for most of the time and if you can get efficiency and functionality easily and cheaply for the majority of the time, that's good enough. Designing for every contingency imaginable is too hard.

The final thing I'll point out is the importance of the tool bar, or spreader, out front. If you simply attach your pull chain to the two skids, it puts a lot of inward pressure to pigeon-toe the structure and rip it apart. By using the spreader pipe out front, that pipe takes all that inward pressure rather than the structure.

This one is designed for 1,000 chickens but certainly it can be made smaller for smaller flocks. We place one nest box per ten hens and make sure they're higher than eye height to reduce loitering and mischief. We found one beneficial surprise with this design over the hoop house: because the chickens have to walk up their chicken ladder and then along the catwalk to get to the nest boxes, the distance from ground to nest box dried out their feet, resulting in much cleaner eggs. When the nest boxes are simply a jump up

photo courtesy of Jessa Howdyshell

from the ground, chickens carry soil moisture and debris on their feet into the nest box, resulting in dirtier eggs.

We don't use roll-away nest boxes because they keep a hen from enjoying her nest building instincts. Every hen comes into the nest and moves things around as she settles in to lay her egg. We think it honors the henness of the hen to let her express her nest building instincts. Just a little mysticism thrown in for good measure to create balance with the engineers among us.

Because the structure gets stressed different ways as we move it, we use bolts on the basic skeleton. Nails or screws eventually work loose on something like this. Although it can't be moved down a public road or routine long distances

because it's on skis, the advantage is that it can't roll and you can never have a flat tire. Its low center of gravity allows you to park it on a hill anywhere and it sits.

The roosts inside and the air flow through it make this an extremely comfortable structure for summer heat. The solid roof protects them from rain; wet hens aren't happy.

This structure can be used for other poultry as well, like turkeys or ducks. When we pick up the birds to bring them into the hoop houses for the winter, we simply move the Millennium Feathernet out of the fencing before dark and then catch the birds once they settle for night. We turn it into a party, enjoying chocolate chip cookies and raw milk when we're done. Enjoy.

Tools

- ❏ Power Drill

- ❏ 8 or 10ft ladder

- ❏ Table Saw

This project requires a tractor or hydraulic loader. We tailored the assembly process to minimize the tractor time needed. If you do not own a tractor, borrowing one is a plausible option because it won't take long to set the prebuilt wooden trusses on top of the welded steel skids.

- ❏ Welder

- ❏ Circular Saw

- ❏ Grinding Wheel

- ❏ Cut-Off Wheel

- ❏ Angle Grinder

- ❏ Socket Driver Set

- ❏ Hand Saw

- ❏ String Line

- ❏ Marking Pencil

- ❏ Tape Measure

- ❏ Speed Square

- ❏ Hammer

- ❏ Driver Bits (for screws)

- ❏ ½" HSS Drill Bit

- ❏ ½" Auger Bit

- ❏ 4ft Level

CHICKENS

Hardware

DECK SCREWS

- ☐ **63x**   1-⅝"

- ☐ **32x**   2"

- ☐ **≈210x**   3"

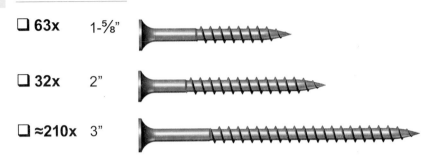

NAILS

- ☐ **≈600x**   20D Galvanized Nails*

- ☐ **4x**   8D Framing Nails

- ☐ **≈300x**   8D Ribbed Nails*

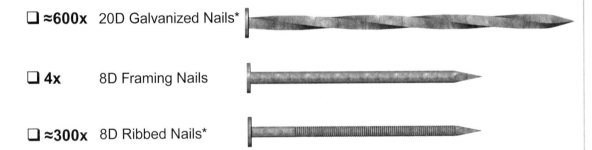

It is recommended that these nails be either spiraled or ribbed for superior holding power.

ROOFING SCREWS

- ☐ **672x**   1-½" Roofing Screws

- ☐ **≈50x**   2-½" Roofing Screws

- ☐ **≈35ft**   ¼" High Test Chain

- ☐ **2x**   ⁵⁄₁₆" Threaded Quick Link Connectors

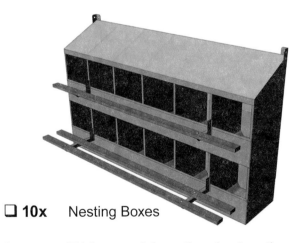

- ☐ **10x**   Nesting Boxes

See page 461 for more information about nesting box selection and setup.

CHICKENS

BOLTS

☐ **15x** ½-13 x 5"
Carriage Bolts

☐ **20x** ½-13 x 6"
Carriage Bolts

☐ **10x** ½-13 x 7"
Carriage Bolts

☐ **2x** ½-13 x 6"
Grade 8 Hex Bolts

☐ **20x** ⅜ x 1-¾"
Hex Head Lag Screws

☐ **49x** ½" Washers

☐ **2x** ½-13 Nylock Nuts

☐ **45x** ½-13 Hex Nuts

☐ **20x** ⅜" Washers

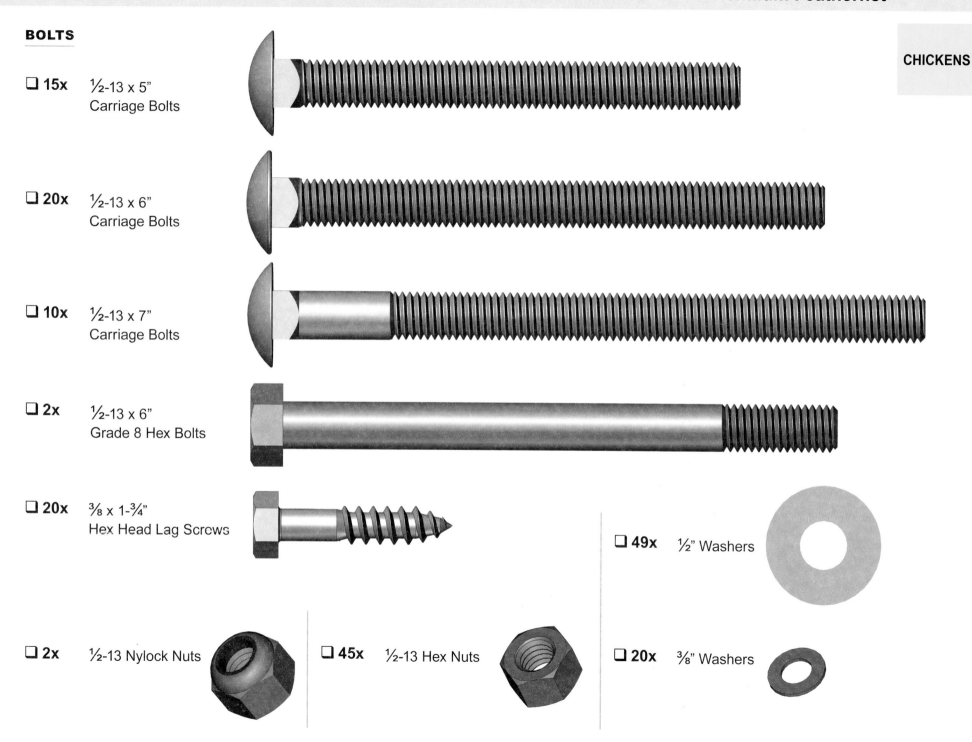

CHICKENS

Wood Cutlist

Wood scraps are highlighted in red.
Do not discard until project is completed.

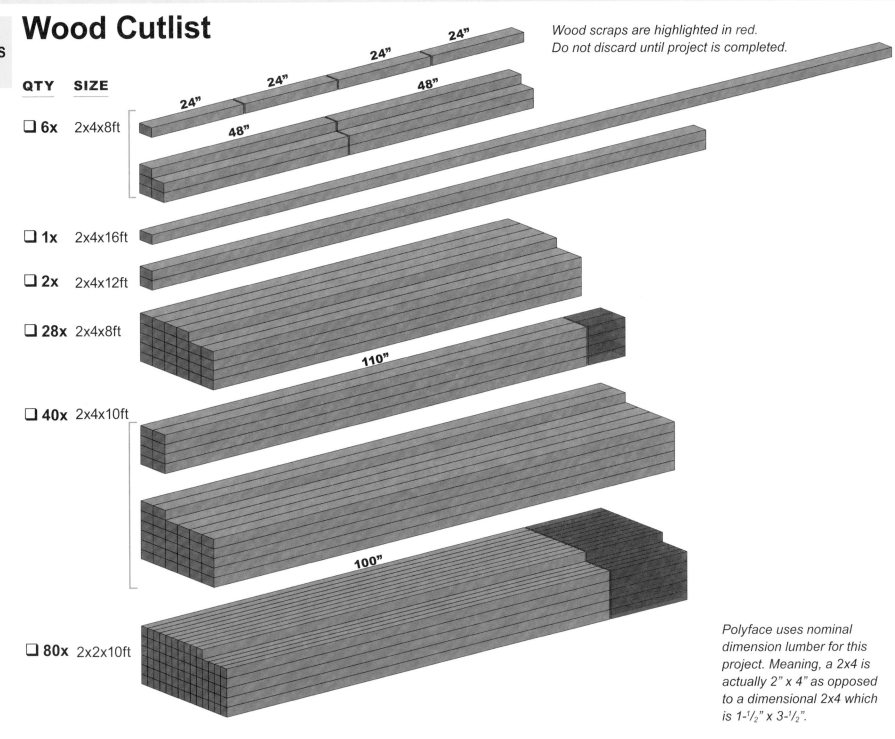

QTY	SIZE
☐ 6x	2x4x8ft
☐ 1x	2x4x16ft
☐ 2x	2x4x12ft
☐ 28x	2x4x8ft
☐ 40x	2x4x10ft
☐ 80x	2x2x10ft

24" 24" 24"
24" 48"
24"
48"

110"

100"

Polyface uses nominal dimension lumber for this project. Meaning, a 2x4 is actually 2" x 4" as opposed to a dimensional 2x4 which is 1-$\frac{1}{2}$" x 3-$\frac{1}{2}$".

CHICKENS

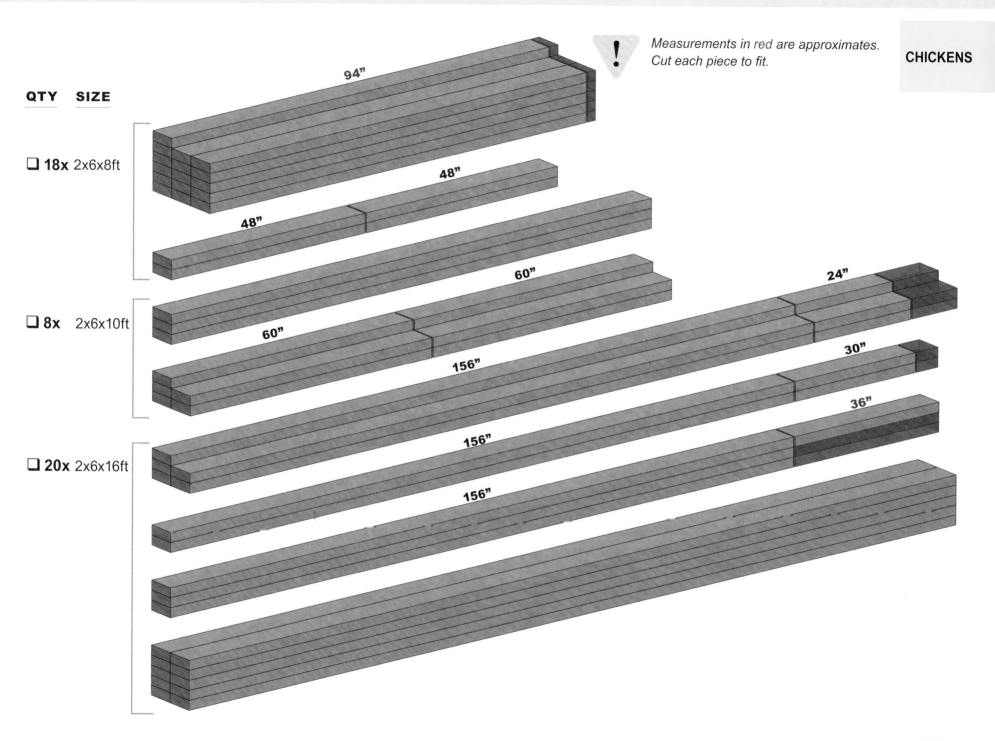

Measurements in *red* are approximates.
Cut each piece to fit.

QTY   SIZE

☐ **18x** 2x6x8ft

☐ **8x** 2x6x10ft

☐ **20x** 2x6x16ft

94"
48"
48"
60"
60"
24"
156"
30"
156"
36"
156"

CHICKENS

Wood Cutlist

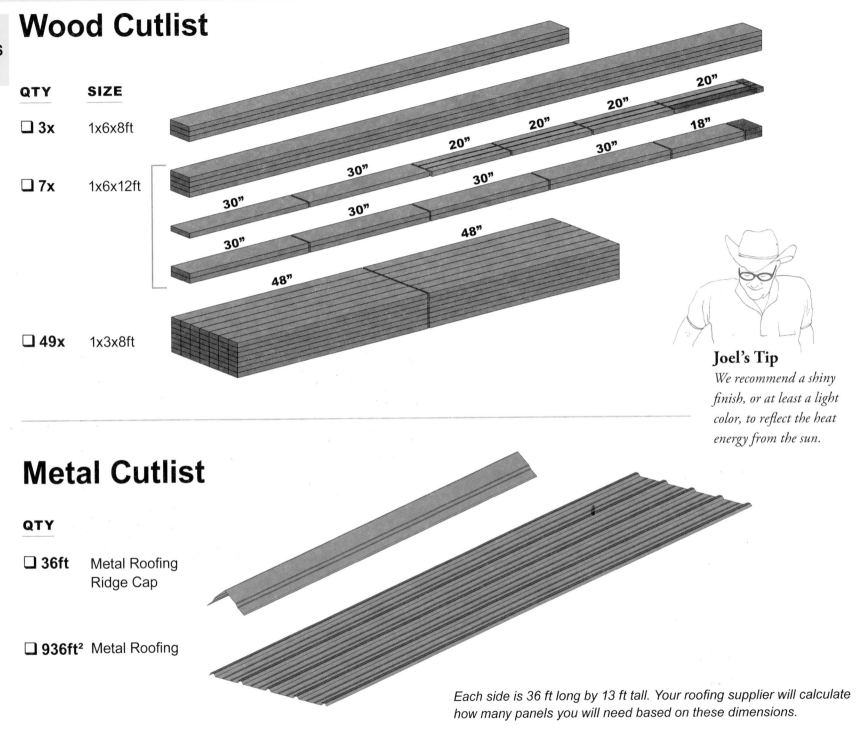

QTY	SIZE
❑ 3x	1x6x8ft
❑ 7x	1x6x12ft
❑ 49x	1x3x8ft

Measurements shown on boards: 20", 20", 20", 18", 20", 30", 30", 30", 30", 30", 30", 48", 48"

Joel's Tip

We recommend a shiny finish, or at least a light color, to reflect the heat energy from the sun.

Metal Cutlist

QTY	
❑ 36ft	Metal Roofing Ridge Cap
❑ 936ft²	Metal Roofing

Each side is 36 ft long by 13 ft tall. Your roofing supplier will calculate how many panels you will need based on these dimensions.

QTY

☐ **40"** C4x7.25 C-Channel

☐ **36"** ¼x3-½" Bar

☐ **64"** 8x4x½" Angle

☐ **2x** 13ft x 3" SCH40 Pipe

☐ **1x** 16ft x 3" SCH40 Pipe

☐ **2x** 20ft x 3" SCH40 Pipe

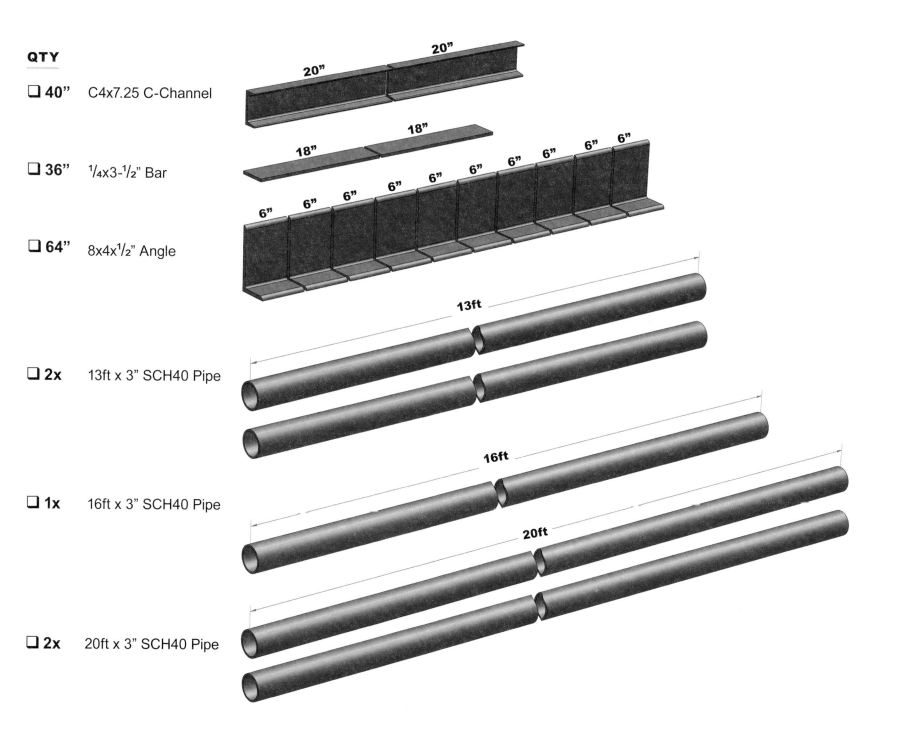

CHICKENS

1

1x   2x6x156"

2

Miter both ends as shown.

54°

54°

3

Draw a line that is parallel with the cuts you just made (54°). Mark both sides.

100"

4

Draw a line that is parallel with the cuts you just made (54°). Mark both sides.

45-¹/₂"

5

Repeat steps 1-4 to create a second side.

6

Stack the two pieces together in the orientation shown, and mark the top 2" wide portion every 24".

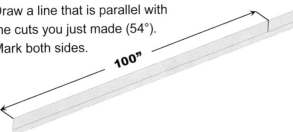

24" **24"** **24"** **24"** **24"** **24"**

These marks will help you align and nail purlins onto the trusses later, as well as with the application of roofing metal.

7

1x   3" Deck Screw

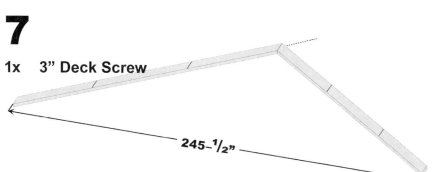

Position the boards so that they meet at the top as shown. The measurement across the bottom should be approximately 245-$\frac{1}{2}$". Drive a screw into the top of the joint just to hold it in place.

8

4x   8D Ribbed Nails
1x   1x6x30"

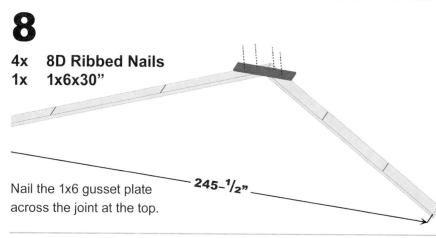

Nail the 1x6 gusset plate across the joint at the top.

9

2x   2" Deck Screws
1x   1x6x144"

This spreader board will only be here temporarily. Its purpose is to maintain the shape of the truss while it is being built.

10

1x   2x6x60"
1x   3" Deck Screw

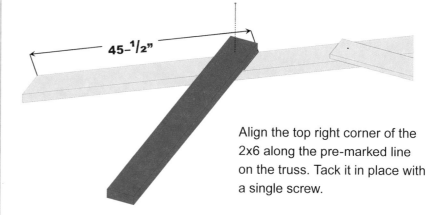

Align the top right corner of the 2x6 along the pre-marked line on the truss. Tack it in place with a single screw.

11

1x   2x6x60"
1x   3" Deck Screw

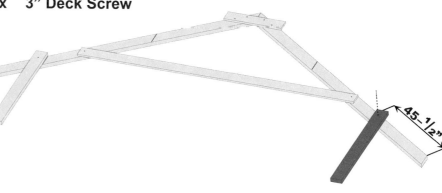

Repeat step 10 on other side.

CHICKENS | **12**

1x   3" Deck Screw
1x   2x6x192"

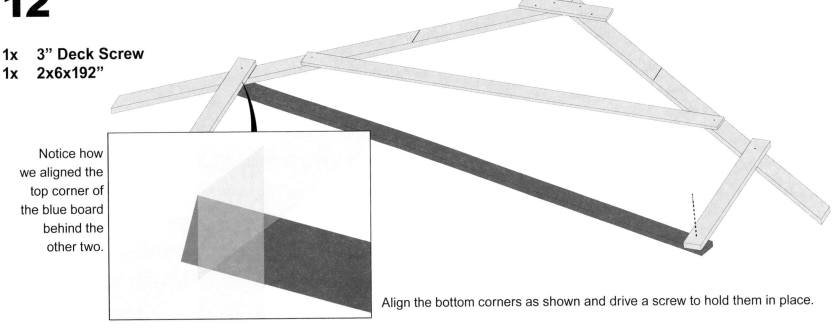

Notice how we aligned the top corner of the blue board behind the other two.

Align the bottom corners as shown and drive a screw to hold them in place.

13

To square this portion of the truss, we need to use some elementary trigonometry. The shaded yellow section in this diagram forms a "right triangle". Therefore, we can apply the Pythagorean Theorem to square our truss

$$A^2 + B^2 = C^2$$

Since we can physically measure the length of the distances marked "A", and "B", we can determine the required value of "C".

If you can't remember how to solve these types of equations don't worry. There are dozens of phone apps and websites on the internet that will solve for these values. Simply type in the corresponding values for "A" and "B" and you will be on your way!

14

1x   3" Deck Screw

With the value of "C" calculated, adjust the cross brace accordingly and screw the board as shown above to hold it in place.

15

1x   3" Deck Screw
1x   2x6x192"

Repeat the process starting at step 12 on the other side.

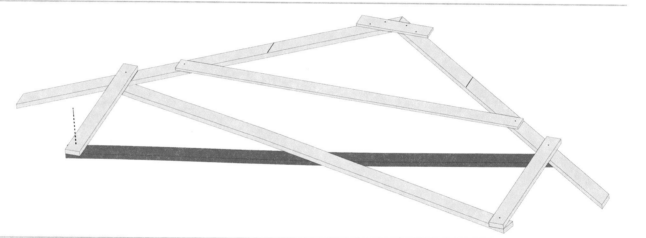

16

This measurement should equal the value of "C" from step 13. Adjust the position of the cross brace accordingly.

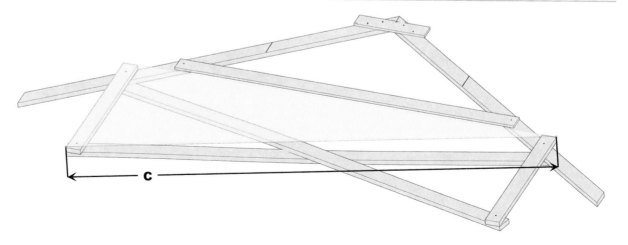

17

1x   3" Deck Screw

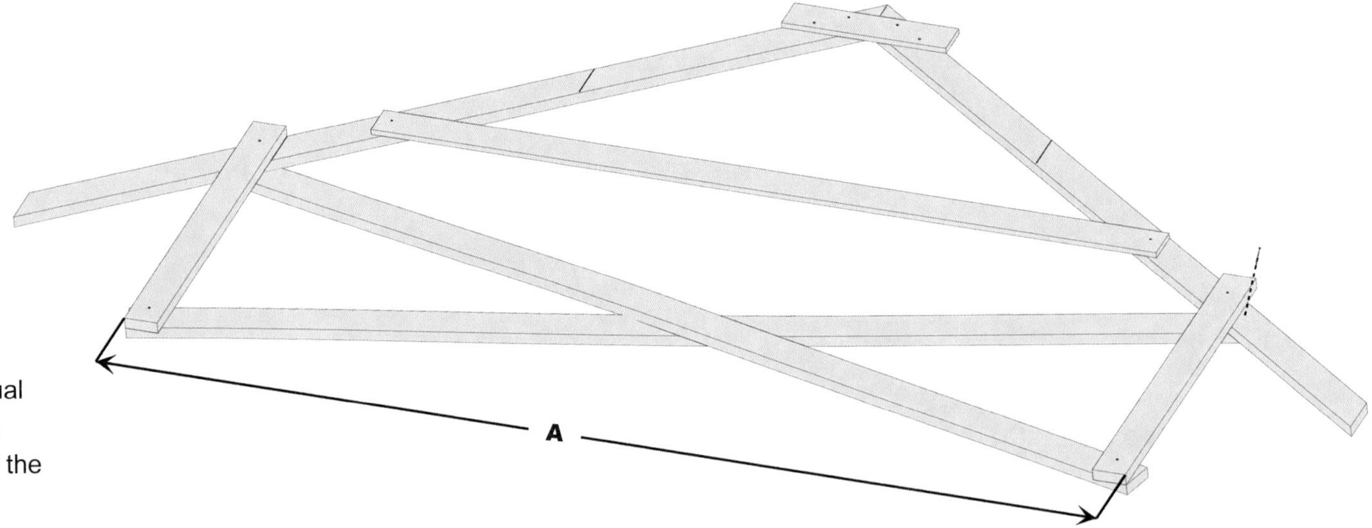

As a check, this measurement should equal the value of "A" from step 13. If it does, then screw the cross brace in place.

A

18

6x   20D Galvanized Nails

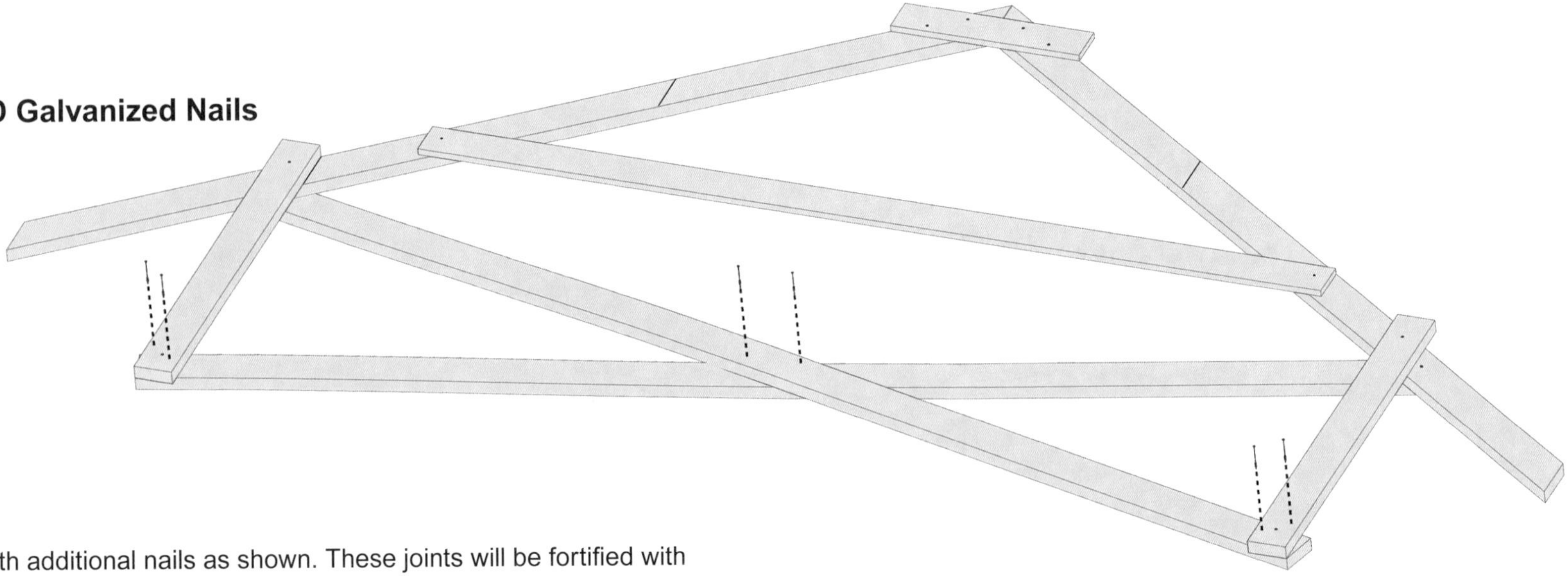

Secure with additional nails as shown. These joints will be fortified with bolts in the following steps.

19

2x ½-13 x 7" Carriage Bolts
2x ½" Washers
2x ½-13 Hex Nuts

Drill holes through all three boards with an auger drill bit. Insert the bolts and fasten with washers and nuts.

Note: It is not uncommon to need a hammer to drive the carriage bolts into the holes. Be sure that the head of the bolt is seated snugly against the wood before attempting to tighten the nuts.

20

1x ½-13 x 5" Carriage Bolt
1x ½" Washer
1x ½-13 Hex Nut

Repeat the process and fasten the center of the cross braces. This is a key joint so do not forget it!

21

1x   **1x6x30"**
2x   **½-13 x 5" Carriage Bolts**
2x   **½" Washers**
2x   **½-13 Hex Nuts**

Place another 1x6 gusset
plate underneath the truss
at the top joint. Drill through
all three pieces and secure
bolts where shown.

*Note: We did not bolt the bottom two joints of the truss. These joints will
be drilled and bolted at the time of installation onto the metal skids.*

22

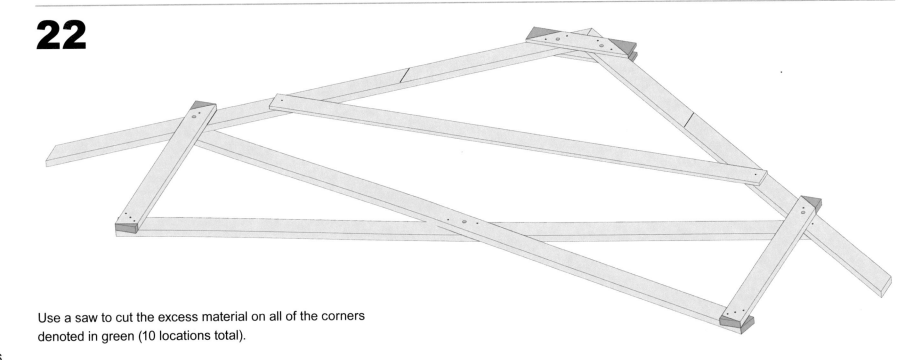

Use a saw to cut the excess material on all of the corners
denoted in green (10 locations total).

23

It is time to remove the temporary spreader. Set it aside for use on the remaining trusses.

24

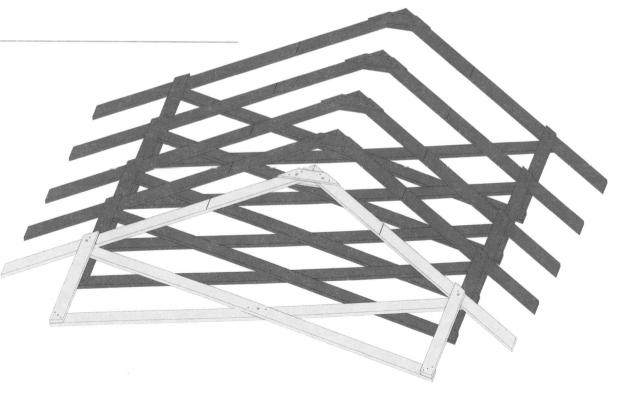

Repeat steps 1-23 to create 5 trusses total.

25

1x   Ø3" x 20ft SCH40 Steel Pipe
1x   Ø3" x 13ft SCH40 Steel Pipe

The method of creating a joint using the mitered cuts as shown is known as a "scarf joint." It creates significantly stronger joints than a typical square ended butt joint. If however, you lack the capability to miter joints accurately, you can butt weld and brace the joints as shown.

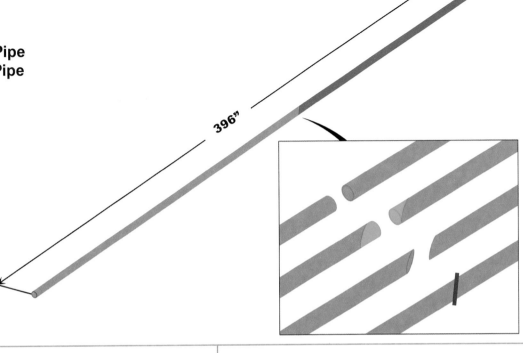

396"

26

1x   C4x7.25 x 20" C-Channel

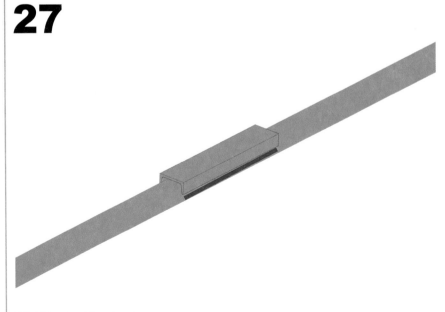

C-channel is used to reinforce the joint.

27

Weld both sides to pipe.

28

1x   8x4x¹/₂ Angle, 6" Long

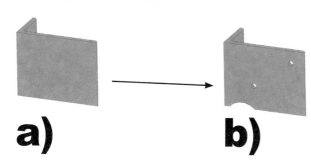

a)   **b)**

See the scale template pattern on page 552 for details on notching and drilling these mounting brackets. Templates can be copied, cut out and traced directly onto the metal.

29

Repeat step 28 to create a total of 10 brackets.

Left   Right

Note: There are 5 left and 5 right orientations. We've provided templates for both on pages 552-553.

30

1x   Mounting Bracket (Left Orientation)

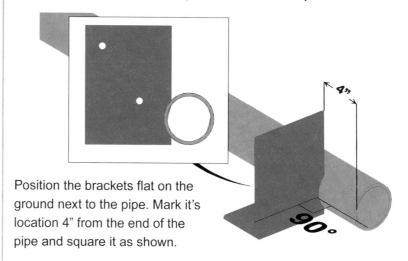

Position the brackets flat on the ground next to the pipe. Mark it's location 4" from the end of the pipe and square it as shown.

31

Weld along the red lines (front, back, and bottom).

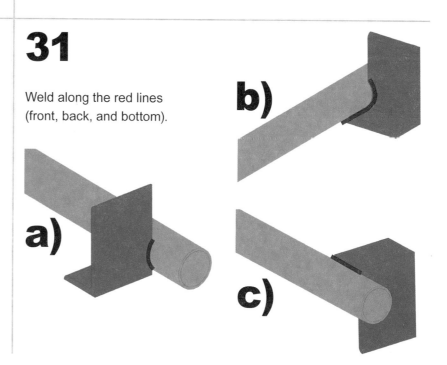

a)   **b)**   **c)**

32

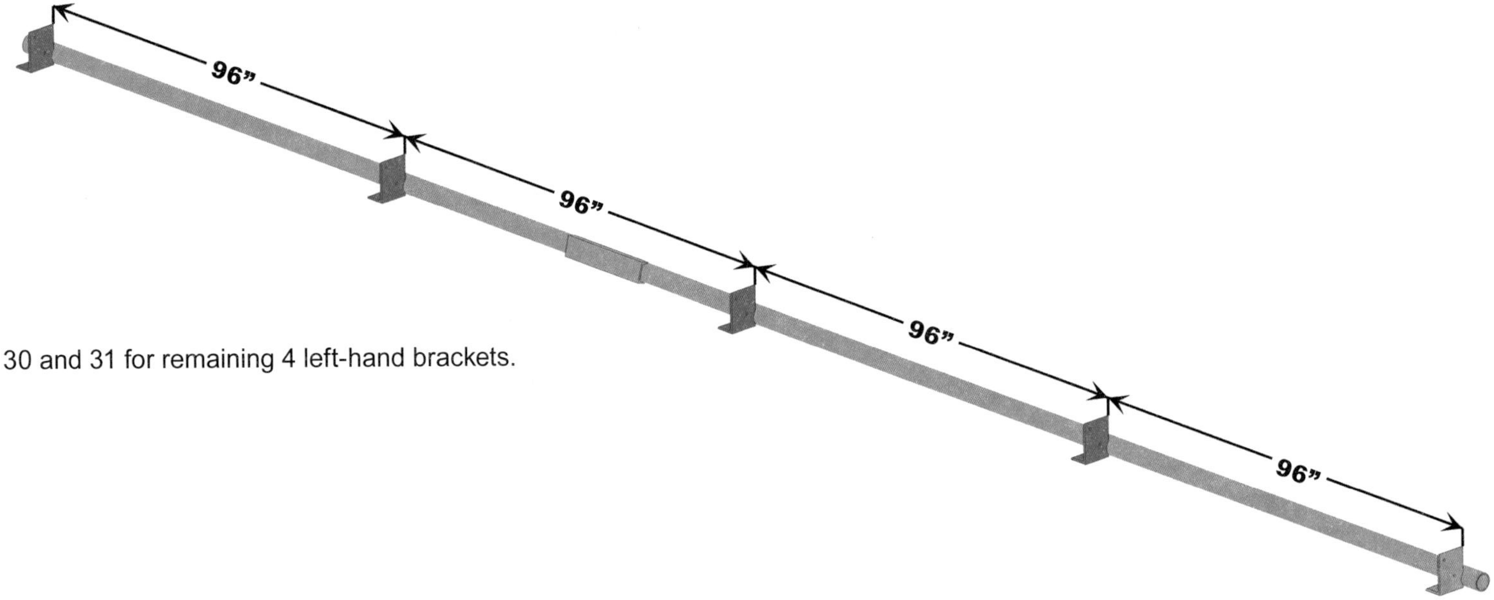

96"

96"

96"

96"

96"

Repeat steps 30 and 31 for remaining 4 left-hand brackets.

33

1x   **¼x3-½x18" Plate Steel**

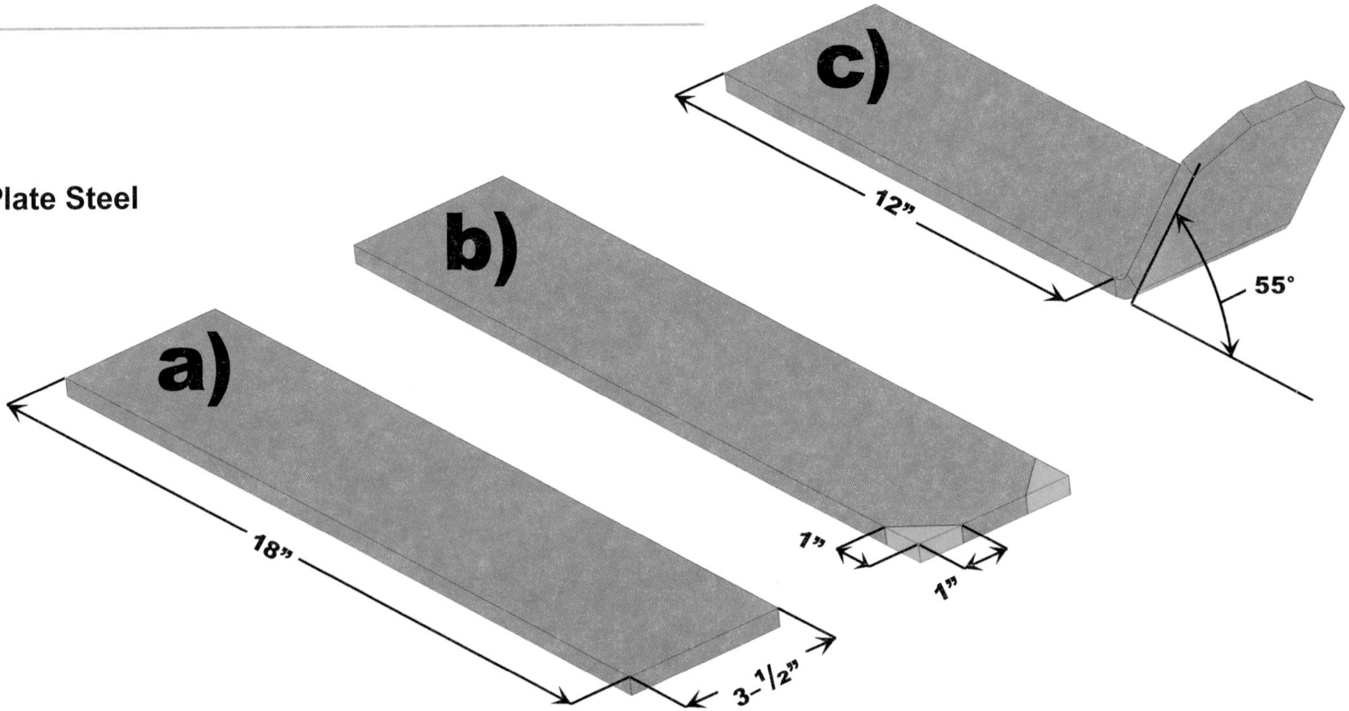

a)

b)

c)

18"

3-½"

12"

55°

1"

1"

In this step, we create the front portion of the skid. After cutting the corners, we bend the front end upwards to help it float across the ground. Imagine the runners on the bottom of a sleigh.

34

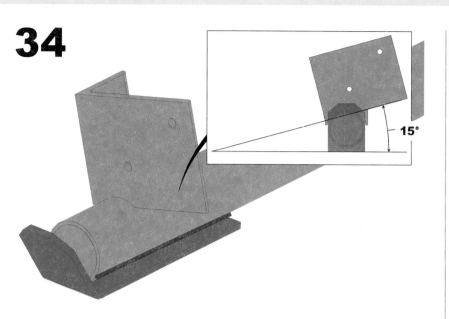

Orient the tube and welded brackets at 15° from horizontal as shown above in red and weld on both sides.

35

2x   8" Length of Chain

a)

b)

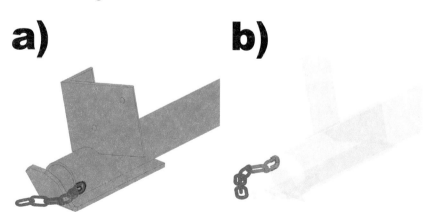

Weld an 8" length of chain to either side of the tube as shown (in red).

36

1x   ⁵⁄₁₆" Threaded Quick Link Connector

Connect the two pieces of chain together at the front of the skid with a threaded quick link.

37

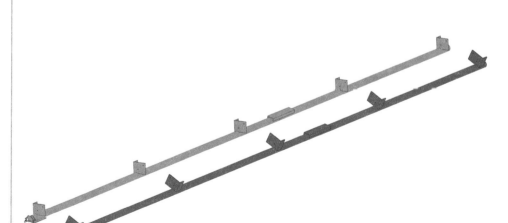

Repeat steps 25-36 but in a mirrored orientation to create the right-hand skid.

38

1x   Premade Truss

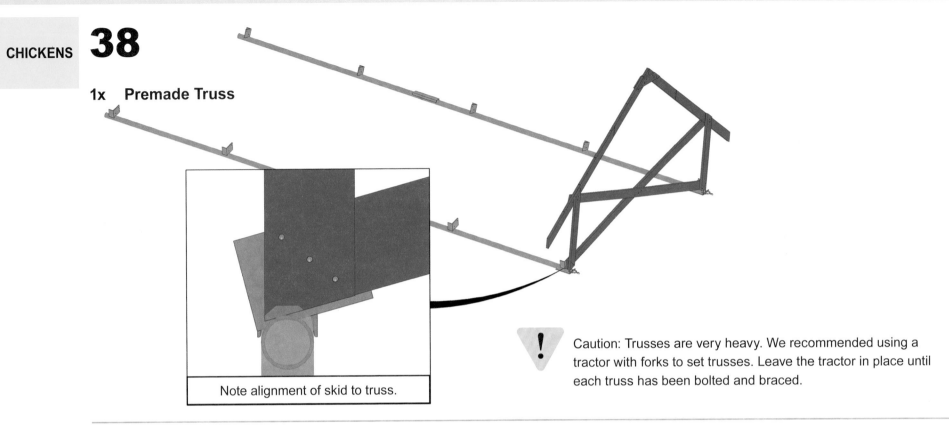

Note alignment of skid to truss.

! Caution: Trusses are very heavy. We recommended using a tractor with forks to set trusses. Leave the tractor in place until each truss has been bolted and braced.

39

4x   ½-13 x 6"Carriage Bolts
4x   ½" Washers
4x   ½-13 Hex Nuts

Using the holes in the brackets as guides, drill through the wood truss and secure the bolts.

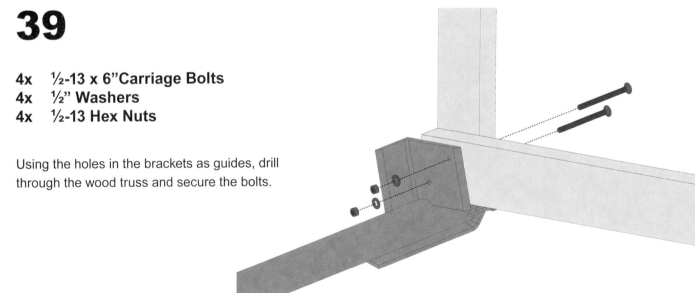

40

2x   3" Deck Screws
1x   2x4x192"

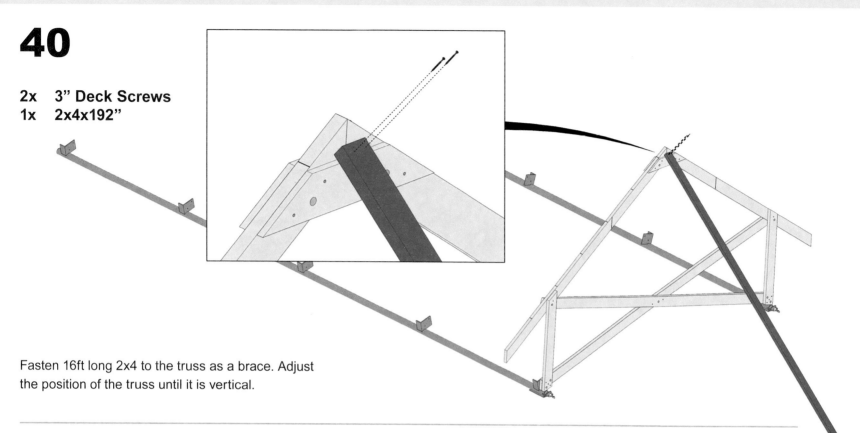

Fasten 16ft long 2x4 to the truss as a brace. Adjust
the position of the truss until it is vertical.

41

1x   3" Deck Screw
1x   2x4" Wooden Stave

Drive a wooden stake into the ground next
to the brace and fasten them together to
stabilize the first truss.

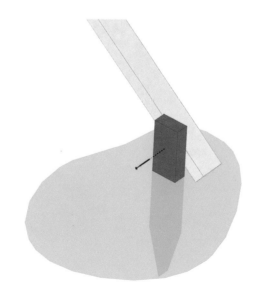

42

1x   2x4x120"

Steps 42-43 are for the creation of temporary "purlin spacers". The idea behind preparing these board in advance is to speed up the process of setting the trusses. Do not worry, none of these boards will go to waste.

96"

Measure and draw a line at 96"

43

2x   3" Deck Screws

Start driving both screws halfway in the locations shown.

44

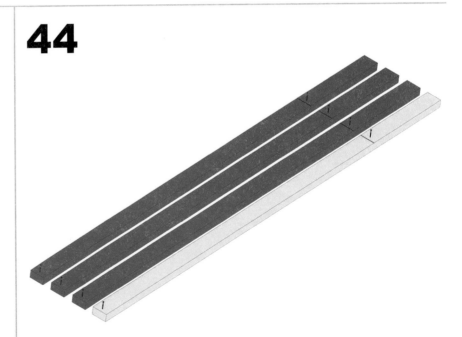

Repeat steps 42-43 to create a total of 4 purlin spacers.

45

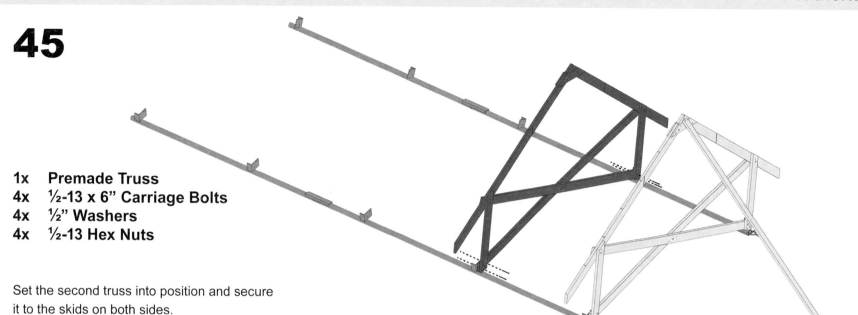

1x Premade Truss
4x ½-13 x 6" Carriage Bolts
4x ½" Washers
4x ½-13 Hex Nuts

Set the second truss into position and secure it to the skids on both sides.

46

1x Premade Purlin Spacer

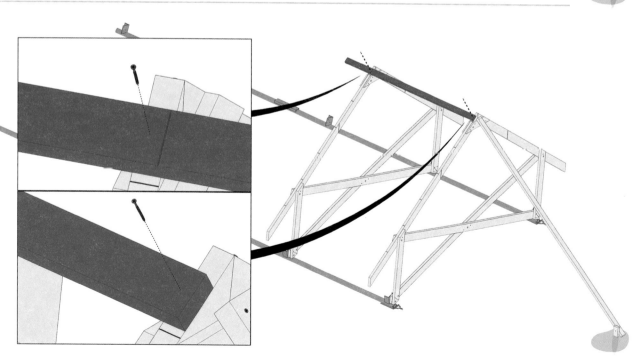

Using a ladder, secure the temporary purlin at the top of the two trusses. This will stabilize the second truss while you brace it further.

47

8x   20D Galvanized Nails
2x   2x4x120"

Trusses should be 96" apart center to center. Contractors use the term "center to center" to clarify that a dimension is measured from the center of one board to the center of another. This ensures that a 96" purlin will overlap half of the truss on either end.

Set purlin in the middle of the truss and drive nails in at an angle as shown.

96"

48

16x 20D Galvanized Nails
4x   2x4x110"

Install cross braces on either side as shown. Make sure that the trusses are square with the skids.

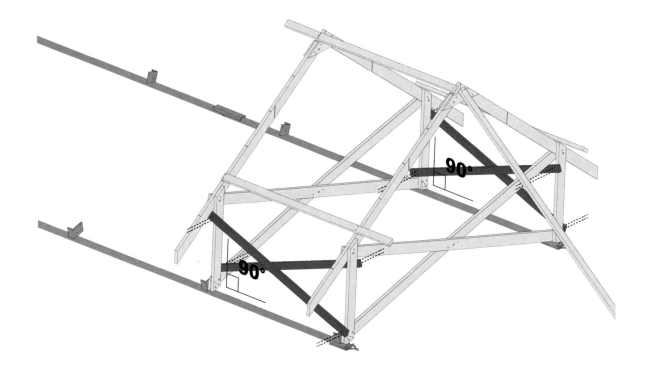

90°

90°

49

16x 2" Deck Screws
2x 1x6x144"

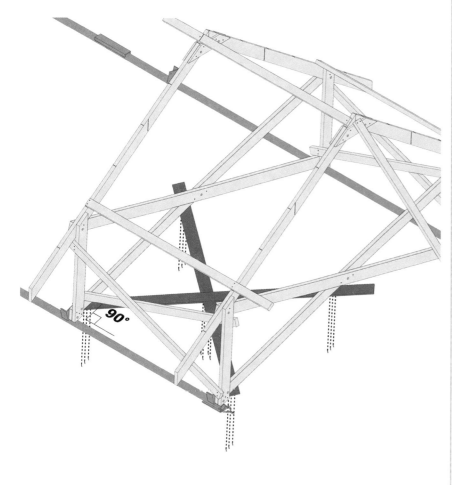

Install additional bracing onto the underside of the cross braces. Ensure that they are square in this direction as well.

50

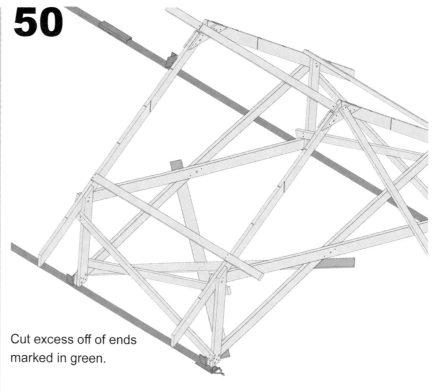

Cut excess off of ends marked in green.

51

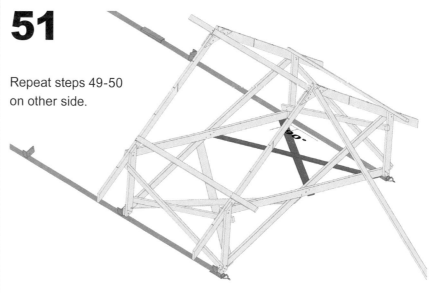

Repeat steps 49-50 on other side.

52

1x   **Premade Truss**
4x   **½-13 x 6" Carriage Bolts**
4x   **½" Washers**
4x   **½-13 Hex Nuts**

Continue the process with setting the third truss.

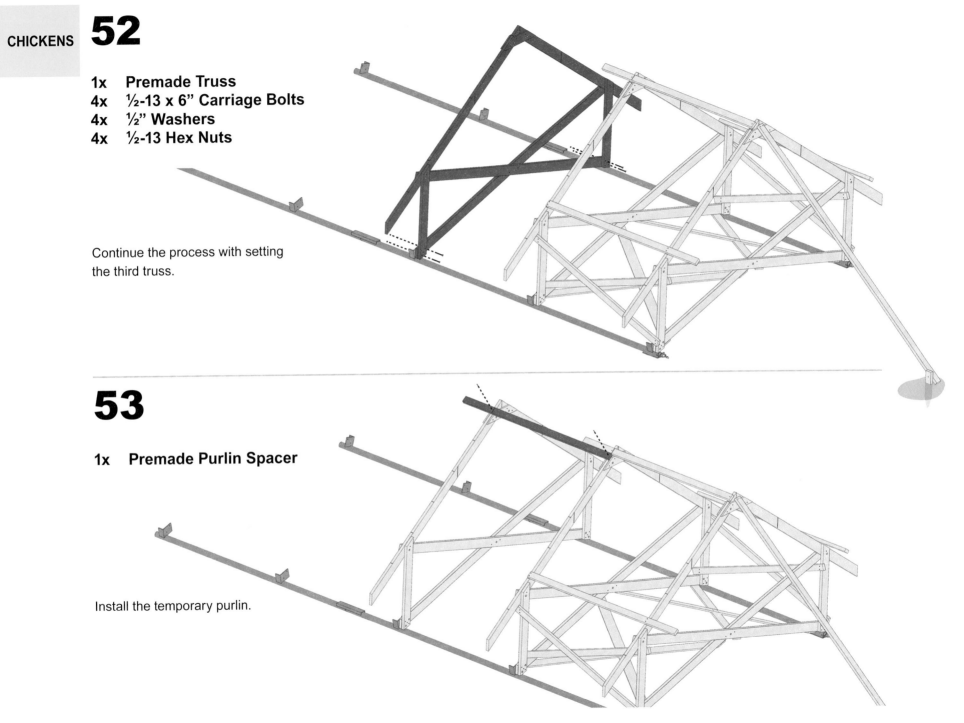

53

1x   **Premade Purlin Spacer**

Install the temporary purlin.

54

8x 20D Galvanized Nails
2x 2x4x96"

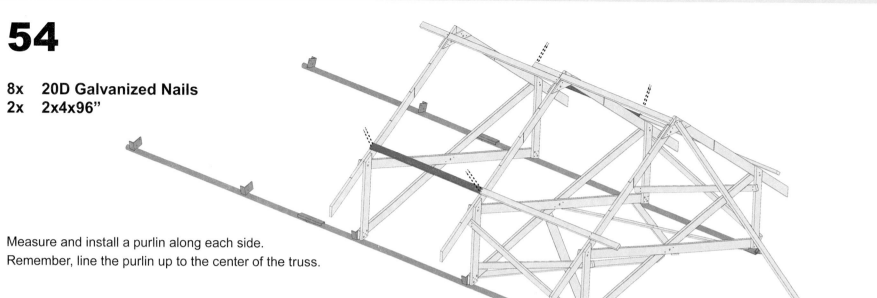

Measure and install a purlin along each side.
Remember, line the purlin up to the center of the truss.

55

1x Premade Truss
1x Premade Purlin Spacer
4x ½-13 x 6" Carriage Bolts
4x ½" Washers
4x ½-13 Hex Nuts
8x 20D Galvanized Nails
2x 2x4x96"

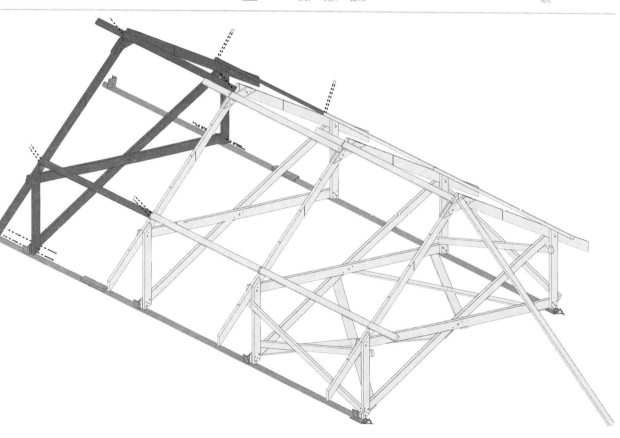

Repeat the process for the fourth truss.

56

1x **Premade Truss**
1x **Premade Purlin Spacer**
4x **$\frac{1}{2}$-13 x 6" Carriage Bolts**
4x **$\frac{1}{2}$" Washers**
4x **$\frac{1}{2}$-13 Hex Nuts**
8x **20D Galvanized Nails**
2x **2x4x120"**

Finally, set the fifth truss.

57

16x **20D Galvanized Nails**
4x **2x4x110"**

Square and brace
the structure just like
in step 48.

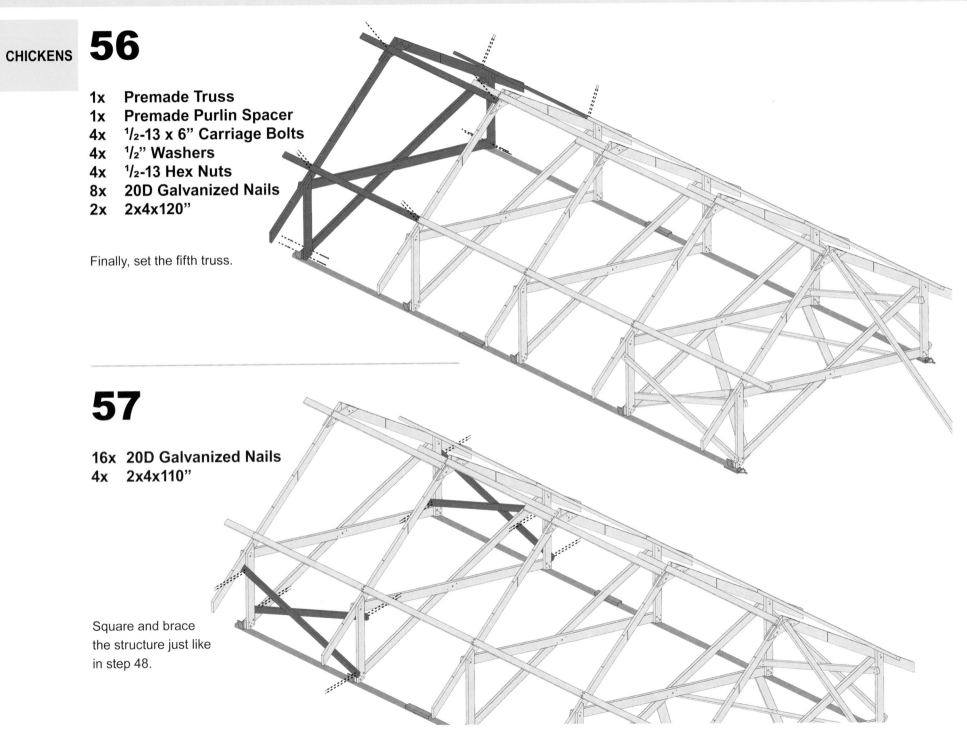

58

48x **20D Galvanized Nails**
6x **2x4x96"**
6x **2x4x120"**

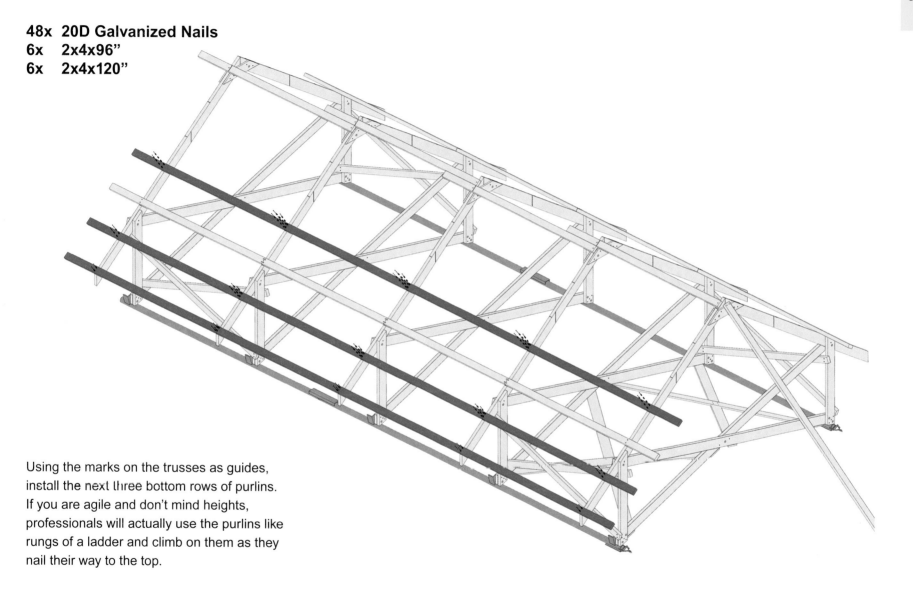

Using the marks on the trusses as guides, install the next three bottom rows of purlins. If you are agile and don't mind heights, professionals will actually use the purlins like rungs of a ladder and climb on them as they nail their way to the top.

59

You may need to measure and cut each 2x6 to fit.

6x **20D Galvanized Nails**
1x **2x6x94"**

Align the 2x6's with the pre-drawn lines and secure as shown. These boards are used as a mounting point for the nesting boxes in a later step. While they can be installed later, we have found this is an opportune point to do it.

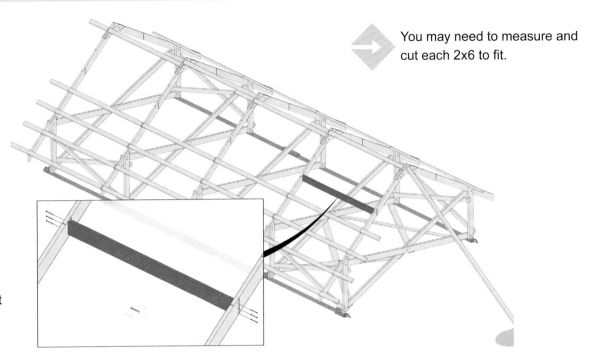

60

8x **20D Galvanized Nails**
1x **2x6x94"**

Notice that you will have to drive nails at an angle on one end.

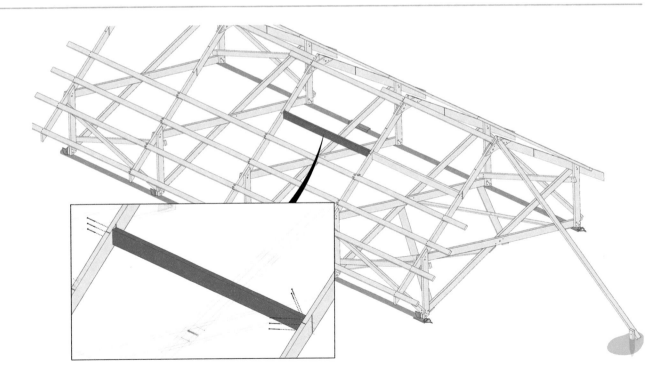

61

16x 20D Galvanized Nails
2x   2x6x94"

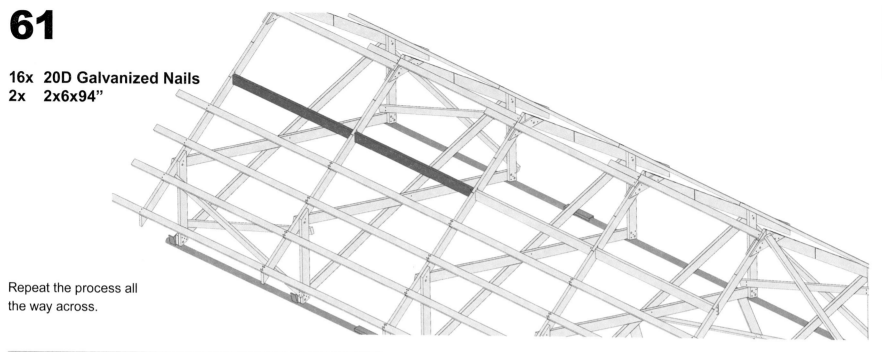

Repeat the process all
the way across.

62

32x 20D Galvanized Nails
4x   2x4x96"
4x   2x4x120"

Nail on the next two
rows of purlins.

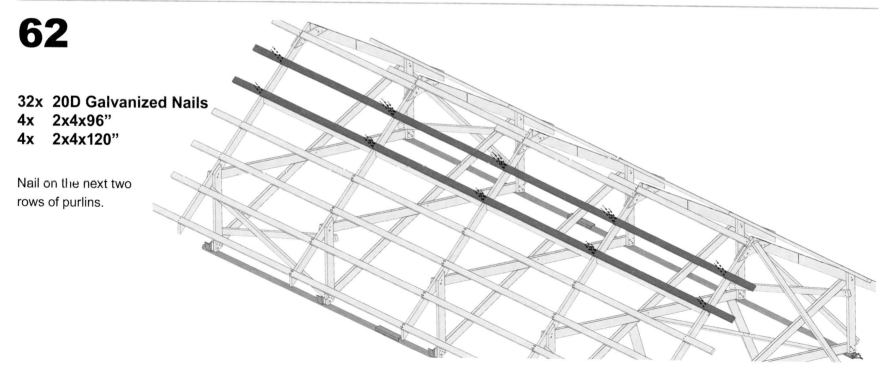

63

At this point, remove the temporary purlin spacers.

64

16x 20D Galvanized Nails
2x   2x4x96"
2x   2x4x120"

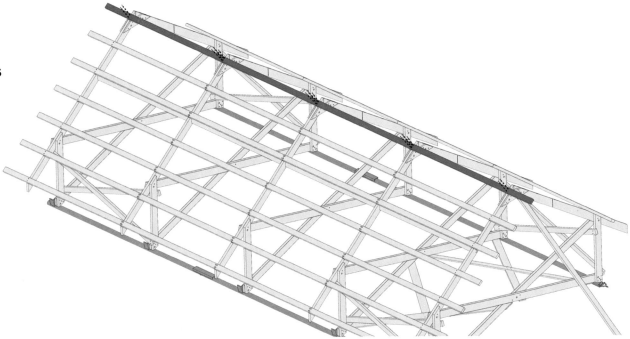

Attach the final row of purlins at the top.

65

Repeat steps 58-64 on other side
of structure.

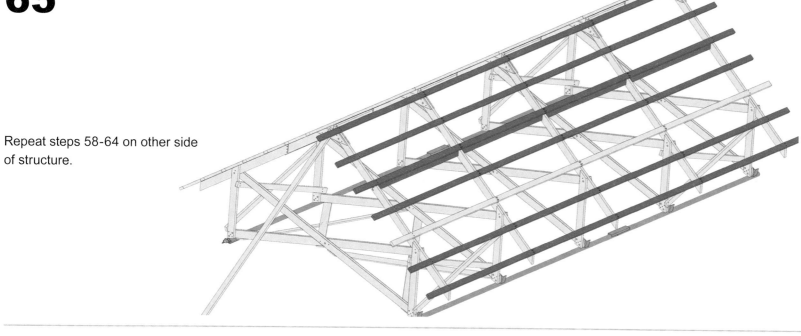

66

1x   Steel Roofing Panel 3x13ft

A professional roofing trick is to pre-mark and
drill all of your roofing panels prior to putting
them up. In this step we are lining the panel up
where it needs to be on the roof line (≈2" above
the top of the topmost purlin) and marking the
drill holes centered on each of the 7 purlins.
This upfront step makes fastening through the
metal easier while you are balancing on the roof
later and it ensures that all of the screw rows
are straight.

*Note: This will only work if all of your purlins are straight and
spaced evenly (hence the pre-marking in step 6).*

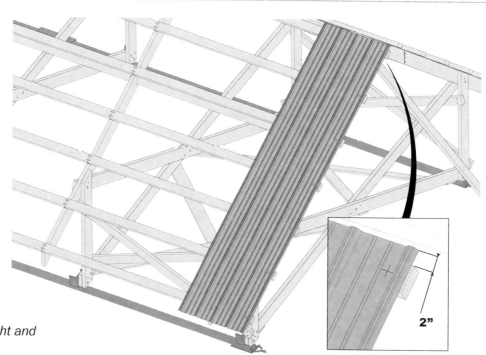

67

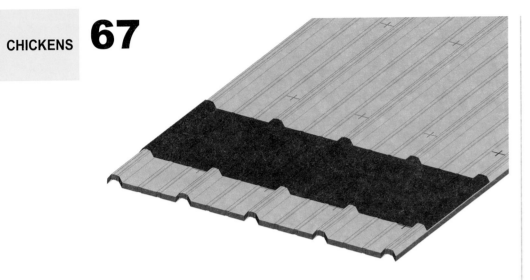

An easy way to transcribe your marks to the other locations across the panel is to take a scrap piece of metal (with a square edge) and slide it along each row of marks.

68

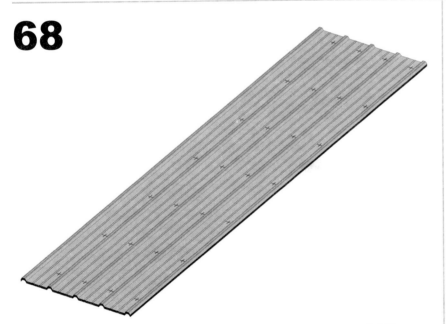

Stack all of the roofing panels for one side of the structure and drill through them all at each marked point.

69

28x 1-½" Roofing Screws

Align panel and drive a screw through each pre-drilled hole.

70

≈308x   1-½" Roofing Screws

Repeat the roofing process for the entire side. Depending on the width of the panels, you may or may not need to trim the end to size.

71

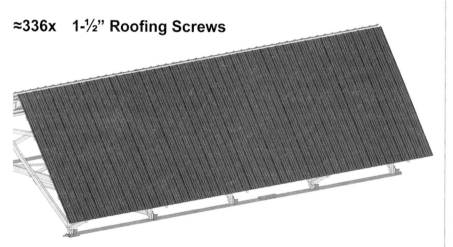

Now that one side of the roof is secured, the temporary brace can be removed. You will notice that the roofing adds a lot of rigidity to the structure.

73

≈12x   2-½" Roofing Screws
1x   Metal Roofing Ridge Cap

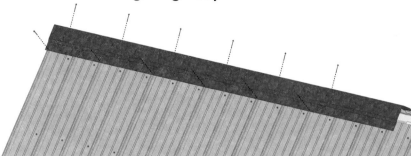

Install the pre-bent roof cap. Use appropriately sized screws per the recommendation of your roofing supplier. Place screws over the ribs of the roofing and make sure you do not over tighten them because they can deform the metal.

72

≈336x   1-½" Roofing Screws

Repeat steps 66-70 for the second side of the roof.

74

≈12x   2-½" Roofing Screws
1x   Metal Roofing Ridge Cap

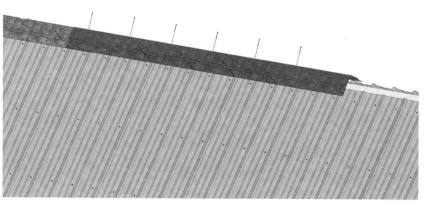

Repeat the process and overlap the previous piece by 10-12".

75

≈12x   2-½" Roofing Screws
1x   Metal Roofing Ridge Cap

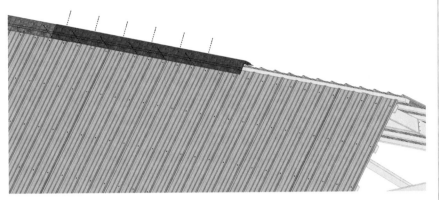

Repeat the process and overlap the previous piece by 10-12".

76

≈12x   2-½" Roofing Screws
1x   Metal Roofing Ridge Cap

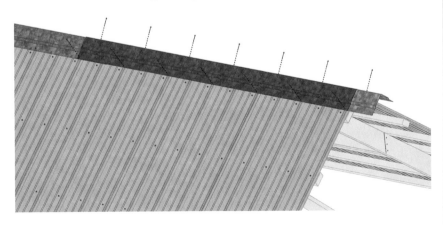

Cut the excess off of the end.

77

For the sake of visibility, we have hidden the entire roof structure from view for the remainder of the build process.

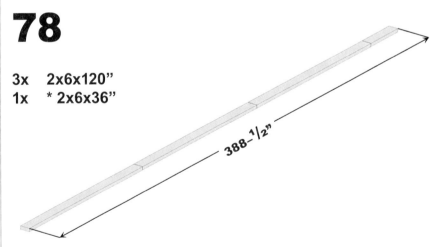

388-½"

The following steps will illustrate the installation of the elevated walkway and its supporting structures. Record the dimension above for use in the next step.

78

3x   2x6x120"
1x   * 2x6x36"

388-½"

*We've purposefully left this piece long with the intent of cutting it down to the exact size needed once everything has been nailed together.

79

24x 20D Galvanized Nails
3x   2x6x24"

Fasten 2x6s together using 24"
long scabs and a liberal number
of nails.

81

4x   20D Galvanized Nails
1x   2x6x24"

Cut and nail a filler piece
as shown.

24"

29°

15°

80

Slide the long piece down
the center of the structure and let
it rest vertically in the crook of the "X"
part of the truss.

82

3x   20D Galvanized Nails
1x   2x6x48"

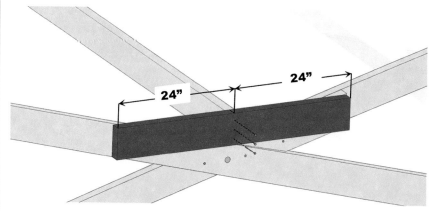

24"   24"

Mark the middle on the 48" long board and center it on the long 2x6 from
step 80. Nail the two together as shown.

83

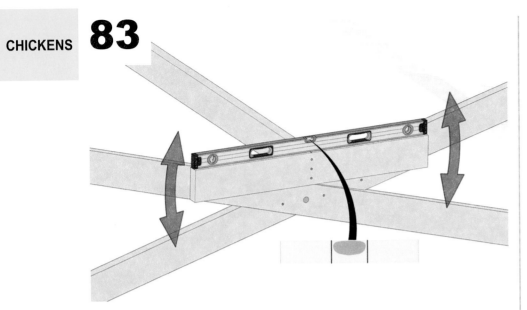

In this step we level what will be the supports for the walkway.

84

6x   20D Galvanized Nails

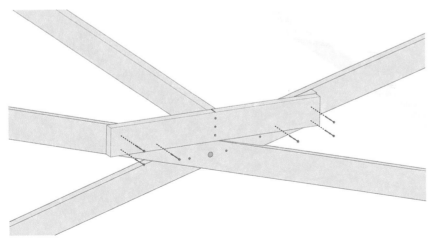

Once structure is level, nail it in place.

85

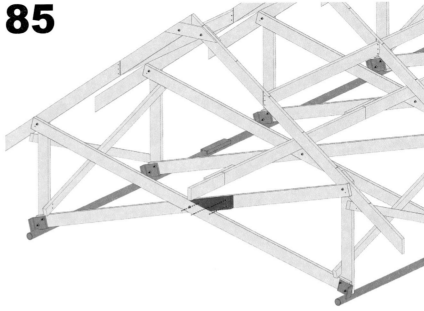

Repeat step 81 on other end.

86

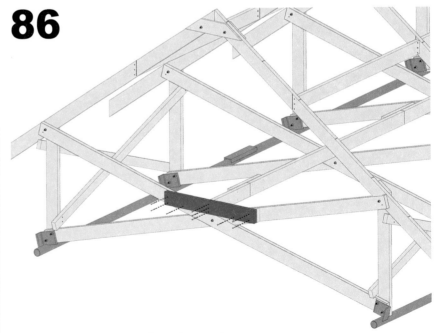

Repeat steps 82-84 on the other end.

87

4x   8D Smooth Nails
1x   String Line

In this step we use string lines to mark and align the supports for the walkway. Drive a nail into each of the 4 corners. Nails should be in line with the outer edge of the 2x6x48" boards. Tighten string lines until they are taut and check that they are 48" apart.

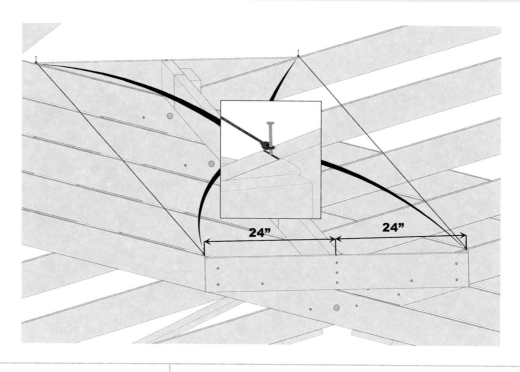

88

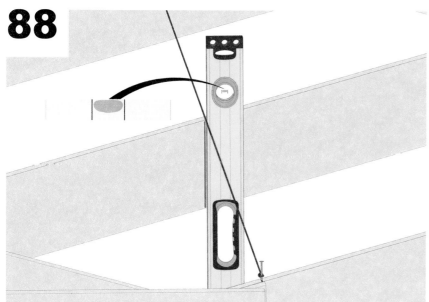

Hold a level against the string line and pencil a line vertically on the truss brace. Repeat this process front and back, left and right on each truss.

89

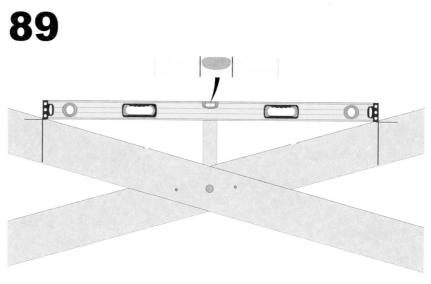

Place level horizontally across the 2x6 center support and pencil lines as shown. Repeat this process front and back on each truss.

90

8x **20D Galvanized Nails**
1x **2x6x94"**

 You may need to measure and cut each 2x6 to fit.

In the following steps we install the supports for the walkway. Each support may need to be trimmed to fit. Line each support up on the inside of the vertical line and below the horizontal lines. Nail in place as shown.

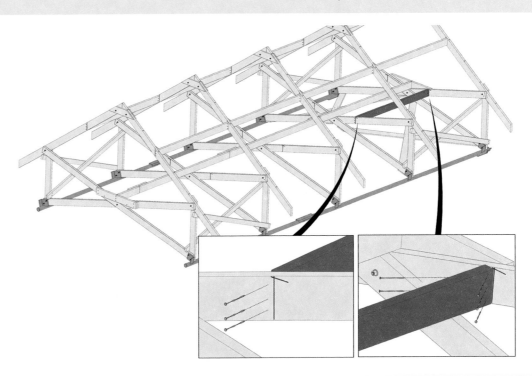

91

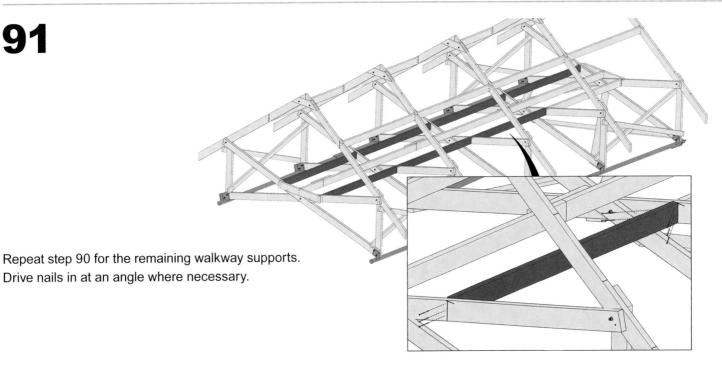

Repeat step 90 for the remaining walkway supports. Drive nails in at an angle where necessary.

92

3x   8D Ribbed Nails
1x   1x3x48"

With the supports all installed, it is time to install the decking. The decking planks consist of 1x3s nailed down with a 1" space in between them to allow manure and debris to fall through. Start the first plank flush with the end and work your way in.

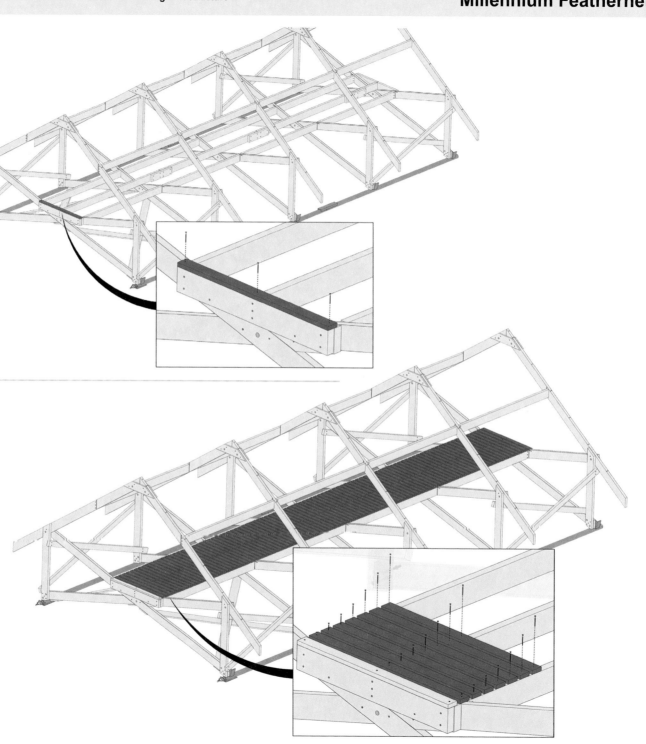

93

≈291x   8D Ribbed Nails
≈97x   1x3x48"

We recommend creating some 1" wide gauge blocks and placing them temporarily between each plank as you nail it in place. This will ultimately speed up the installation process and provide you with consistent spacing throughout.

94

40x　20D Galvanized Nails
10x　2x4x48"

These vertical supports will be utilized for the installation of the nesting boxes in a future step. Use a level as needed to ensure that they are vertical.

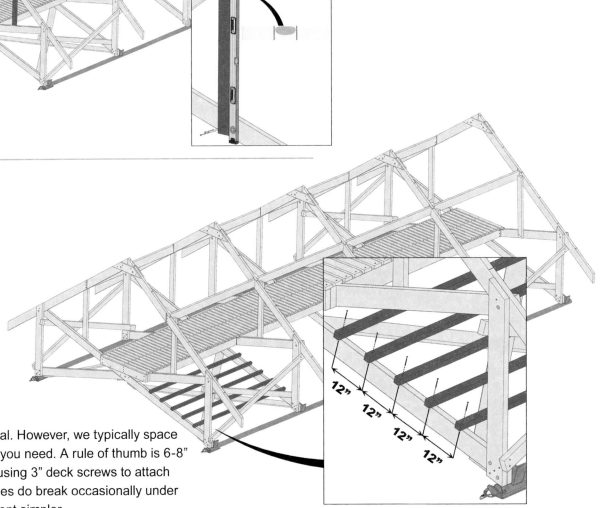

95

10x　3" Deck Screws
5x　2x2x100"

We are ready to install perch bars throughout the entire structure. The beauty of the X-braced truss design is the availability to place plenty of roosting space under the protection of the structure. These roosting locations are sheltered from the elements and also from airborne predators.

Exact placement of these perches is not crucial. However, we typically space them 10-12" apart. Add as few or as many as you need. A rule of thumb is 6-8" per bird, depending on size. We recommend using 3" deck screws to attach them (as opposed to nails) because the perches do break occasionally under the weight of the birds and it makes replacement simpler.

96

30x 3" Deck Screws
15x 2x2x100"

Continue installing perches across structure as shown. Notice how the perches are offset above and below each other.

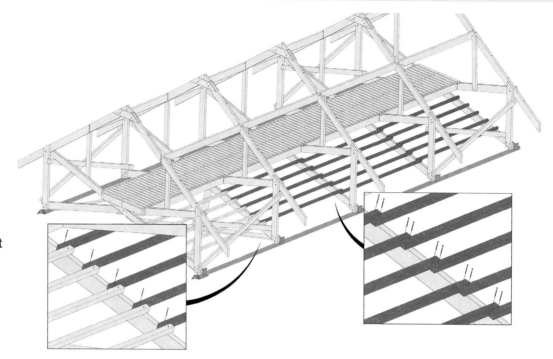

97

10x 3" Deck Screws
5x   2x2x100"

Again, exact placement of these perches is not critical but we typically space them 10-12" apart.

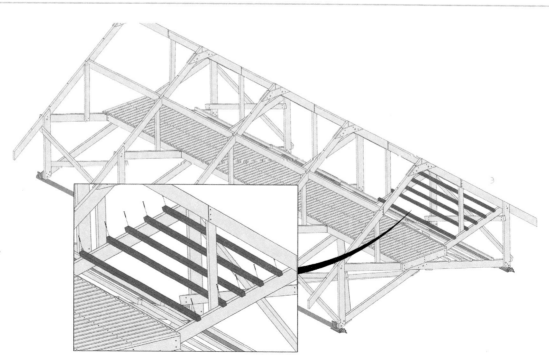

98

30x 3" Deck Screws
15x 2x2x100"

Finish installing perches across the upper section, overlapping above and below each other as shown.

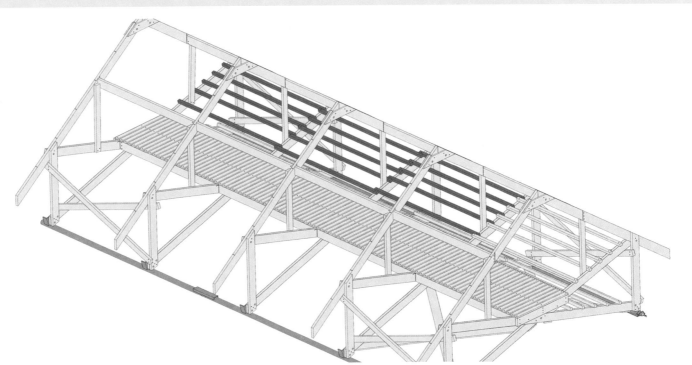

99

80x 3" Deck Screws
40x 2x2x100"

Repeat steps 95-98 on the opposite side to finish the perch installation.

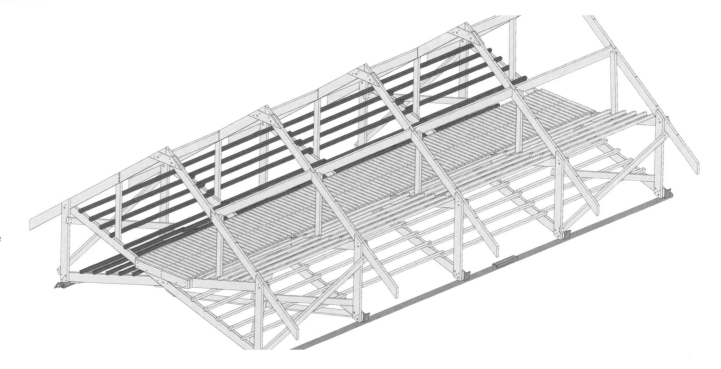

100

1x   2x4x144"
2x   2x4x120"

Line up several boards to achieve a total length of 384".

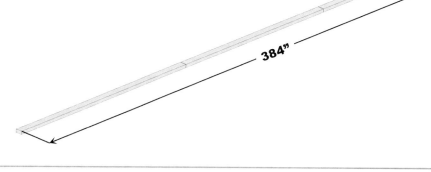

384"

101

8x   20D Galvanized Nails
2x   2x4x24"

Use some scrap pieces of lumber as scabs and nail everything together.

102

Repeat steps 100-101 to create a second 384" long piece.

CHICKENS # 103

20x 20D Galvanized Nails

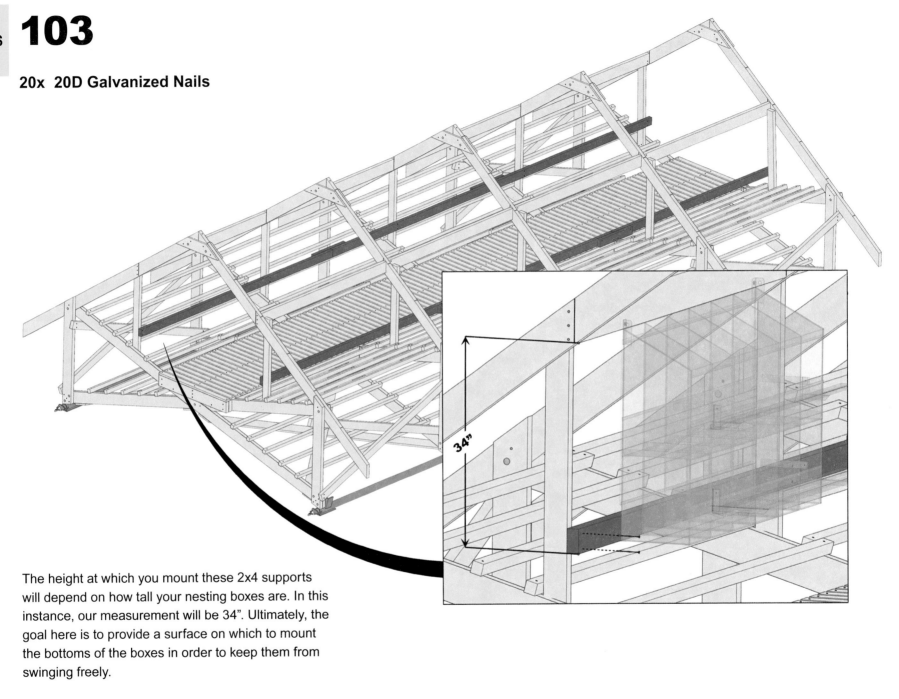

34"

The height at which you mount these 2x4 supports will depend on how tall your nesting boxes are. In this instance, our measurement will be 34". Ultimately, the goal here is to provide a surface on which to mount the bottoms of the boxes in order to keep them from swinging freely.

104

2x ⅜ x 1-¾" Hex Head Lag Screw
2x ⅜" Washers

We cannot tell you where exactly to mount your nesting boxes. It depends on how many, and how big they are. There are a few things to consider when making your decisions:
Hens tend to favor the nests closer to the inside. Boxes near the outside are often brighter and more subject to the elements. Therefore, we recommend placing your boxes as far from the ends as you can.

Pre-drill holes prior to inserting lag bolts. Choose a drill bit diameter that is slightly smaller than the threads of the lag bolts. Use some left-over roofing screws to fasten the bottom of the boxes to the 2x4 support to prevent it from swinging freely.

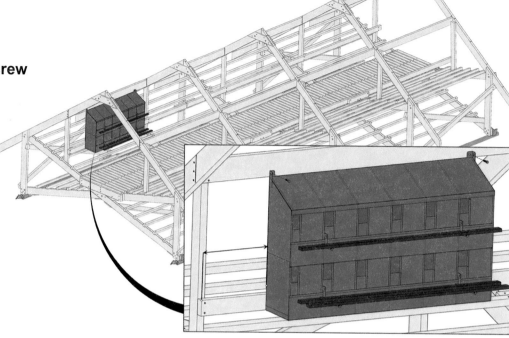

105

Repeat mounting process for remaining nesting boxes. Our example shows 10 boxes for a total 120 nesting boxes. We often close off several units as egg production slows over the course of the season.

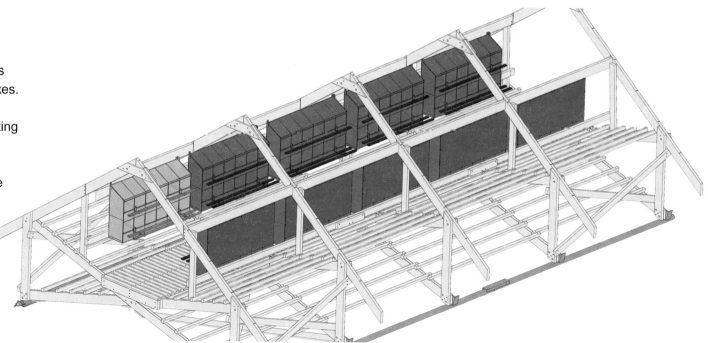

106

≈35ft   ¼" High Test Chain
2x   ½-13 x 6" Grade 8 Hex Bolts
4x   ½" Washers
2x   ½" Nylock Nuts
16ft   Ø3" SCH40 Steel Pipe

There is engineering behind the rigging shown here. The 16ft pipe serves as a "spreader bar". Without it, the skids have a tendency to pull together towards the center. This causes undo stress on the structure which may lead to failure. Simply put, use a spreader bar – trust us.

The straight lengths of your chain can be 4ft or 12ft instead of 8, it really doesn't matter. But, keep in mind that we attach a grain buggy to the front of this structure and you want ample room to work between the two.

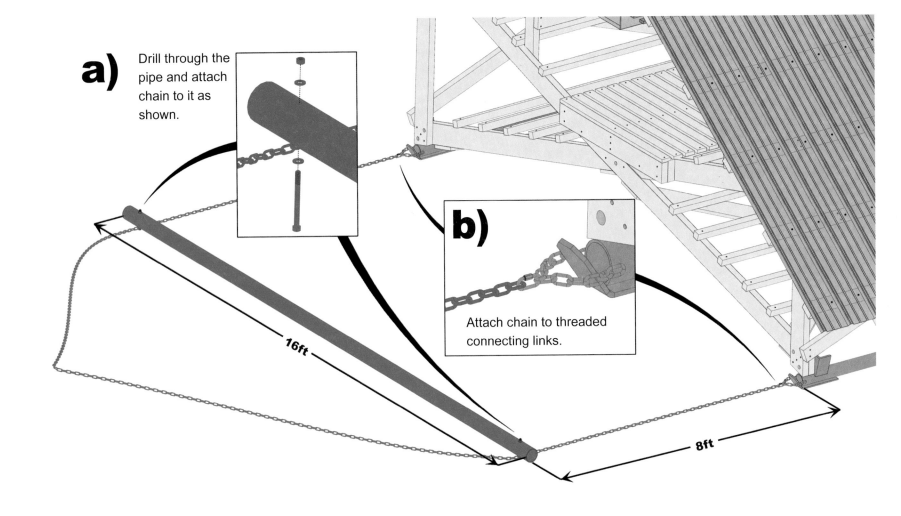

a) Drill through the pipe and attach chain to it as shown.

b) Attach chain to threaded connecting links.

16ft

8ft

Build A Ramp

This ramp is a standard design we use on both the Feathernet and the Gobbledygo structures. It is 8ft long and 18" wide. The 1x2 rungs are crucial for the birds to safely travel up and down the ramp. We have found that the size of the rungs and spacing between them really does matter. Generously use screws during the assembly. These ramps take a lot of abuse and need to be durable.

108

2x   1x6x18"

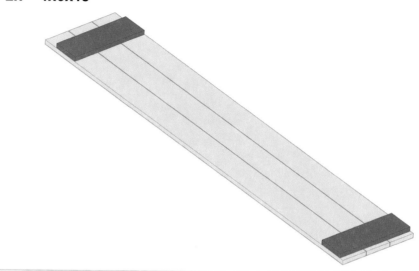

107

3x   1x6x96"

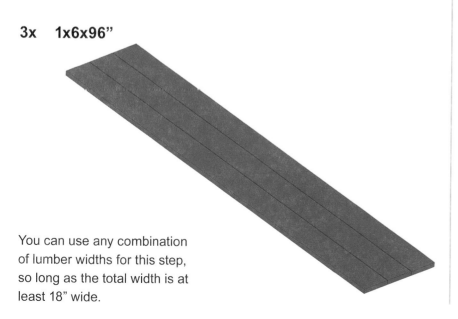

You can use any combination of lumber widths for this step, so long as the total width is at least 18" wide.

109

24x 1-⅝" Deck Screws

After fastening these screws, flip the ramp over and check that none of them punctured through the other side. Grind or file down any sharp points that may be protruding in order to protect the chicken's feet.

CHICKENS

110

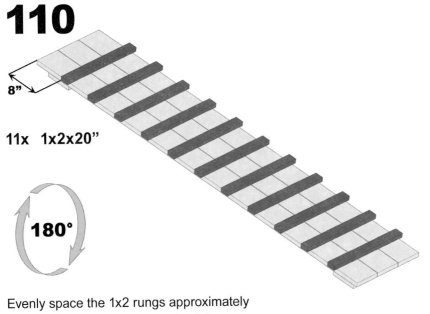

8"

11x 1x2x20"

180°

Evenly space the 1x2 rungs approximately 8" apart along the entire length of the ramp.

112

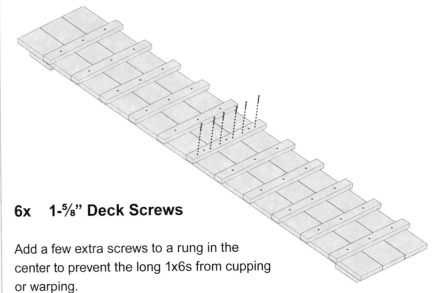

6x 1-⅝" Deck Screws

Add a few extra screws to a rung in the center to prevent the long 1x6s from cupping or warping.

111

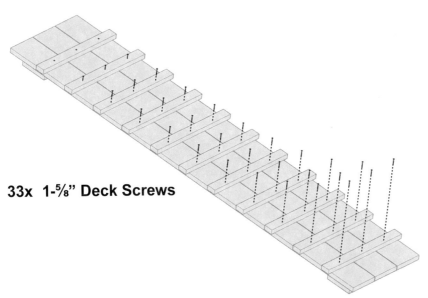

33x 1-⅝" Deck Screws

113

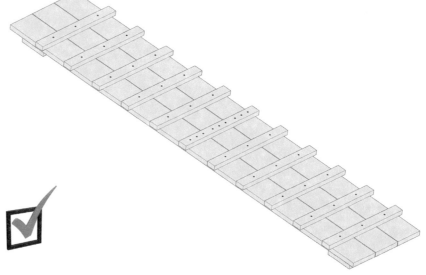

The ramp is complete!

Build A Ladder

Build a ladder to step up into the Feathernet structure. This same design is also used on the mini brooder when it is mounted on a trailer. We also use these steps on our feed barrels that we make from old oil tanks sawed in half. They can also be used to create steps for the mezzanine structures in winter poultry hoop houses. You can easily modify this design to add more steps as needed—it is truly a simplistic yet versatile design!

115

Cut a 30° miter for the foot of the step stringer.

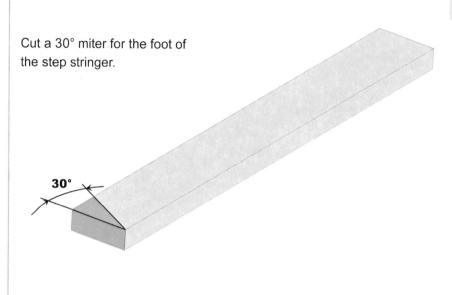

30°

114

1x   1x6x48"

116

Cut the top end as shown.

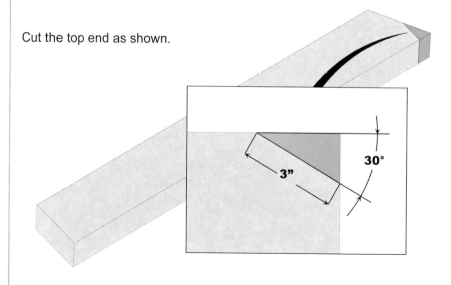

3"

30°

CHICKENS

117

Make a second cut to complete the top end of the stringer.

119

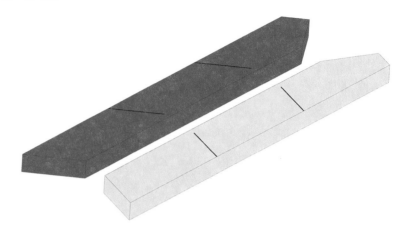

Repeat steps 114-118 to create an opposite hand step stringer as shown.

118

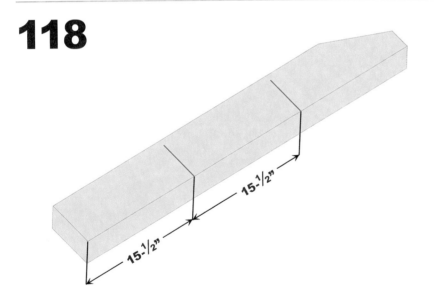

Next, measure and mark two lines for the step locations. These lines should be parallel with the 30° miter you cut in step 115.

120

6x　3" Deck Screws
2x　2x6x30"

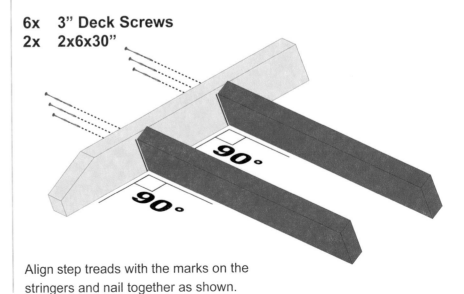

Align step treads with the marks on the stringers and nail together as shown.

121

6x   3" Deck Screws

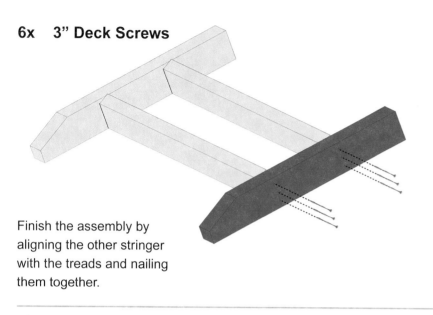

Finish the assembly by aligning the other stringer with the treads and nailing them together.

122

The ladder is complete!

123

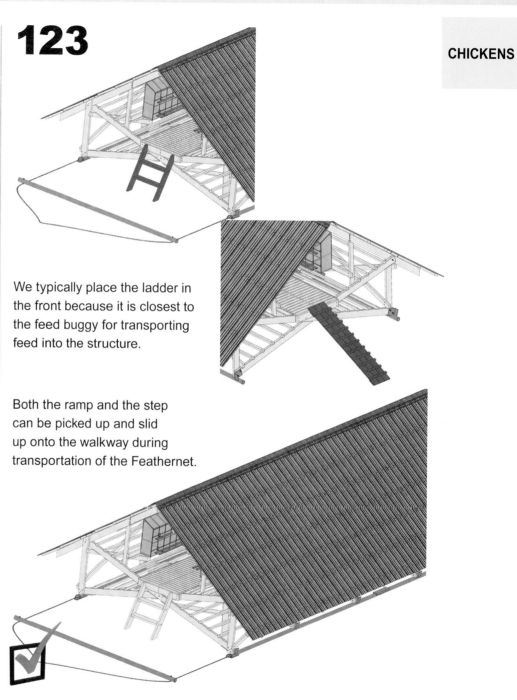

We typically place the ladder in the front because it is closest to the feed buggy for transporting feed into the structure.

Both the ramp and the step can be picked up and slid up onto the walkway during transportation of the Feathernet.

Congratulations, you have completed your Millennium Feathernet!

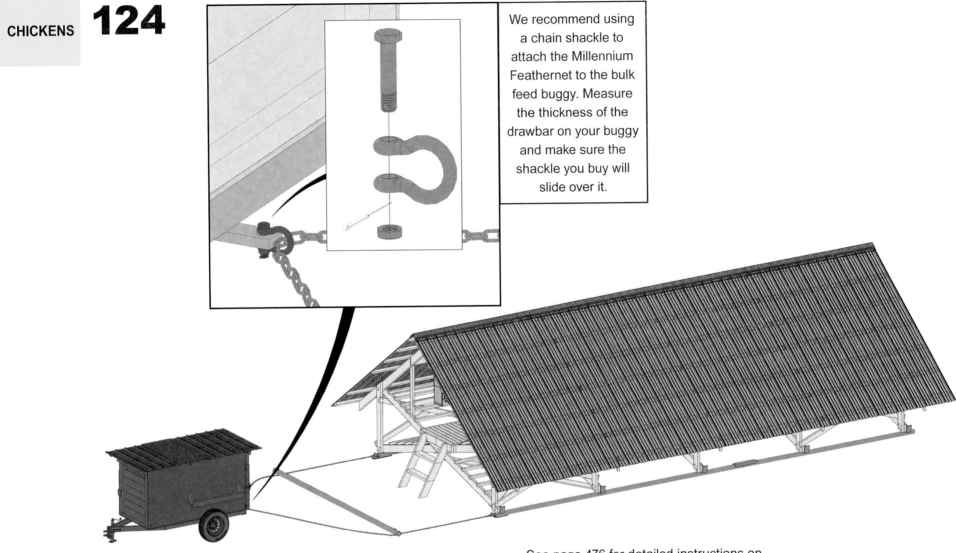

We recommend using a chain shackle to attach the Millennium Feathernet to the bulk feed buggy. Measure the thickness of the drawbar on your buggy and make sure the shackle you buy will slide over it.

See page 476 for detailed instructions on the construction of a bulk feed buggy.

The bulk feed buggy is the perfect compliment to the Millennium Feathernet system. By attaching the two together it consolidates chore tasks and makes the moves easier and more efficient. However, disconnect them when traveling over long distances or over rough terrain.

photo courtesy of Jessa Howdyshell

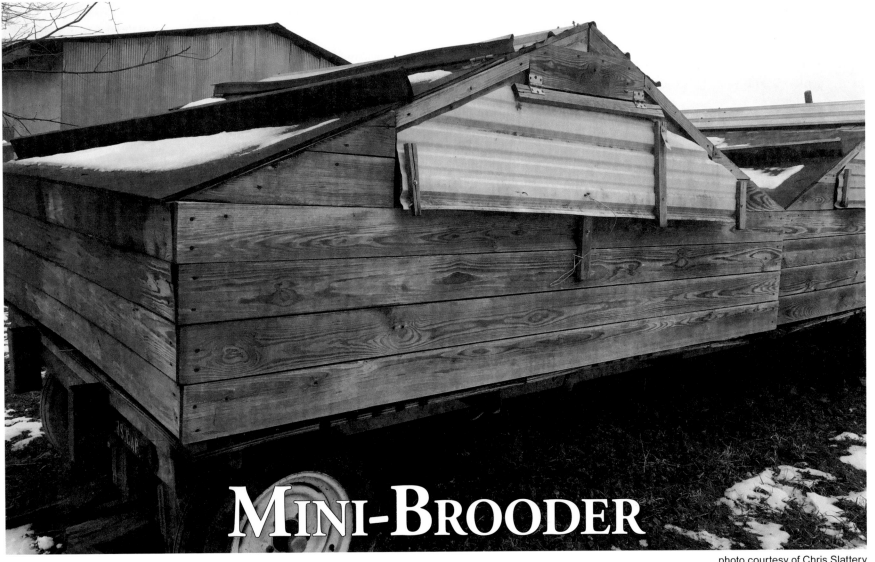

MINI-BROODER

photo courtesy of Chris Slattery

Broilers go from infancy to almost maturity in 60 days. If we equate that to human development, it's equivalent to going from birth to 16 years old in 60 days. As if that weren't acceleration enough, realize that the first couple years of human development set the stage for the rest of life; in other words, each year is not weighted equally in the big scheme of things.

The point I'm getting to is that every day of the chick is a quarter year of equivalent human development, which means the first 8 days is equivalent to two years. If you mess up one day, that's like messing up an infant for 3 months. I cannot stress enough the importance of getting that brooder right; like getting the nursery period right.

That means we need an environment that can be kept as close as possible to what living underneath the mother hen would be like. The most important element is warmth. Regardless of heat source, that can be extremely expensive, so the smaller the area you have to heat, the better off you are.

You might be surprised that in our design, we don't have insulation; after all, insulation would seem to make heating easier. But in my lifetime of brooding chicks, the single biggest threat, after temperature, is predation from rats. If you start building an insulated brooder, you'll have rats living in the walls. And if you have a dirt floor, they will eventually tunnel in from underneath. A single hefty sow rat in one night can take more than 50 chicks. I've seen it.

So this structure is small to reduce cubic heating space, not insulated to thwart rats, and mounted on a hay wagon up off the ground so the rats can't dig underneath. If you have a couple of cats they will help patrol at night and keep rats from making that vulnerable dash from ground to chassis to brooder.

Chicks need 25 square feet per 100 birds up to 3 weeks, so 8ft x 8ft offers 64 square feet and is big enough to brood 250 chicks. If you're going to go beyond 3 weeks, increase the space 30 percent. At 8 batches a season, that's 2,000 birds in this little brooder. The bottom box is deep enough to accommodate deep litter. Perhaps the most important thing we can do to ensure healthy birds is to put them on vibrantly decomposing litter, like compost. Our favorite material is wood shavings--the curly shavings from furniture and cabinet shops. Because they're bouncy, they don't pack down. And because they're friable,

photo courtesy of Chris Slattery

the chicks scratch into them, quite deep actually, injecting oxygen and mixing manure with the carbon.

In order to create a vibrantly decomposing environment, this bedding must be at least a foot deep. A little skiff of bedding might be sterile, but it doesn't grow good bugs to beat out the bad bugs. Few mediums protect housed animals from disease like a compost-type bedding. We all know compost needs mass in order to function. A couple of inches won't house good bugs.

By making the entire roof a door, you can access the entire structure without crawling inside. That would be pretty confining if you had to do it. With both doors fully open, you can step

in and catch birds easily by using chicken crates on edge as a corral.

While heat is a big factor, so is ventilation. By putting the ventilation ports up high in the peak , it offers fresh air without draft. Chickens need fresh air, but they don't want to be in a draft. Roof peak type ventilation is superior to chick-height ventilation.

Benefits of propane over electric: Heat can be supplied a number of different ways. The most poor-boy is heat lamps, but they are difficult to regulate and often don't offer enough heat capacity on a very cold night. Another system is hot water run through pipes on the floor. The chicks can get as close or as far away as they want. A pipe grid on 12 inch centers with 1 inch PVC is the most preferred design. A thermocouple suspended a foot off the floor to turn the water on and off regulates the heat.

Our preference is propane because when you need heat, you have it right now and plenty of it. These ceramic hoovers are not expensive and are industry standard, meaning they're just about fool proof. They're mounted high enough off the floor to not be a fire hazard (unlike heat lamps) and their heat settings can be adjusted easily. To get the same level of heating power from heat lamps is far more costly than propane.

These little structure are light enough that you can move them around. Two strong people can do it but four people (one on each corner) makes picking it up simple. Invite some folks over for cake and ice cream and have a brooder moving party.

I can hear someone say "if I'm brooding

CHICKENS

Joel's Tip

Chicks need 25 square feet per 100 birds up to 3 weeks, so 8ft x 8ft offers 64 square feet and is big enough to brood 250 chicks.

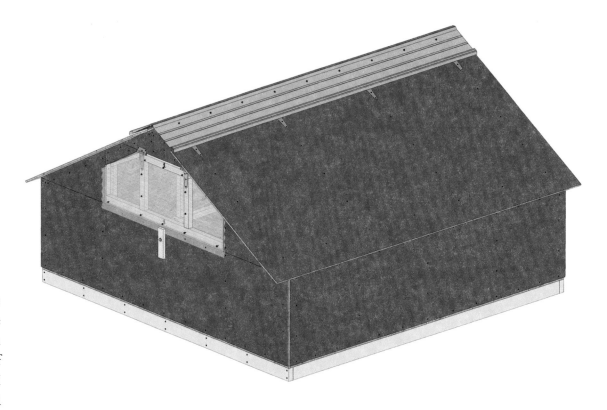

more than 2,000, I'll make a bigger one." That's fine, but as soon as you make it bigger, it'll be heavier, harder to move, harder to mount on a hay wagon bed, and harder to make the roof one sheet of plywood. When I outgrew one, I made a second one. That put me up to 4,000 bird capacity for the season.

Another advantage in modular duplication is that if something goes wrong in one, it might not go wrong in the other. You cut your risk a bit by having multiple units rather than one big one.

To be sure, now that we're brooding up to 30,000 chicks a year, we don't use these original mini-brooders anymore and have upgraded to a 1,200 square foot concrete based permanent structure. But these little brooders will stand you in good stead for a long time before you need to upgrade. They're cheap and easy to build; goodness, you can throw this together in a day. And if you set it on an existing hay wagon deck, you won't even need to build a floor structure. The beauty of this is in the clean out; you just lift

it off the bedding at the end of the season. How efficient is that?

One other note: modular brooders facilitate different types of poultry. Perhaps you want to do some turkeys or ducks. You can have a separate mini-brooder for each kind of bird, with the customized temperature and environment each prefers. Brooders don't have to be elaborate. They need temperature control, rat prevention, and deep bedding. Other than clean water and a good ration, that's about all you need to be successful.

Over the years we've created various renditions of these mini-brooders. Some changes were improvements, others weren't, but we've

learned something every time. For this book, we wanted to incorporate the "best-of's" all into one design for you, the reader, to build the best structure possible. We've incorporated some improvements in the form of more modern building materials. In the days of our first builds, there weren't many options for weather resistant materials. Today, however, we have advanced plywood sheathing available to us that allows the inside to breath by allowing moisture to exit, while simultaneously repelling water on the outside. All-in-all this is the best mini-brooder we've come up with yet, and are excited to share it with you.

Tools

❑ Hand Saw

❑ Power Drill

❑ Circular Saw

❑ Table Saw

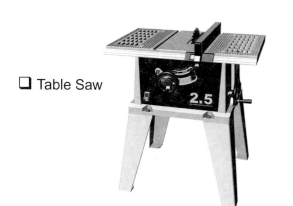

❑ Stapler

❑ ½" Staples

❑ Socket Driver Set

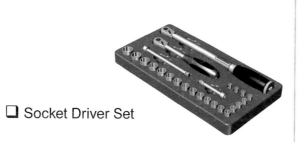

❑ Hammer

❑ Speed Square

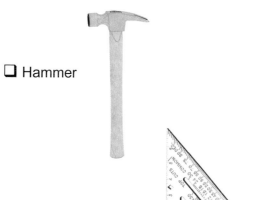

❑ Marking Pencil

❑ Tape Measure

❑ Driver Bits (for screws)

❑ 8" Tin Snips

❑ 4ft Level

Hardware

DECK SCREWS

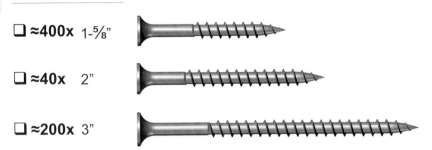

☐ **≈400x** 1-⅝"

☐ **≈40x** 2"

☐ **≈200x** 3"

BOLTS

☐ **1x*** ⅜ -16 x 2-½" Hex Head Bolts

☐ **2x** ⅜ x 3" Hex Head Lag Screws

☐ **≈22x** ⅜" Washers

☐ **1x*** ⅜ -16 Hex Nut

Bolt hardware is used to secure chain to hang brood lamp. Size of hardware may vary based on size of chain. Make sure bolt can pass through center of chain link.

☐ **≈40x** 1-½" Roofing Screws

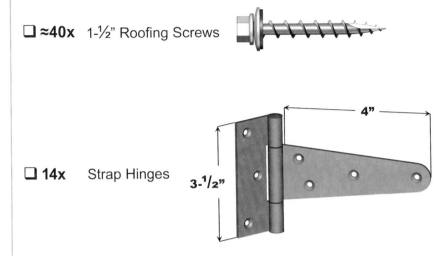

☐ **14x** Strap Hinges

4"

3-¹/₂"

Materials

☐ **≈8ft** Roll of 24" Chicken Wire

☐ **≈8"** Lightweight Chain

☐ **1x** S-Hook

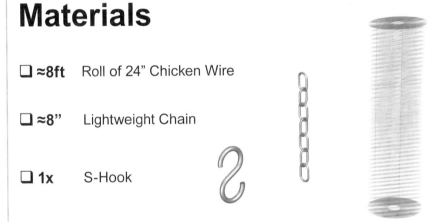

☐ **1x** Brooder Lamp

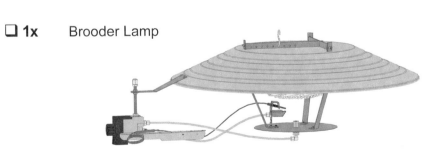

We typically use the Shenandoah® ShenVEC™ system with a 34" canopy and a heat energy output of 17-31,000 BTUs. Any comparable unit or smaller will suffice for this small brooder.

Wood Cutlist

Wood scraps are highlighted in red.
Do not discard until project is completed.

QTY	SIZE
☐ **2x**	1x2x8
☐ **21x**	2x4x8ft
☐ **1x**	2x4x10ft
☐ **2x**	2x6x10ft
☐ **1x**	2x6x8ft

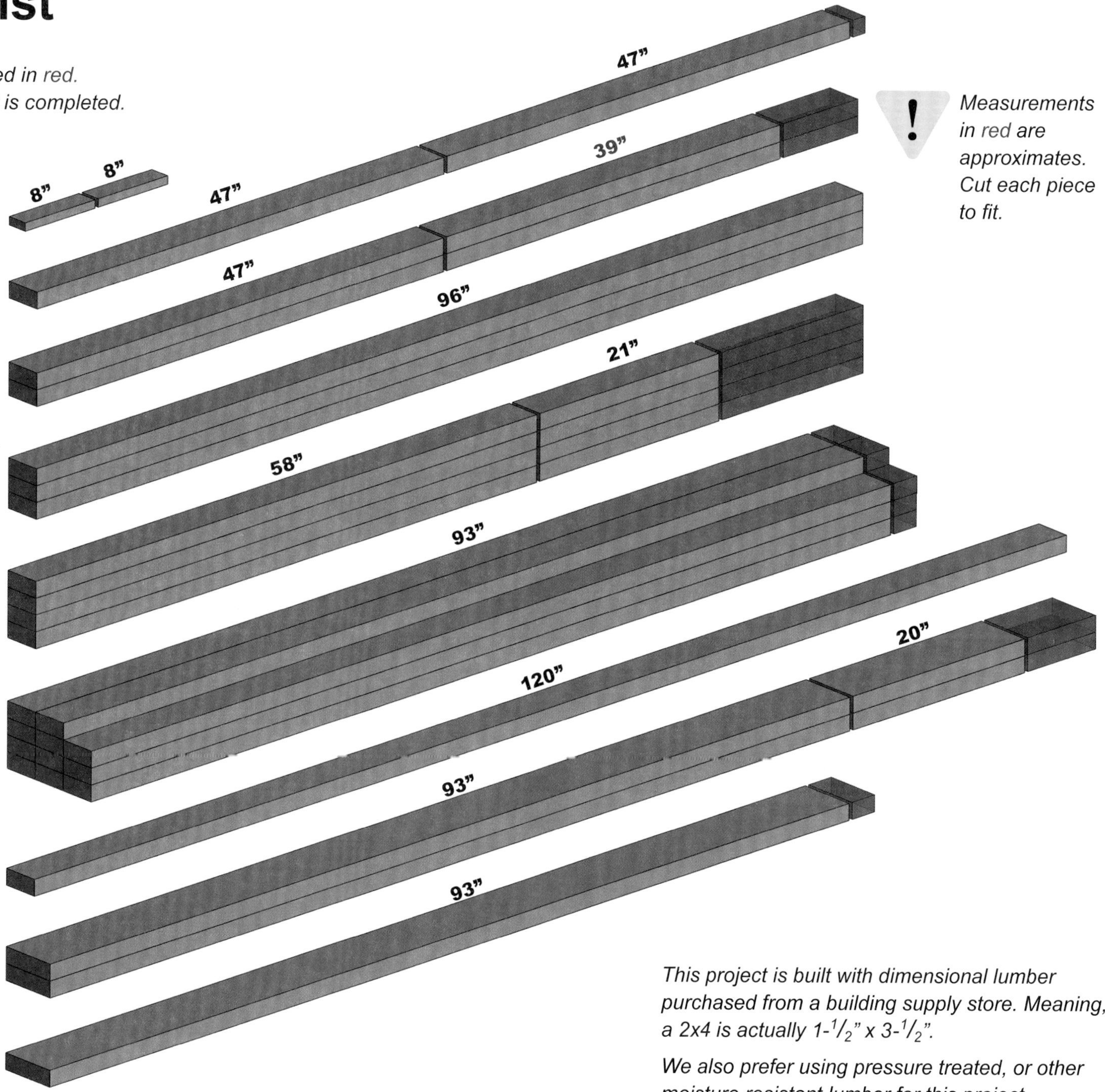

! *Measurements in red are approximates. Cut each piece to fit.*

47"

8" 8" 47" 39"

47" 96"

21"

58"

93"

120" 20"

93"

93"

This project is built with dimensional lumber purchased from a building supply store. Meaning, a 2x4 is actually 1-$\frac{1}{2}$" x 3-$\frac{1}{2}$".

We also prefer using pressure treated, or other moisture resistant lumber for this project.

Wood Cutlist

QTY   SIZE

☐ **7x** Pressure Treated 2x4s 8 feet long (cut in half lengthwise)

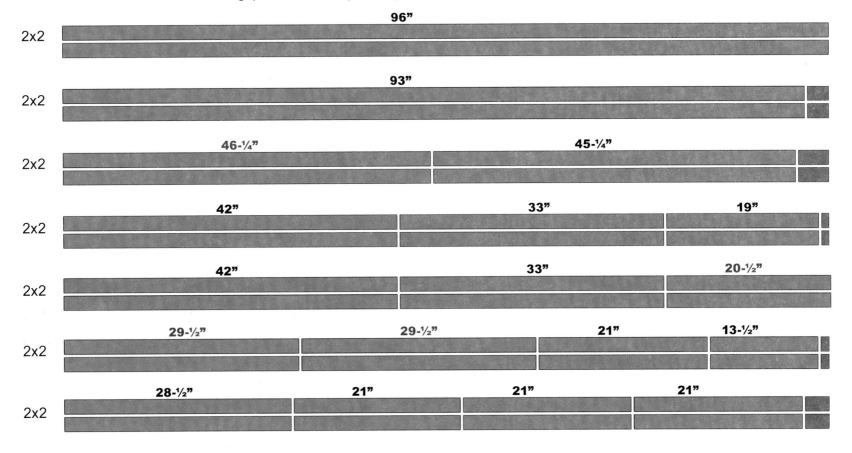

☐ **1x** Pressure Treated 2x4 10 feet long (cut in half lengthwise)

  Measurements in red are approximates.
Cut each piece to fit.

Wood scraps are highlighted in red.
Do not discard until project is completed.

QTY	SIZE

☐ **2x** 4x8ftx²³⁄₃₂ Flooring Panels

We recommend using AdvanTech® Flooring Panels by HUBER Engineered Woods because they have an excellent resistance to moisture.

☐ **2x** 4x8ftx⁷⁄₁₆ Sheathing Panels

We recommend using ZIP System® Roof and Wall Sheathing by HUBER Engineered Woods because they have an excellent resistance to moisture. They are relatively stable even when wet. The sheathing is designed to be exposed to weather for extended periods of time. It is also designed to be impermeable by water but still breath.

☐ **1x** 4x8ftx⁷⁄₁₆ Sheathing Panel

Smaller pieces from this panel will bc cut to fit.

☐ **2x** 4x9' 1-⅛"x⁷⁄₁₆ Sheathing Panels

ZIP System® also makes "Long Length" Panels in 9 and 10ft lengths, making it great for the lids to extend over the edges.

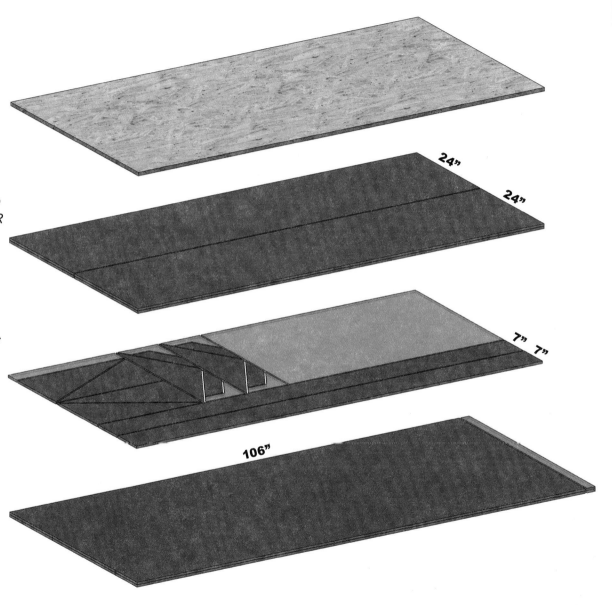

175

Mini-Brooder

Materials

QTYX	SIZE
☐ 2	Clear Polycarbonate Roofing Panels
☐ 1x	Galvanized Steel Roofing Panel

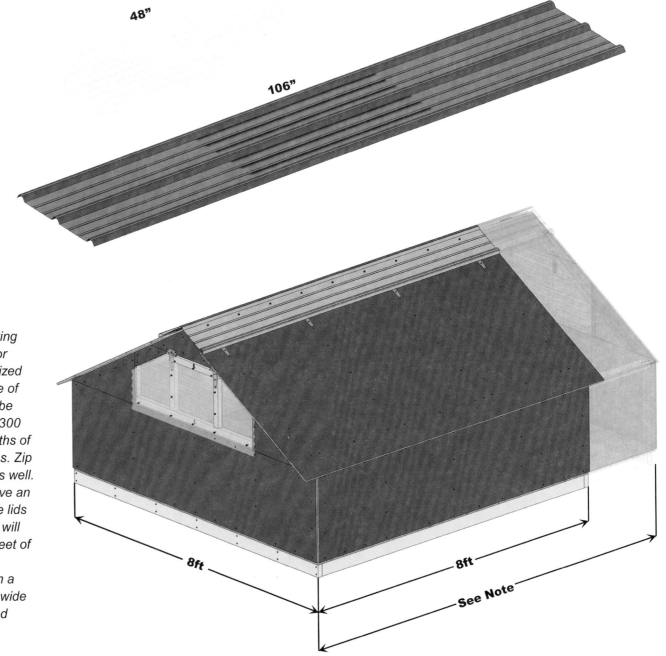

48"

106"

8ft

8ft

See Note

Note: This brooder is 8ft by 8ft, giving you roughly 64ft² space (enough for 250 birds for 3 weeks). While we sized this mini-brooder 8ft by 8ft for ease of portability, this brooder can easily be modified to be 8ft by 10ft (80ft² or 300 chicks). Simply purchase 10ft lengths of lumber where we call for 8ft lengths. Zip panels also come in 10ft lengths as well. With a 10ft brooder you will not have an "eave" overhang off the ends of the lids but you can still make it work. You will also need to purchase an extra sheet of the Advantech flooring material.

If you plan to mount this directly on a hay wagon, your deck may not be wide enough to accommodate the added length. Plan accordingly.

Build The Floor

1

2x   2x4x96"

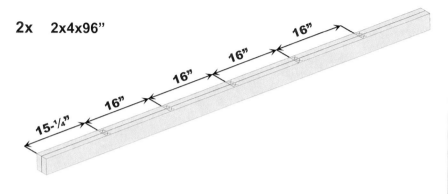

If you have a pre-existing floor to work off of, such as a hay wagon deck, then you can skip this section and proceed directly to step 6.

3

1x   48x96" Flooring Panel
35x 1-⅝" Deck Screws

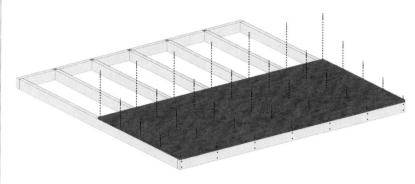

Fasten flooring every 12-16".

2

7x   2x4x93"
28x 3" Deck Screws

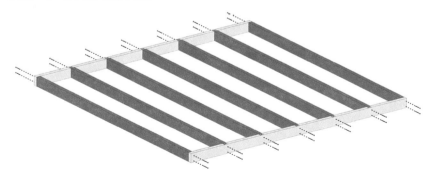

Align floor joists to the marks created in step 1 and fasten in place.

4

1x   48x96" Flooring Panel
35x 1-⅝" Deck Screws

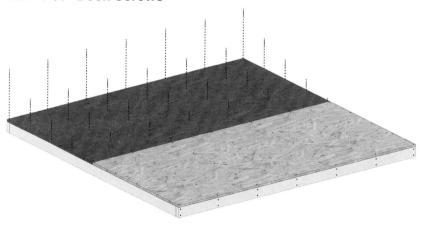

5

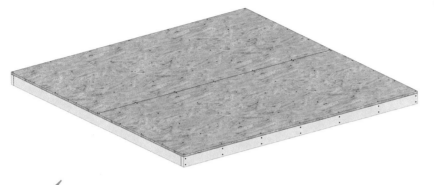

The floor is complete!

Build The End Wall

6

1x   2x2x96"

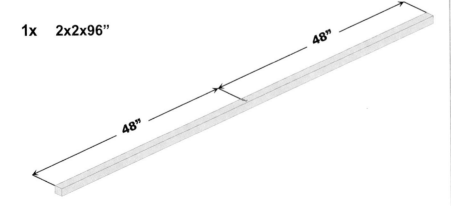

48"

48"

Mark the centerline.

7

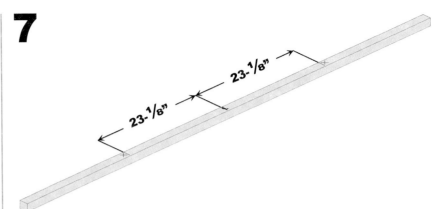

23-¹/₈"

23-¹/₈"

Mark stud locations to either side. Note which side of the line studs will appear.

8

2x   2x4x21"

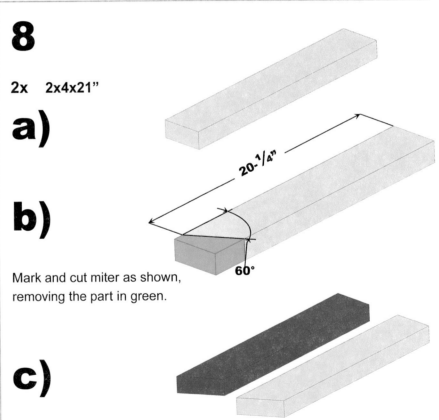

a)

b)

20-¹/₄"

60°

Mark and cut miter as shown, removing the part in green.

c)

Repeat process to create two identical pieces.

9

1x   Piece From Step 8
2x   3" Deck Screws

Fasten to end of 2x2 from step 7.

10

1x   Piece From Step 8
2x   3" Deck Screws

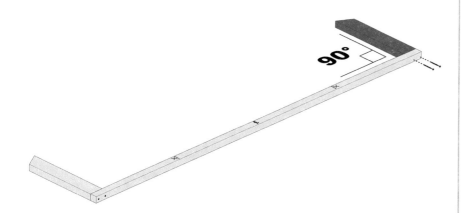

Repeat on the other side.

11

2x   2x4x58"

a)

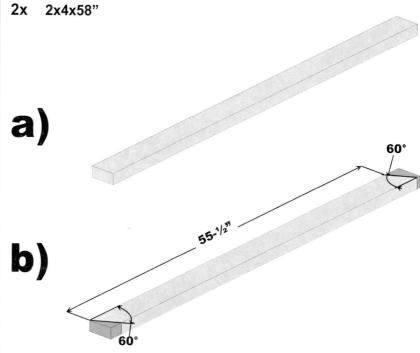

b)

Mark and cut miter as shown, removing parts in green.

c)

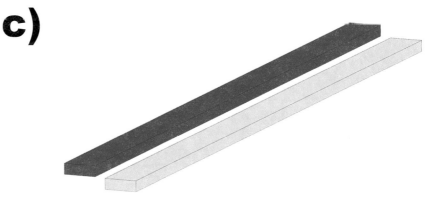

Repeat process to create two identical pieces.

12

1x   **Piece From Step 11**
2x   **3" Deck Screws**

Fasten with screws from both directions.

13

1x   **Piece From Step 11**
3x   **3" Deck Screws**

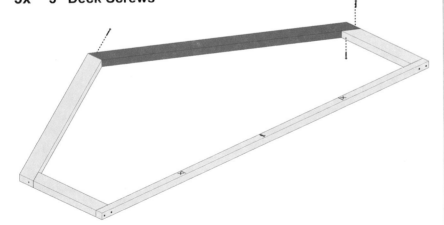

Fasten other side with screws from both directions.

14

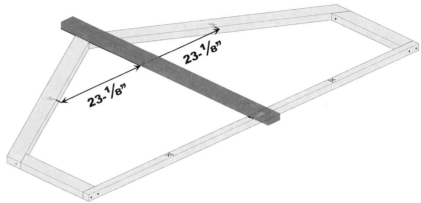

23-1/8"   23-1/8"

Using a piece of lumber as a straight edge, line up the centerlines of the wall. Measure and mark on either side as shown.

15

1x   **2x2x33"**

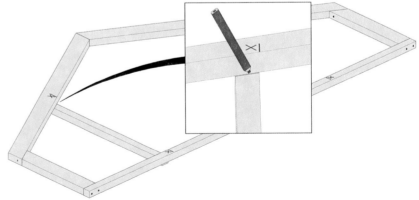

Hold the bottom end flush and scribe the mitered end with a pencil as shown, being sure the board is on the correct side of the marked lines.

16

a)

b)

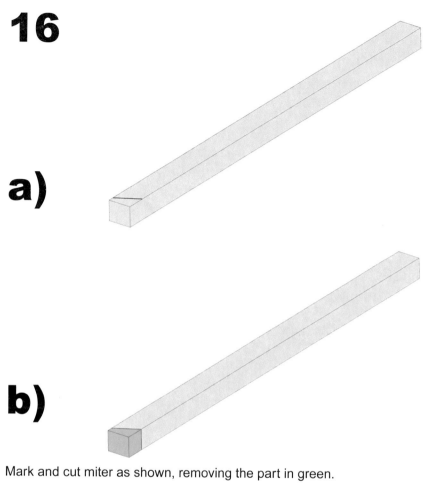

Mark and cut miter as shown, removing the part in green.

c)

17

1x   **Piece From Step 16**
2x   **3" Deck Screws**

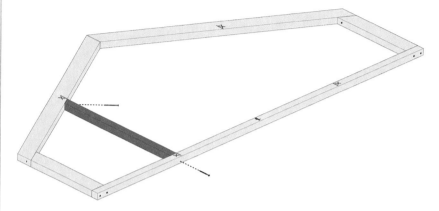

Line up with marks and fasten in place.

18

Repeat steps 15-17 on other side.

19

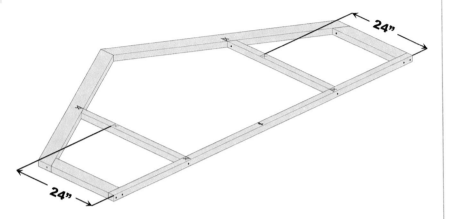

Mark in two locations.

20

1x **2x2x46-¼"**
2x **3" Deck Screws**

Cut 2x2 to fit and fasten from both ends.

21

1x **2x2x20-½"**
2x **3" Deck Screws**

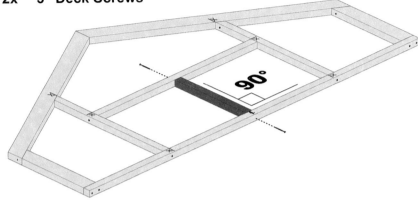

Cut 2x2 to fit, then center 2x2 along centerline and fasten.

22

1x **2x6x20"**

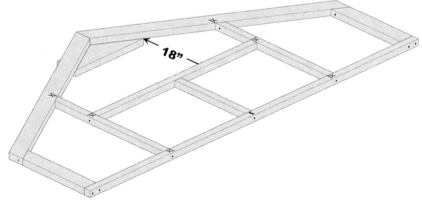

Position the 2x6 18" from the bottom of the window opening. Make sure it is even all the way across. Scribe cutlines with pencil.

23

a)

b)

Cut along the scribed marks, removing the green highlighted portion.

c)

24

1x   **Piece From Step 23**
2x   **3" Deck Screws**

Fasten as shown.

25

1x   **24x96x$^7/_{16}$" Sheathing Panel**
30x   **1-$^5/_8$" Deck Screws**

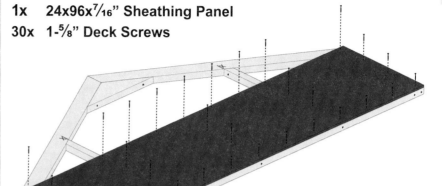

Screw sheathing every ≈8" along each stud.

26

2x   **$^7/_{16}$" Sheathing Panels** (Measured to fit)
≈12x   **1-$^5/_8$" Deck Screws**

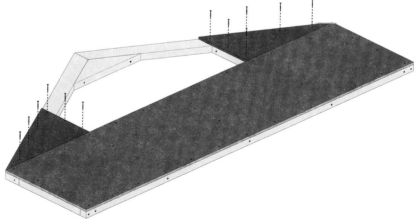

It is best to measure and cut each piece to fit. Fasten accordingly.

CHICKENS

27

1x ⁷⁄₁₆" **Sheathing Panel** (Measured to Fit)
≈6x 1-⁵⁄₈" **Deck Screws**

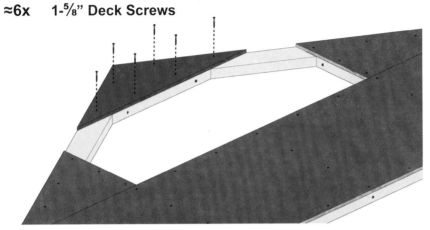

It is best to measure and cut each piece to fit. Fasten accordingly.

28

2x ⁷⁄₁₆" **Sheathing Panels** (Measured to Fit)
≈8x 1-⁵⁄₈" **Deck Screws**

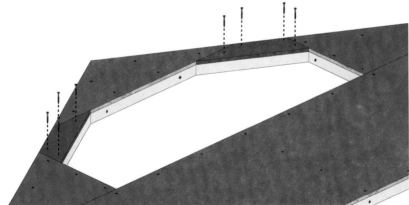

It is best to measure and cut each piece to fit. Fasten accordingly.

29

≈4x 24" **Roll of Chicken Wire**

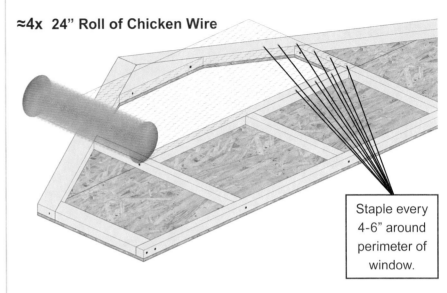

Staple every 4-6" around perimeter of window.

30

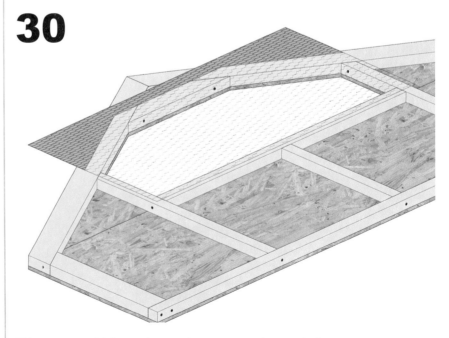

Trim excess chicken wire as shown, removing parts in green.

Build The Window Frame

31

1x 2x2x45-¼"

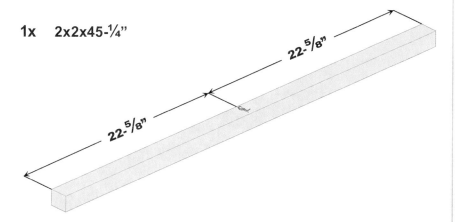

Mark centerline.

33

1x 2x2x13-½"
1x 3" Deck Screw

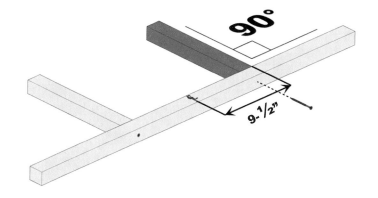

32

1x 2x2x13-½"
1x 3" Deck Screw

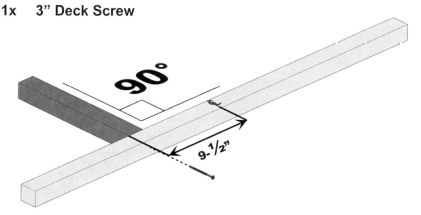

34

1x 2x2x19"
2x 3" Deck Screws

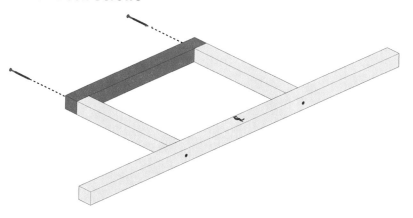

35

2x **Strap Hinges**
≈8x **1-⅝" Deck Screws**

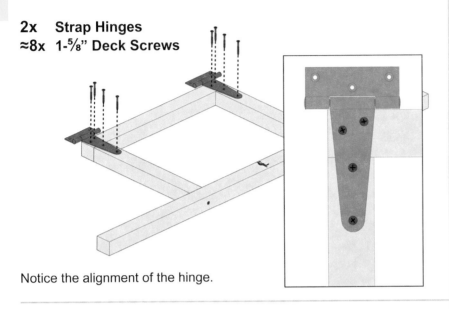

Notice the alignment of the hinge.

36

≈6x **2" Deck Screws**

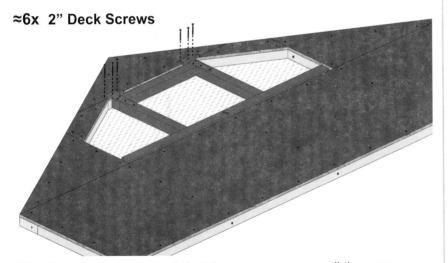

Align the window frame such that there are even gaps all the way around, and screw hinges in place.

37

1x **Clear Polycarbonate Roofing Panel, 48" Long**

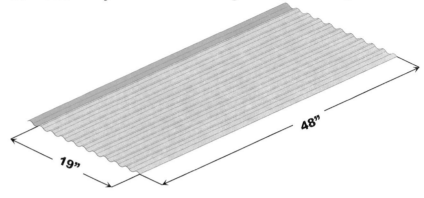

You may need to trim the polycarbonate depending how wide it is when purchased.

38

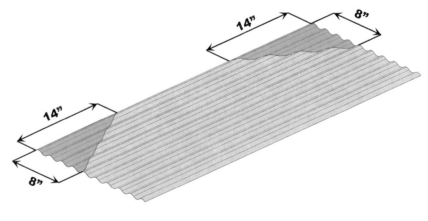

These dimensions are estimations only. Place panel in place over window and mark cuts for the best fit.

39

≈10x   1-½" Roofing Screws

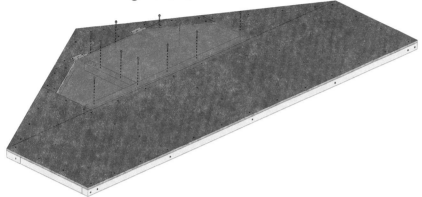

Window should overlap on the bottom end to prevent water intrusion. Fasten to frame with roofing screws.

41

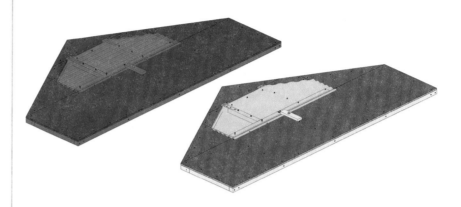

Repeat steps 6-40 to create a second end piece.

40

1x   1x2x8"
1x   ⅜ x 3" Hex Head Lag Screw
≈9x   ⅜" Washers

This step illustrates the assembly of a latch mechanism for the window. It consists of a wooden catch that swivels on a lag bolt. By off-setting the hole in the wood catch, gravity will always make sure it is oriented in the vertical position which happens to latch the window shut.

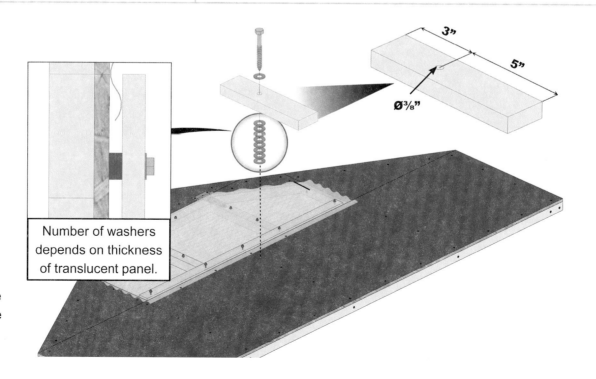

Number of washers depends on thickness of translucent panel.

3"

5"

Ø⅜"

Attach The Walls

42

1x   2x2x93"
1x   2x4x93"

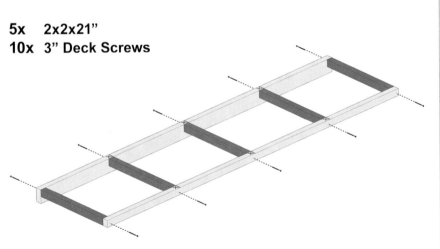

Mark stud locations on both pieces.

44

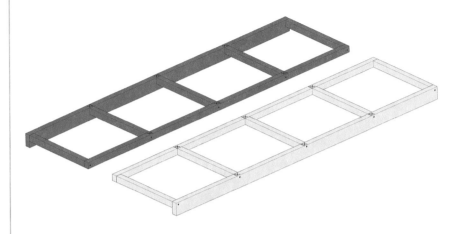

Repeat steps 42 and 43 to create a second side.

43

5x   2x2x21"
10x   3" Deck Screws

45

5x   3" Deck Screws

Fasten endwall to the floor.

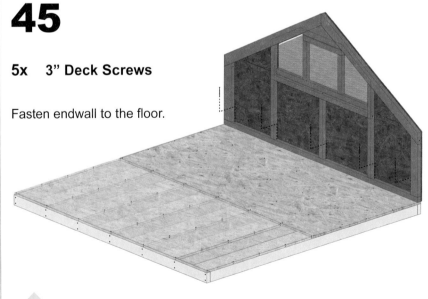

Be aware of the orientation of the floor joists. This will be important if you plan to mount this on a wagon.

46

9x   3" Deck Screws

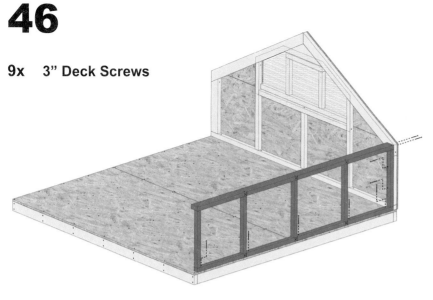

Fasten side to floor and endwall.

48

13x   3" Deck Screws

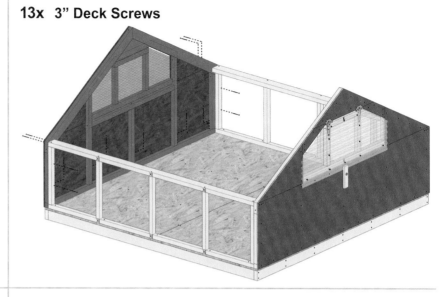

47

9x   3" Deck Screws

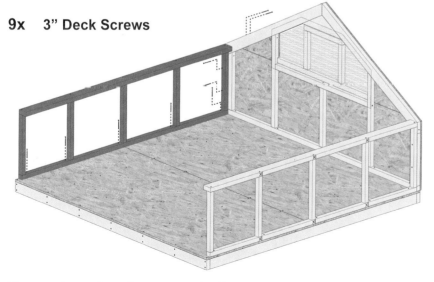

Fasten other side to floor and endwall.

49

1x   24x96x⁷⁄₁₆" Sheathing Panel
36x 1-⁵⁄₈" Deck Screws

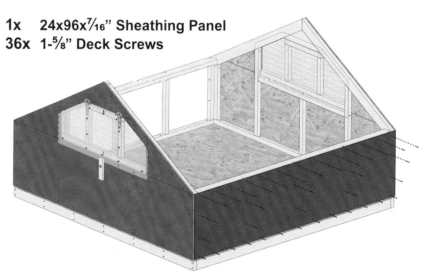

Fasten the sheeting to open sides with screws every 8".

50

1x   24x96x$^{7}/_{16}$" Sheathing Panel
36x   1-$^{5}/_{8}$" Deck Screws

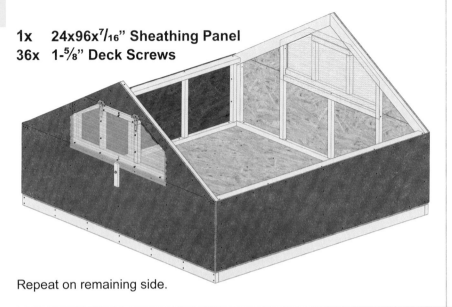

Repeat on remaining side.

51

1x   2x4x93"
4x   3" Deck Screws

8"

8"

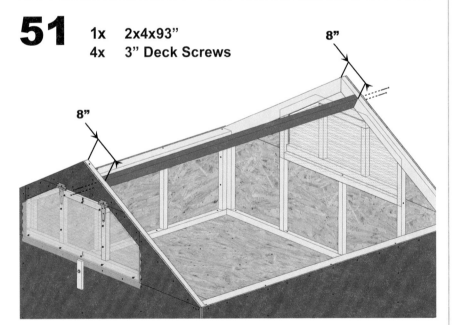

Measure down 8" from the peak on one side and fasten a 2x4 cut to fit.

52

1x   2x6x93"
9x   3" Deck Screws

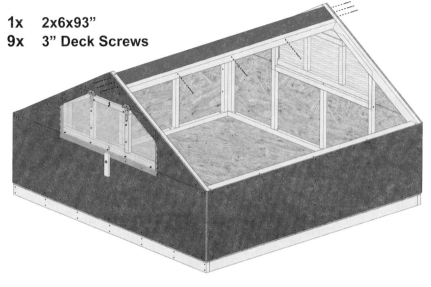

53

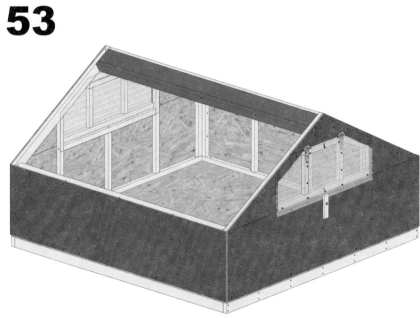

Repeat steps 51 and 52 on the other side.

54

1x   **2x6x93"**
2x   **⅜" Washers**
1x   **⅜" Nut**
8"   **Lightweight Chain**
1x   **⅜-16 x 2-½" Hex Head Bolt**

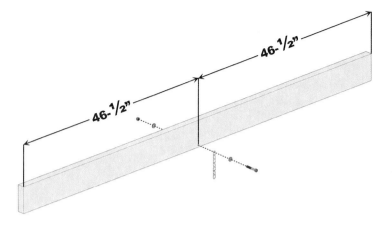

Prior to installing the board, it is easiest to drill and mount the chain.
This chain will be used to hang the brood lamp in a future step.

55

12x 3" Deck Screws

If you Increase the size of the brooder to 8x10ft, we suggest replacing this 2x6 with a 2x8 for added support.

For added ventilation, you can drill some ½" holes at an angle through the top of this structure.

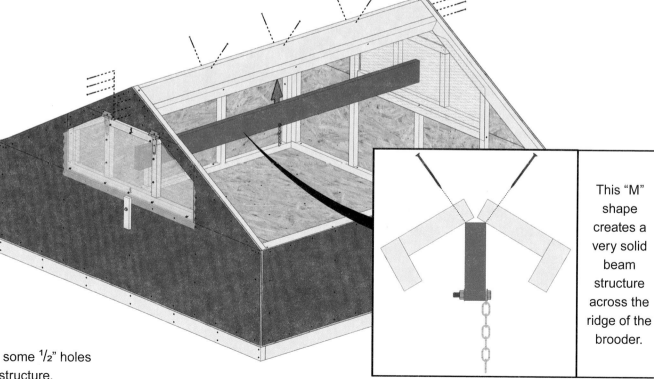

This "M" shape creates a very solid beam structure across the ridge of the brooder.

56

1x   2x4x96"
1x   2x4x120"

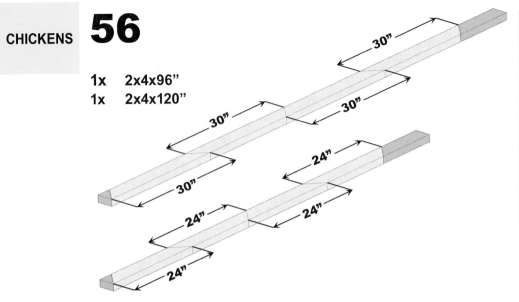

This may look confusing, but it is a simple way to maximize the usage of lumber and make less cuts. All miters are 45°.

57

1x   30" Brace From Step 56
3x   3" Deck Screws

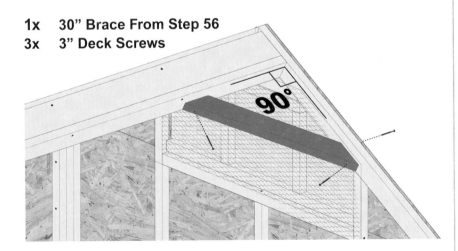

Proceed to install the 30" long braces at the top of the openings as shown. Square the corners before fastening.

58

1x   24" Brace From Step 56
4x   3" Deck Screws

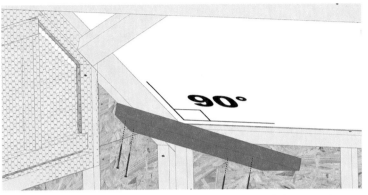

Now install the 24" long braces at the bottom of the openings as shown. Square the corners before fastening.

59

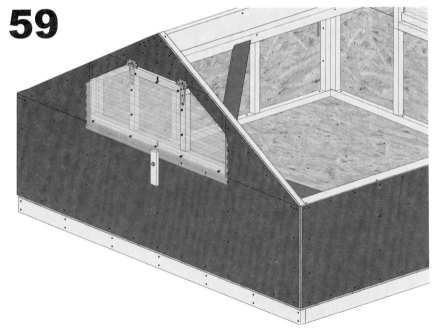

Repeat steps 57 and 58 on the left side of the opening.

60

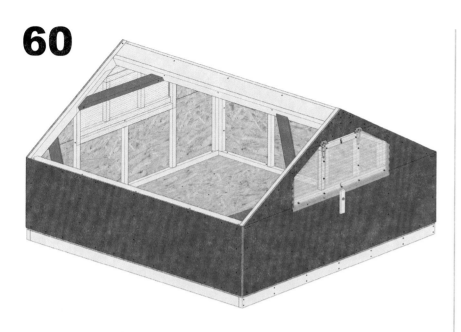

Repeat steps 57-59 to secure the 4 braces on the other roof opening.

61

1x   48x106x⁷⁄₁₆"
Sheathing Panel

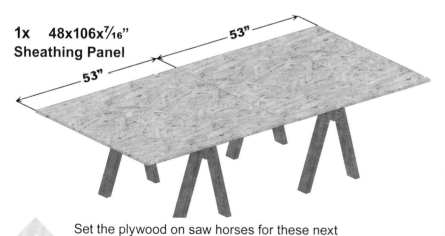

 Set the plywood on saw horses for these next steps because you will need to fasten pieces from below.

Mark centerline of plywood sheet.

62

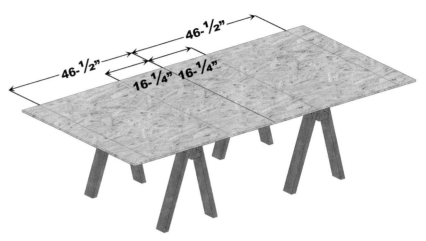

Mark the following 4 lines off of the centerline.

63

Lastly, mark lengthwise.

CHICKENS

64

2x   2x4x47"

a)

b)

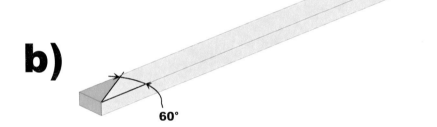

60°

Mark and cut miter as shown, removing the part in green.

c)

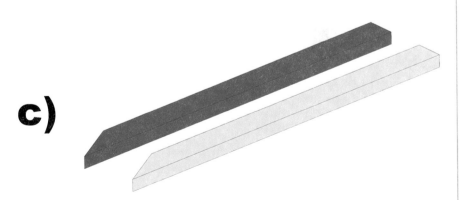

Repeat process to create two identical pieces.

65

2x   Piece From Step 64
8x   1-⅝" Deck Screws

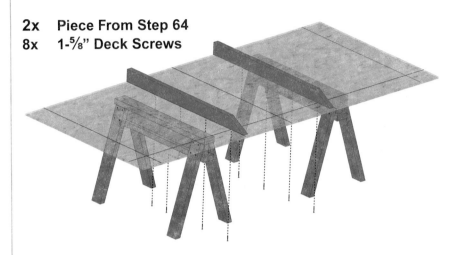

Boards should be aligned on the inside of the marks. Fasten from below.

66

2x   2x2x42
8x   1-⅝" Deck Screws

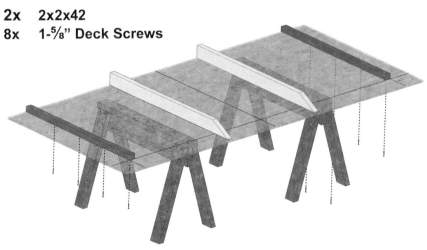

Align boards on the inside of the marks. Fasten from below.

67

2x   2x2x29-½"
6x   1-⅝" Deck Screws

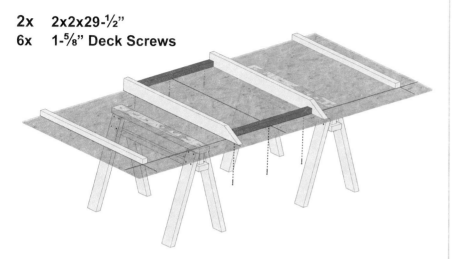

Cut 2x2's to fit and fasten from below.

69

4x   Strap Hinges
≈16x   1-⅝" Deck Screws

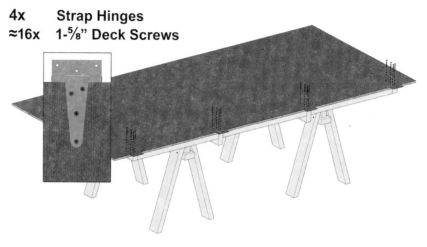

Notice the alignment of the hinges.

68

4x   2x2x28-½"
12x   1-⅝" Deck Screws

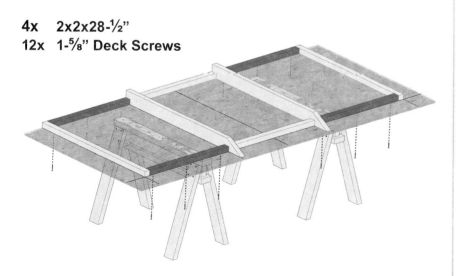

Cut 2x2's to fit and fasten from below.

70

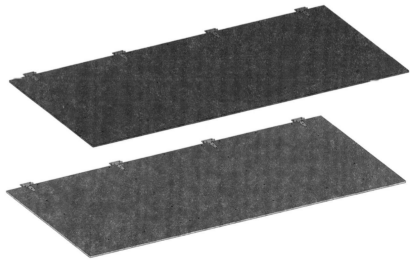

Repeat steps 61-69 to create a second lid.

71

1x   7x96x⁷/₁₆" Sheathing Panel
5x   1-⁵/₈" Deck Screws

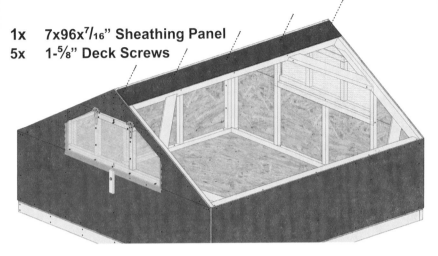

Cut and fasten sheathing to cover top. Just a few screws along the top will suffice for now.

73

3x   1-⁵/₈" Deck Screws

≈19"

Fasten from underneath.

Prop stick should be able to rest on the top angle brace.

72

1x   2x4x39"
1x   Strap Hinge
4x   1-⁵/₈" Deck Screws

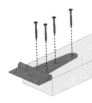

Create a prop stick to hold open the lid while working. Dimensions may vary slightly from what we've listed. It's best to leave it a little long in length and trim as needed.

74

12x   2" Deck Screws

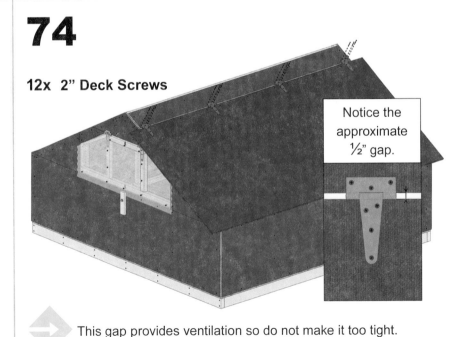

Notice the approximate ½" gap.

This gap provides ventilation so do not make it too tight.

75

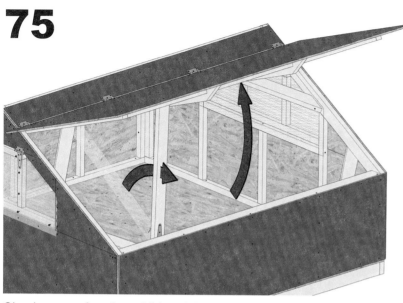

Check proper function of lid and the prop stick. Make adjustments as needed.

77

1x Galvanized Steel Roofing Panel, 106" Long

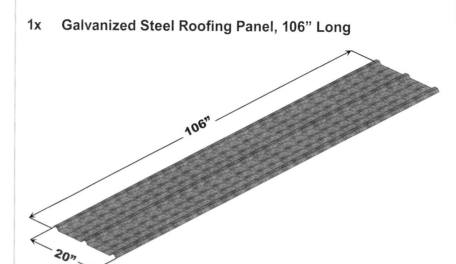

106"

20"

76

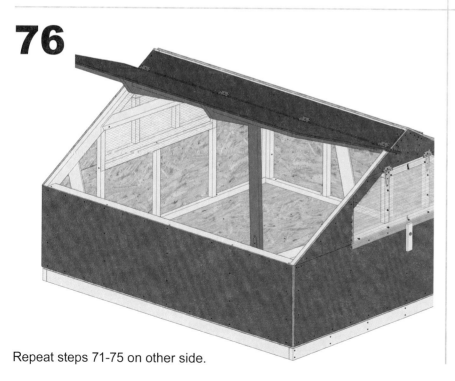

Repeat steps 71-75 on other side.

78

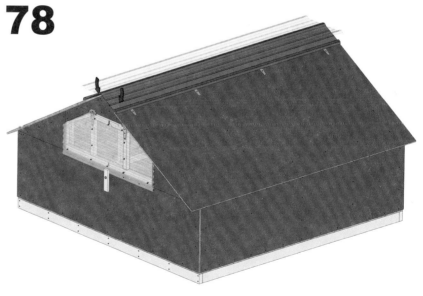

Center on ridge and fold down either side. Metal should cover the gap between the hinges to prevent water from seeping in.

Mini-Brooder

79

≈18x   1-½" Roofing Screws

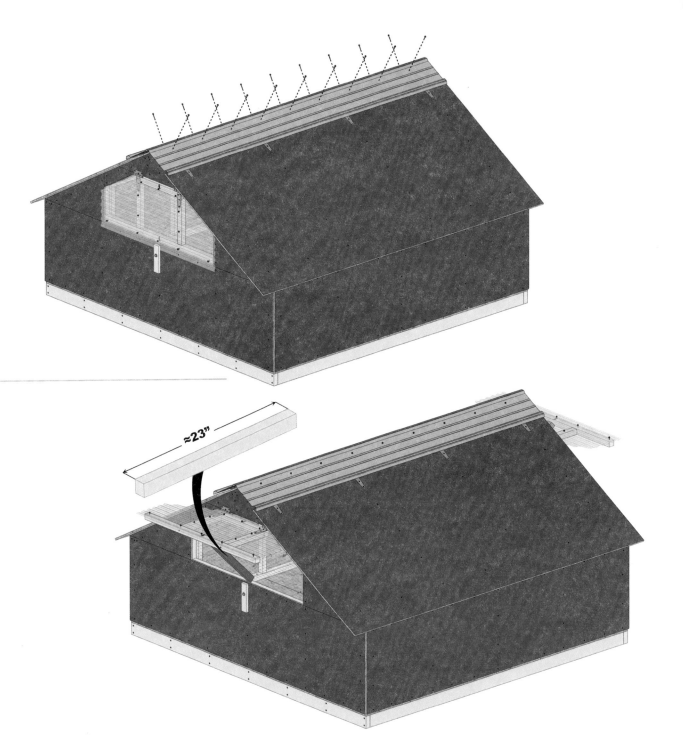

Fasten with screws only along the top. You want the bottom to flex so that it doesn't hinder the motion of the lid and hinges.

80

2x   Scrap Pieces of Wood

≈23"

Cut prop stick to length. Ideally the window should rest loosely against the overhanging portion of the lid without binding. Repeat process on other side.

81

1x   **Brooder Lamp**
1x   **S-Hook**

Be sure to follow the installation instructions for the brood lamp you choose. There is typically a recommended height from the top of the bedding.

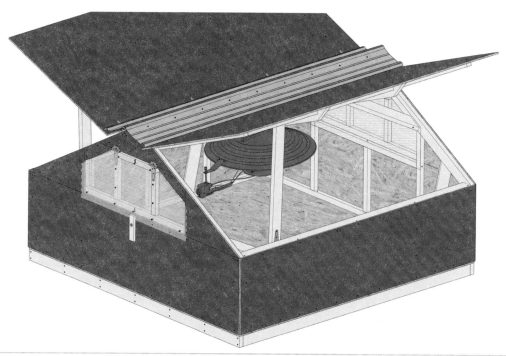

82

If your brooder is propane powered, we recommend hiring a professional to install your gas line and check for leaks. Some propane providers will not even service you unless it has been professionally installed. Check with local regulations and propane utilities before starting.

The mini-brooder is complete!

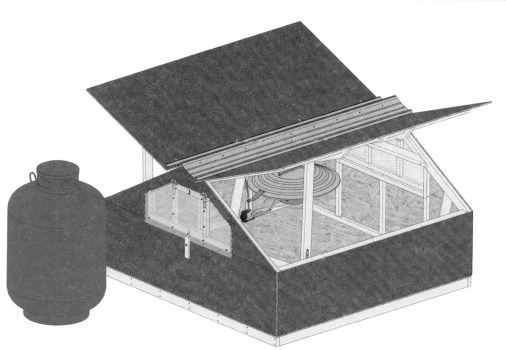

Mounting Options

Concrete Blocks

Concrete blocks can work, but I suggest putting it at least two or three blocks high to make it more difficult for rats to climb up and get in. Rats can climb up concrete blocks easily. Rats can get in holes you can barely stick your finger through; unless you're a first class builder, you'll have a hole that size somewhere around the brooder. Rats can come in from the floor, the roof, the side. They're not picky. The main deterrent is to keep the area clean. Mow the grass short in a 15 ft. perimeter; don't stack any junk around. And keep a couple of good hunting cats.

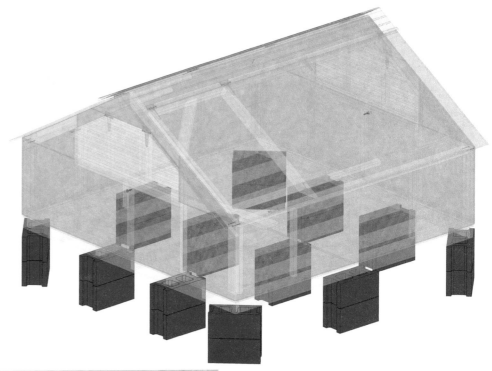

Skids

Ditto everything I said above about the blocks, except realize that the skid option creates even more vulnerability to rats. Over the decades, rats have taken more chickens than any other predator at Polyface. Fortunately, they seldom attack a chicken older than 4 weeks, but when they attack little chicks, a couple of hard working rats can take 100 chicks in a night. I've seen it.

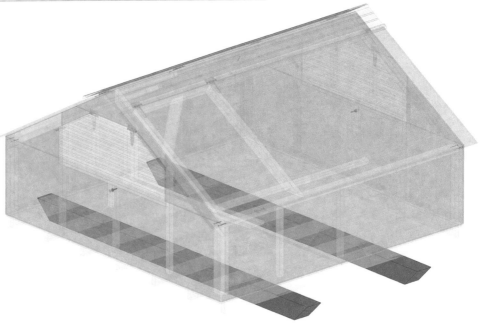

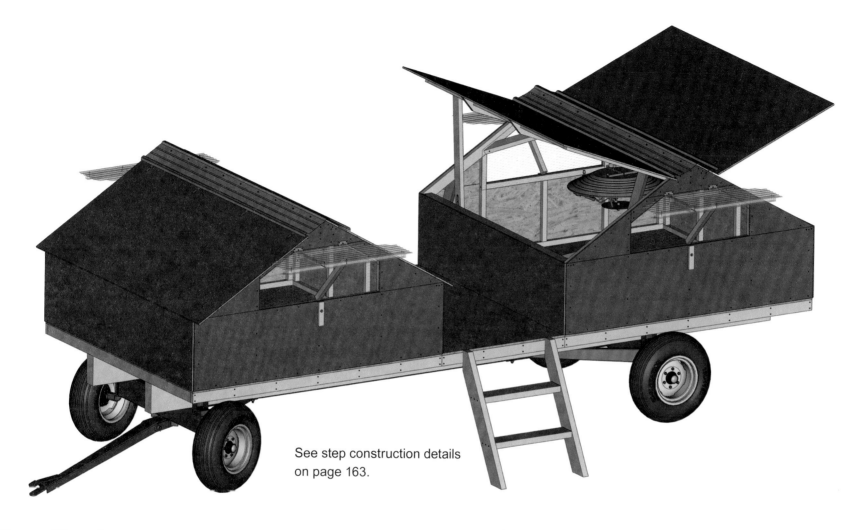

See step construction details
on page 163.

Wagon Chassis

While the hay wagon may be more costly than blocks or skids, what are 100 or 200 chicks worth? A few losses in a season can easily pay for a wagon. While a rat can climb up the tires and get onto the wagon, it's not as easy. To my knowledge, we've never had a rat attack on a hay wagon chassis. This is also the most portable option. And, if you procure a wagon large enough, you can install two brooders on it.

That splits the cost in half per unit and makes that initial expenditure more palatable.

One of the beauties of the wagon is that if it already has a deck, then you don't have to build a floor under the brooder. You can just set it on the wagon. If the deck has gaps, then fasten a few sheets of AdvanTech® flooring to it and then place the brooder on top.

If all you have is a bare running gear,

procure two large timbers and install them as beams along the length of the chassis, and simply mount the brooder directly on top. If you are feeling fancy you can add a deck in between the two brooders for easier access. The Millennium Feathernet step design works perfectly on this setup too. Just be sure to slide the steps up onto the deck at night to prevent rats from climbing up.

COWS

SHADEMOBILE

In my view, the triumvirate of forage management is electric fence, water pipe, and mobile shade. When we came to our farm in 1961 (I was only 4) and Dad began moving the cows around in paddocks, he quickly saw the value in mobile shade.

Most people think of it as most valuable for animal comfort, but it's perhaps more valuable as a fertility engine. While it is true that heat stressed cattle do not eat much and therefore don't gain much, where they lounge and drop their manure and urine is equally important.

While I appreciate silvo-pasture and agro-forestry type designs to use trees as shade, nothing beats mobile shade. Anyone who disagrees, in my opinion, has never tried mobile shade. One of the problems with trees in pasture is that they complicate hay making. We do not have designated hay fields and want to be able to mow anything. Trees drop dead branches that can damage hay equipment and their shade keeps hay from drying, creating wet spots prone to mold and heat. From a land management efficiency standpoint, I like open fields bounded by trees. Too much diversity can reduce functionality.

Dad designed and together we built our first Shademobile from a burned out mobile home frame back in 1969. In those days, the farm's impoverished fertility responded to Shademobile spots with dramatic dark green squares. Visible for 7 years, these squares played a significant role in our fertility rehabilitation program.

I'm convinced that one of the reasons the Shademobile spots respond so dramatically is because they offer shade to the urine. The lion's share of the nitrogen is in urine; on a hot, cloudless day a lot of that can evaporate. But under the Shademobile, it's protected for at least several hours, which gives it time to percolate into the ground and even get metabolized by soil life. As a vegetation management tool, too, the Shademobile has no equal. If you have a patch of weeds or brambles, you can park that Shademobile on top of them and in one day literally change a spot from thin grass and brambles to thick grass and clover. It's miraculous. Because they pack in so tightly, a mixed herd will average only 10 sq. ft. per animal. Our 600 sq. ft. Shademobiles, therefore, can handle 60 mixed head. Four hooked together can handle 240 head.

University of Kentucky several years ago concluded that on a hot day a stocker calf will gain an additional .8 pounds if it has shade. At $1.40 a pound, that .8 pound is worth $1.12. If you have 100 under two of these Shademobiles, that's an additional $112 a day, for an additional move time of about 10 minutes. I don't know about you, but I can sure work for $100 per 10 minutes.

While you can get the shade benefit with trees, you can't get any of the fertility benefits. Who wants to lose all the pasture-generated manure and urine under trees? Not only is it in the wrong place, it's essentially a campsite that builds up toxic levels of manure in one place while the pasture goes begging. A Shademobile allows you to place the pasture droppings in the exact spot where they'll do the most good. And the halo extends several years.

The main idea in design is to minimize obstructions where the animals stand. You don't want to box them out of any area, which is why the H chassis allows access to every spot. Fortunately, cows enjoy getting wet in a rain, so

the nursery shade cloth on top (we use 80 percent light exclusion) works great. A solid roof tends to catch too much wind and blow over. Because shade cloth is permeable, even hurricane force winds have no effect except a slight rippling of the fabric.

In the winter, we take off the shade cloths and fold them up like bed linens to reduce weathering and tearing. If folded methodically, They can be quickly installed with minimal labor. (Check out the appendix for a step-by-step on folding and unfolding shade clothes). Because the purlins are not permanently fastened, but lie in notches, they can be removed quickly, stacked, and tied to the tongue for transport down a public road. The whole idea is to be able to use it in multiple locations with your herd on land that you do not own.

While the superstructure certainly does not have to be as strong as it is to hold up something as light as the shade cloth, it has to be strong enough to withstand cows rubbing on it. The roof needs to be beyond their reach so they don't tear up the shade cloth, but the support skeleton needs to be strong.

The original one that Dad built was 20ft. wide and 50ft. long, all metal, with a metal roof, on three legs (a tricycle). That made the massive thing maneuverable, but it was too wide to take down a public road. It blew over twice until we abandoned that and went to shade cloth. Our current design was an effort to build something we could legally take down a public road.

If you had sat around our kitchen table during the years we wrestled with designs, you

would have been extremely entertained. At one point I wondered if we could put together a metal lattice of conduit with a helium weather balloon in each square. We would tie it to a cow and wherever she went, the shade would go. Can you imagine?

As with most good designs, what we finally came up with was extremely simple and cheap. Initially we didn't want to fool with all those 4-wheel chassis and imagined trailers all hooked together, but the problem with trailers is that going down a steep hill, they could start pushing sideways on the first one if you had 4 hooked together. I've been scared to death on a steep hill pulling a little square baler and hay wagon when the hay wagon pushed the baler sideways. Not a good feeling.

By going to a 4-wheel chassis, we solved that problem because in a turn, all the weight of the Shademobiles would not be against a single axle. The turning front wheels take a lot of that pressure away.

Because the residual effect of these Shademobiles is long and often you can move them more than once a day, they can maintain fertility on nearly half your pasture acreage. That's no small coverage. While it's true that on soft ground right after a rain the animals can pug up a square, it'll look terrible for a year but the following year it'll be the most beautiful spot in the field. We've done this for 40 years so have plenty of experience to back this up. Ideally, someday someone will invent self-propelled Shademobiles that you can move with a remote.

COWS

Tools

COWS

- [ ] Hand Saw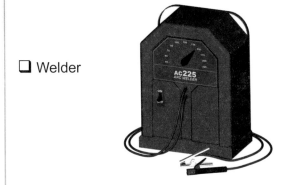

- [ ] Power Drill

- [ ] 6 and/or 8ft ladder

- [ ] Hammer

- [ ] Table Saw

This project requires a tractor or hydraulic loader. We tailored the assembly process to minimize the tractor time needed. If you do not own a tractor, borrowing one is a plausible option because it won't take long to set the prebuilt wooden structure on top of the wagon chassis.

- [ ] Welder

- [ ] Circular Saw

- [ ] Grinding Wheel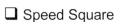

- [ ] Cut-Off Wheel

- [ ] Angle Grinder

- [ ] Socket Driver Set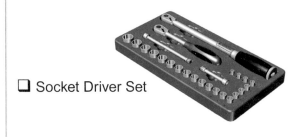

- [ ] Marking Pencil

- [ ] Tape Measure

- [ ] Speed Square

- [ ] ½" HSS Drill Bit

- [ ] Driver Bits (for screws)

- [ ] ½" Auger Bit

- [ ] 4ft Level

Hardware

DECK SCREWS

☐ **48x**   2-½"

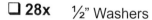

☐ **94x**   3"

NAILS

☐ **20x**   20D Galvanized Nails

BOLTS

☐ **28x**   ½" -13 Hex Nuts

☐ **28x**   ½" Washers

☐ **8x**   ½-13 x 3"
Carriage Bolts

☐ **20x**   ½-13 x 5" Carriage Bolts

☐ **1x**   Shade Cloth 20'x36',
80% Shade

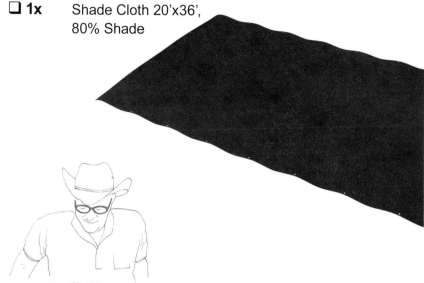

COWS

Joel's Tip

We use 80% shade cloth because that seems to most closely approximate the variegated shade of a tree. If you look at shade under a tree, it's generally not opaque, but mottled, or kind of blotched because every particle of sunlight is not blocked by a leaf. That actually is more enjoyable for the animals. Less than 80 percent is not shady enough. The shade cloth is permeable but cows don't mind getting wet; they kind of like it on a hot day. The main benefit of the permeable cloth is to let the wind through. Even in extremely high winds (80 mph) the cloth catches no wind and simply flutters a bit.

Grommets strategically placed down the edges and corners offer strong fastening points. You can also purchase grommet kits so you can install them yourself. Any major horticultural supplier has shade cloth. Most will custom cut it for you, but it's always cheaper to find a dimension they sell standard. The more you buy, the cheaper it is. We always fold them up and store them in a shed over the winter to reduce ultraviolet breakdown and weathering.

Wood Cutlist

Wood scraps are highlighted in red.
Do not discard until project is completed.

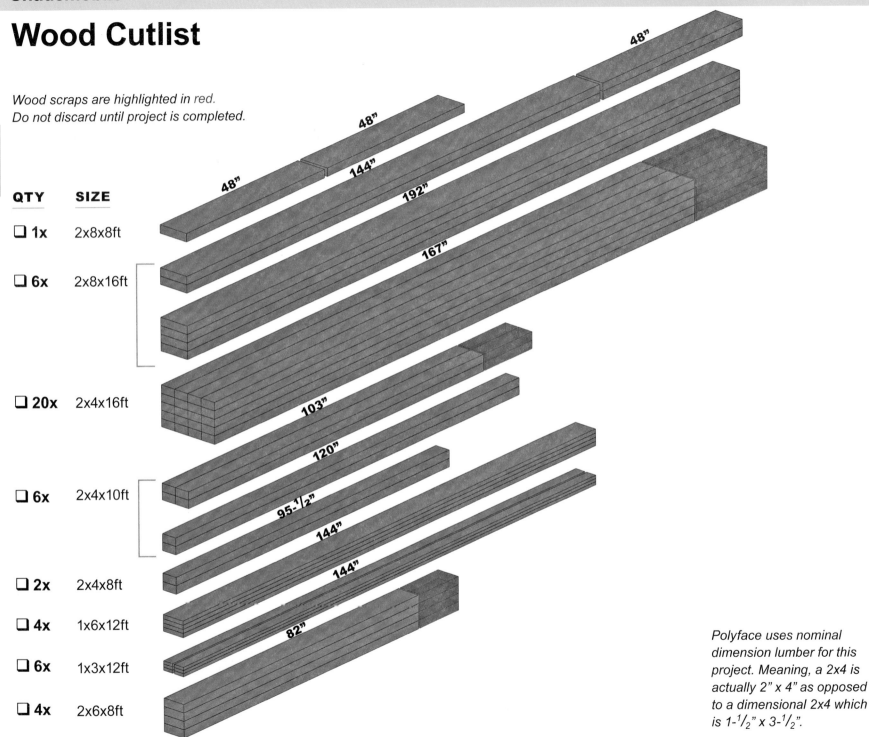

QTY	SIZE
☐ 1x	2x8x8ft
☐ 6x	2x8x16ft
☐ 20x	2x4x16ft
☐ 6x	2x4x10ft
☐ 2x	2x4x8ft
☐ 4x	1x6x12ft
☐ 6x	1x3x12ft
☐ 4x	2x6x8ft

Polyface uses nominal dimension lumber for this project. Meaning, a 2x4 is actually 2" x 4" as opposed to a dimensional 2x4 which is 1-$\frac{1}{2}$" x 3-$\frac{1}{2}$".

Metal Cutlist

! *Measurements in red are approximates.*
See step 7 prior to cutting to length.

COWS

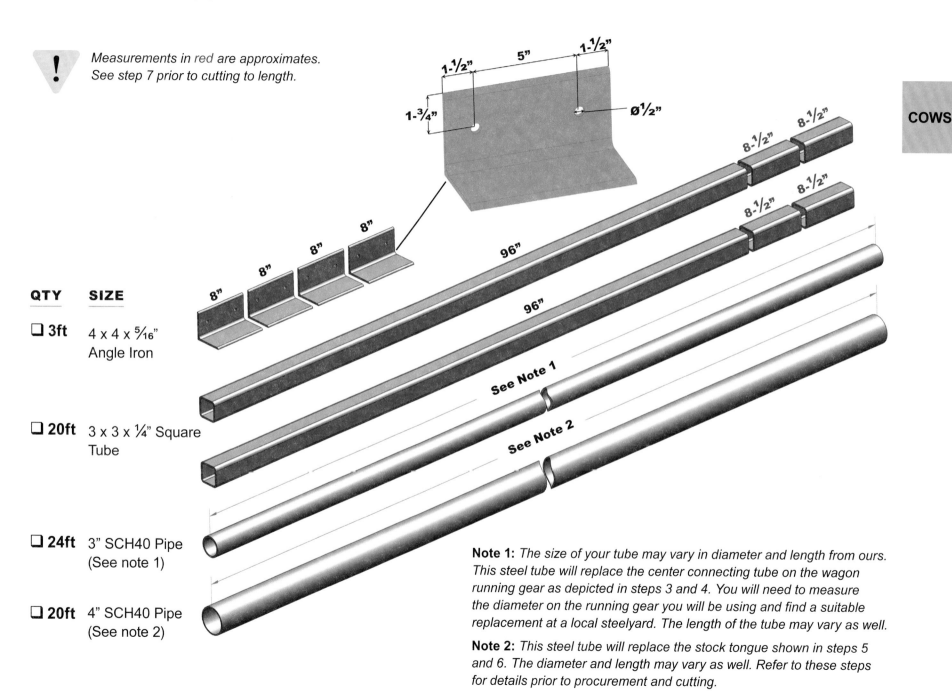

QTY	SIZE	
☐ **3ft**	4 x 4 x ⁵⁄₁₆" Angle Iron	
☐ **20ft**	3 x 3 x ¼" Square Tube	
☐ **24ft**	3" SCH40 Pipe (See note 1)	
☐ **20ft**	4" SCH40 Pipe (See note 2)	

Note 1: *The size of your tube may vary in diameter and length from ours. This steel tube will replace the center connecting tube on the wagon running gear as depicted in steps 3 and 4. You will need to measure the diameter on the running gear you will be using and find a suitable replacement at a local steelyard. The length of the tube may vary as well.*

Note 2: *This steel tube will replace the stock tongue shown in steps 5 and 6. The diameter and length may vary as well. Refer to these steps for details prior to procurement and cutting.*

1

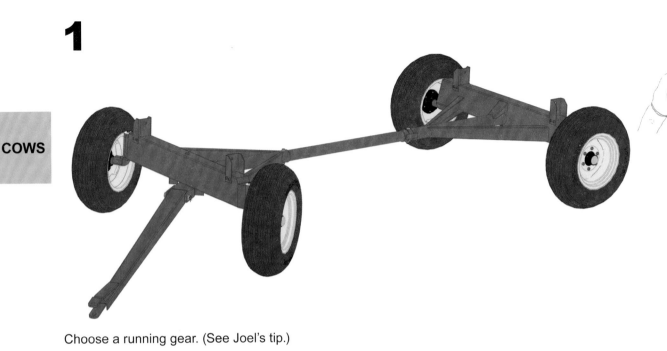

COWS

Choose a running gear. (See Joel's tip.)

2

If your running gear is equipped with "bolster stakes", then they will need to be removed. Unbolt or cut them off as needed. Save them as they are invaluable to have on hand. Inevitably you will procure a running gear in the future that is missing one or all of them.

Joel's Tip
Choosing a Running Gear:

Because these Shademobiles are extremely lightweight, you don't need a heavy chassis. Most farm auctions include several hay wagons in various states of disrepair. Of course, you don't need a bed, so if it's a raw running gear, so much the better. Make sure all the wheels spin. If the tires are rotted off, no problem. But you do want to check the rims for rust; often buying a matched rim can cost as much as the whole running gear itself. Pick up the tongue and move it from side to side to make sure the tie rods aren't frozen. Lots of times at farm auctions a hay wagon running gear may have been on the proverbial bone pile for 20 years; make sure everything that's supposed to move does in fact move.

Tires are easy to replace; don't worry about them. Finally, check what I call the wobble factor. How tight is the front end? You don't need it tight enough to go 50 miles an hour down the highway, but you also don't want it so wonky and wobbly that it won't track through the field or through a gate. Remember that if you're going to hook several together, your extended tongue will exacerbate any wobbliness.

3

Measure the diameter of the center chassis tube and try to find something similar at your local steelyard in a longer length.

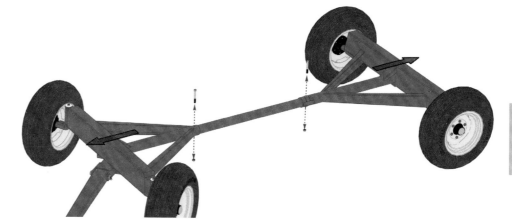

Running Gear Length

We anticipate the most common design change to be made will be the length. While a 36ft shade with a 24ft long running gear is advantageous for maximizing the amount of square footage, it is more difficult to build and is cumbersome to move around. If your landscape necessitates a smaller shade structure, then you will need to modify this design. Here are a few simple suggestions:

Keep It Low

Taller shades are more top heavy and less stable in wind and on hills. You do need it tall enough to go over gate posts unless you make fence access points 24ft wide. You also want it tall enough that big animals can't lick and rub the purlins and shade cloth.

Width

The width of the shades is dictated by the length of the purlins. It's a cheap way to gain square footage so maximize it if you can. Keep in mind the width of roads, crossings, etc. on your farm and make it narrow enough to fit everywhere. Wider than 20ft becomes problematic, especially in windy or hilly locations. We've worked with different dimensions before, but now use 20ft wide as standard, adjusting the length only. The reason is that 20ft is about as wide as you can go with lightweight purlins without additional bracing

from the main stringers. As soon as you need stabilizers, you complicate the structure. The 20ft allows for an 8ft width between stringers and 6ft purlin overhangs on either side; that's the limit of lightweight material.

The reason you want it wide rather than long and skinny is because the chassis takes up room and is an obstacle for the animals. Width offers the maximum amount of unencumbered loafing area versus narrower and longer.

Length

Your length may be dictated by a need for maneuverability in tight spaces. The longer the running gear is, the larger the turning radius will be. It may also reduce maneuverability on a public road if you're towing it from property to property. Material might also be a limiting factor. Your local steelyard may not carry pipe lengths longer than 20ft. Check availability before beginning this project. You can adjust the length easily from 20ft up to 36ft. Our standard is 36ft and it will accommodate 60 animals of mixed sizes.

Adding length off of the front or the back of the chassis is doable, but remember that this

will affect your tongue lengths. Having 5-6ft of overhang on the front and back seems to work best for us. If you never anticipate chaining multiple shades together, you could potentially add significantly more length off of the back end (bracing as needed) and gain a substantial amount of area that way. You need at least 6ft between Shademobiles when hooking them together so when you turn, the corners don't collide as the inside of the turn pinches them.

We have chosen to standardize our design so that Shademobiles can be linked or unlinked as herd sizes grow and we know that any combination of shades will link together without interference or the need for adjusting tongue lengths.

Unlike the Eggmobiles, in this case we want a hay wagon chassis so they are not trailers. The reason is that if you hook 4 trailers together, for example, and head down a hill, the back ones can push the front ones. When the front trailer starts pushing sideways, the whole shebang piles up in a mess that is both destructive and dangerous. Wagons, on the other hand, track behind each other so you never get that direct push from behind.

Adjust Chassis Tube

4

≈24ft   3" SCH40 Pipe

COWS

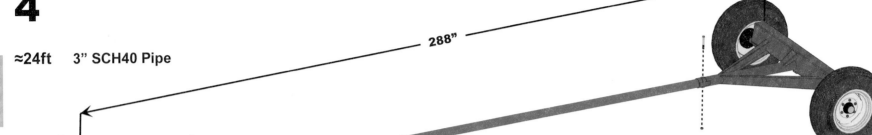

288"

In this example, the desired wheelbase of this chassis is 24ft (288"). The center chassis tube should be cut, drilled, and fastened accordingly.

5

Remove the hardware that connects the original tongue to the running gear.

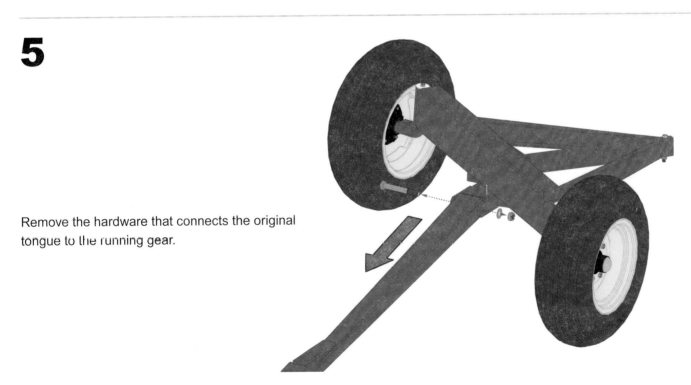

Determine Tongue Length

6

≈20ft   4" SCH40 Pipe

Determining tongue length depends on several factors.

1. The overall dimensions of your shade (are they the same as ours?)
2. Will the shade ever be daisy chained onto another Shademobile (as shown below).
3. Will it just be pulled by a truck or tractor?

A tongue length of 20ft for a daisy chain and 12ft for pulling behind a tractor are safe bets for this design. If you alter the dimensions of the shade, these lengths may vary. When in doubt leave it long and you can always shorten it later.

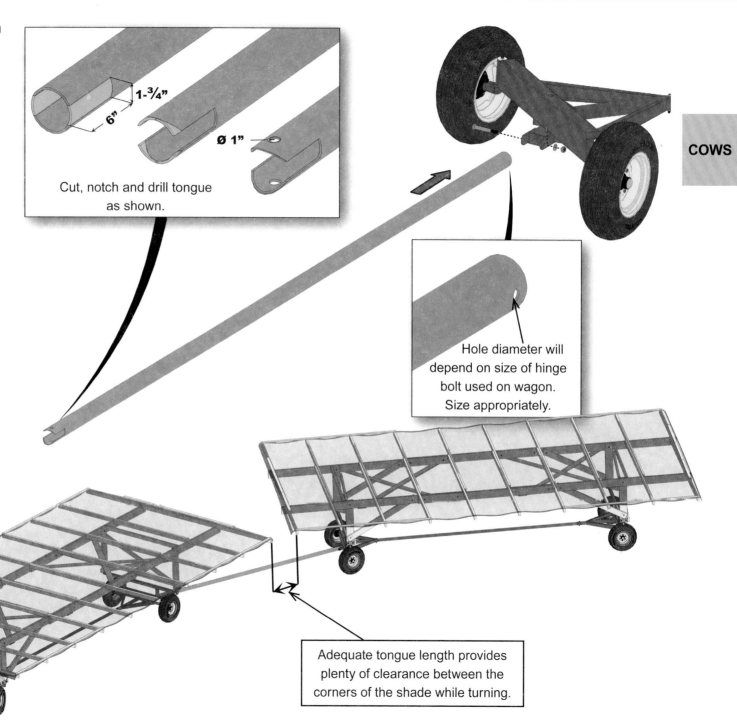

1-¾"

6"

Ø 1"

Cut, notch and drill tongue as shown.

COWS

Hole diameter will depend on size of hinge bolt used on wagon. Size appropriately.

Adequate tongue length provides plenty of clearance between the corners of the shade while turning.

7

COWS

2x   3x3x¼" Square Tube, ≈8-½" Long
1x   3x3x¼" Square Tube, 96" Long

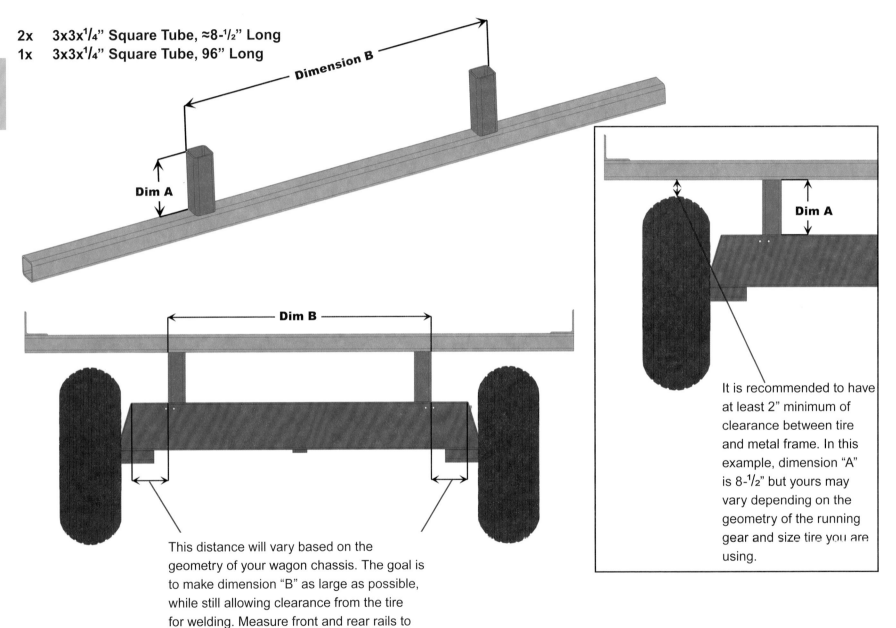

Dimension B

Dim A

Dim B

Dim A

This distance will vary based on the geometry of your wagon chassis. The goal is to make dimension "B" as large as possible, while still allowing clearance from the tire for welding. Measure front and rear rails to be sure they are the same. In the example shown, dimension "B" is 46".

It is recommended to have at least 2" minimum of clearance between tire and metal frame. In this example, dimension "A" is 8-½" but yours may vary depending on the geometry of the running gear and size tire you are using.

8

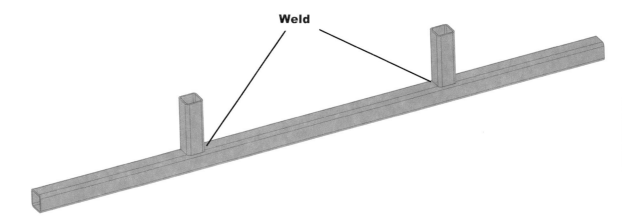

Weld

Weld posts on all four sides.

9

2x   **4x4x⁵⁄₁₆" Angle Iron, 8" Long**

Weld

Angle iron should be aligned flush with the ends of the square tubing. Weld joints on both.

10

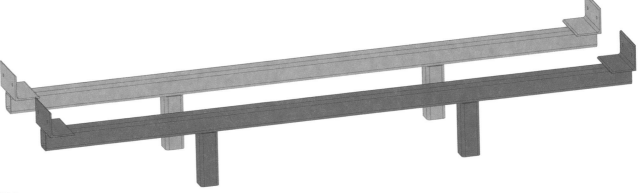

Repeat steps 7-9 to create second piece.

COWS

COWS

11

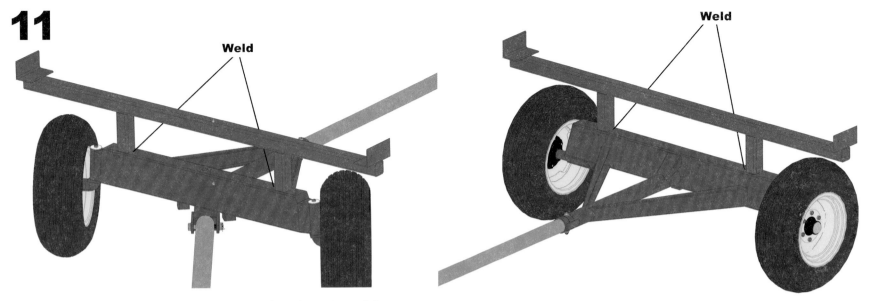

Center supports on the running gear and weld each post on all four sides. Repeat on the other end.

12

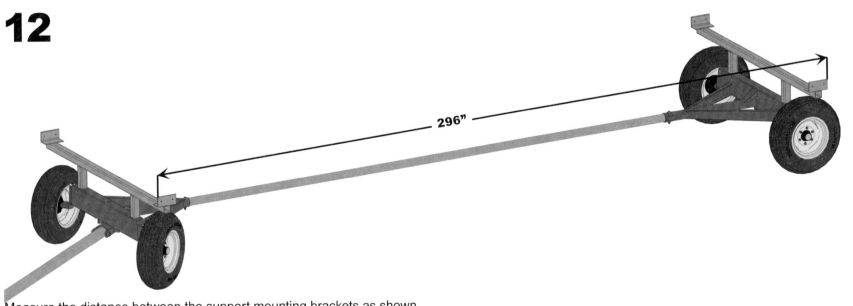

Measure the distance between the support mounting brackets as shown and record the dimension for use in step 20. (If your chassis wheel base varies from ours, then this value will also change).

13

✓

The chassis is complete!

Build The Frame

14

2x   2x8x192"

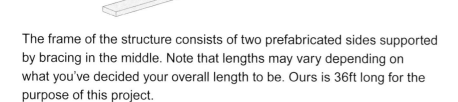

The frame of the structure consists of two prefabricated sides supported by bracing in the middle. Note that lengths may vary depending on what you've decided your overall length to be. Ours is 36ft long for the purpose of this project.

COWS

15

1x   2x8x144"

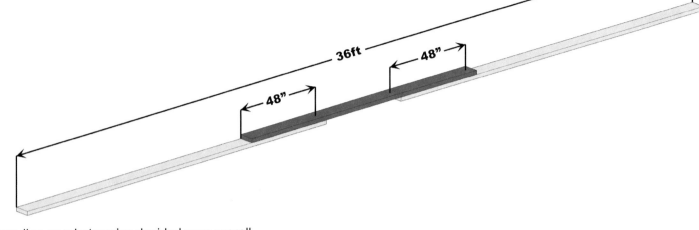

This length may vary depending on what you've decided your overall length to be. You can either overlap these pieces more to achieve the desired overall length or procure shorter pieces as needed.

16

4x   20D Galvanized Nails

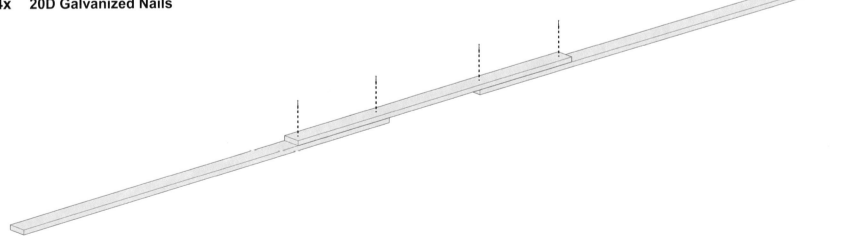

Tack the members in place with nails.

17

1x 2x8x48"
1x 20D Galvanized Nail

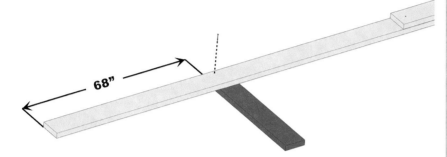

This 68" is the overhang measurement off of the front of the Shademobile but it may vary in your application. Refer to step 6 for an in depth discussion on this topic.

19

2x 20D Galvanized Nails

COWS

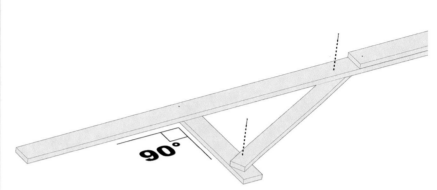

Tack the angle brace in place with some nails.

18

1x 2x8x82"

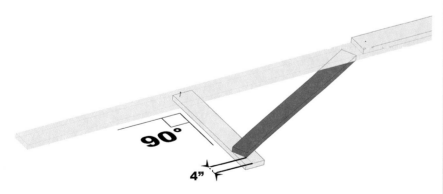

Square the upright post and then brace it as shown. Leave at least 4" of room at the bottom to attach to the angle iron mounting brackets.

20

1x 2x8x48"
1x 2x8x82"
3x 20D Galvanized Nails

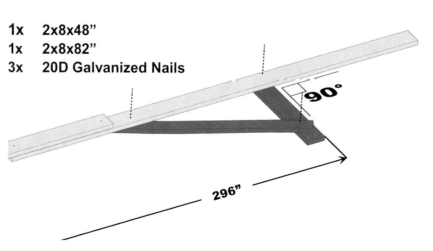

Measure from the left side of the 2x8x48" installed in step 17 to the right side of the brace shown. This measurement should equal 296" from step 12. Square and tack everything in place.

COWS

21

10x ½-13 x 5" Carriage Bolts
10x ½" Washers
10x ½-13 Hex Nuts

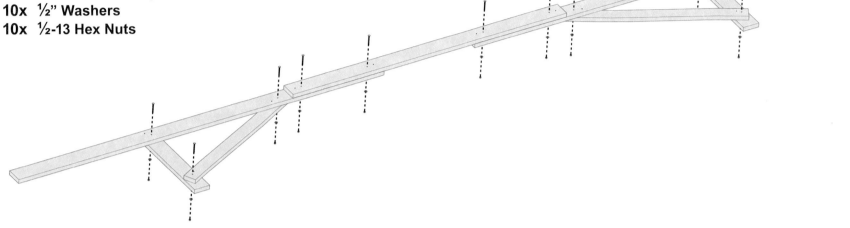

Drill and insert fastener hardware everywhere shown.

22

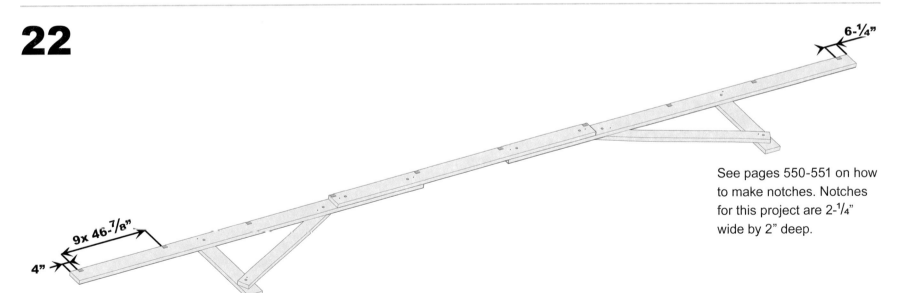

6-¼"

See pages 550-551 on how to make notches. Notches for this project are 2-¼" wide by 2" deep.

9x 46-⅞"

4"

Evenly space notches on the top of the frame no more than 48" apart. These notches will coincide with notches will hold the purlins in place.

23

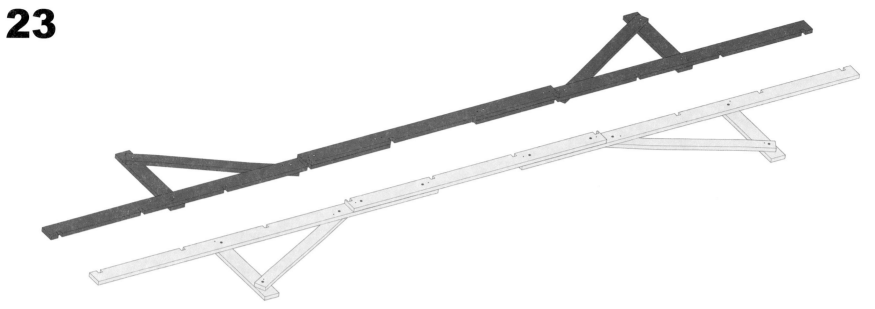

Repeat steps 14-22 to create the opposite hand version.

24

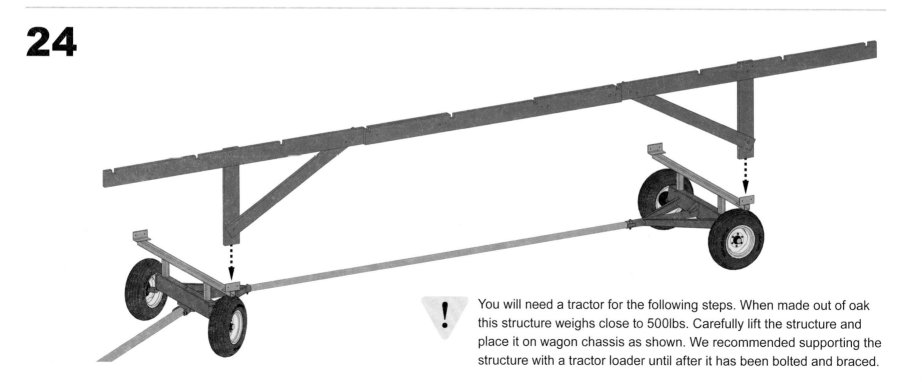

You will need a tractor for the following steps. When made out of oak this structure weighs close to 500lbs. Carefully lift the structure and place it on wagon chassis as shown. We recommended supporting the structure with a tractor loader until after it has been bolted and braced.

COWS

25

2x ½-13 x 3" Carriage Bolts
2x ½" Washers
2x ½ -13 Hex Nuts

Using the predrilled holes in the steel as a guide, drill through the wood members and insert the carriage bolts.

26

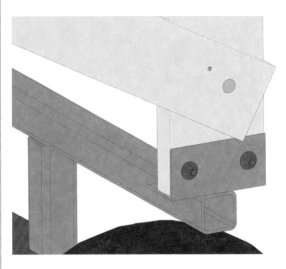

Repeat step 25 on the rear mounting location.

27

4x 3" Deck Screws
2x 2x4x120" *(or something similar that can be used temporarily as a brace)*

Attach the brace posts as shown.

28

Place structure on other side.

29

4x ½-13 x 3" Carriage Bolts
4x ½" Washers
4x ½-13 Hex Nuts

Drill and fasten the frame to the angle iron brackets.

30

4x 3" Deck Screws
1x 2x4x103"

90°

Square the frame and install the first angle support as shown.

31

4x 3" Deck Screws
1x 2x4x95-½"

Install the horizontal brace directly above it.

32

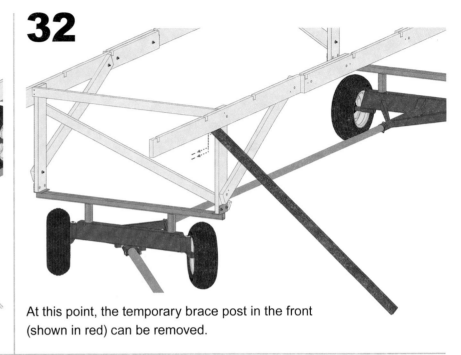

At this point, the temporary brace post in the front
(shown in red) can be removed.

33

4x 3" Deck Screws
1x 2x4x103"

Install the final angle support as shown.

34

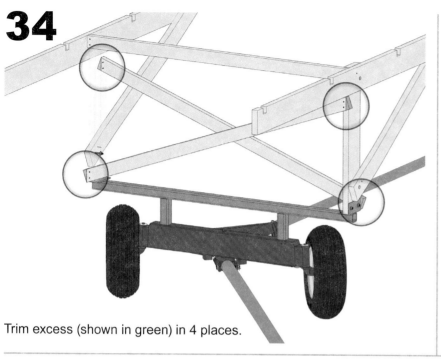

Trim excess (shown in green) in 4 places.

35

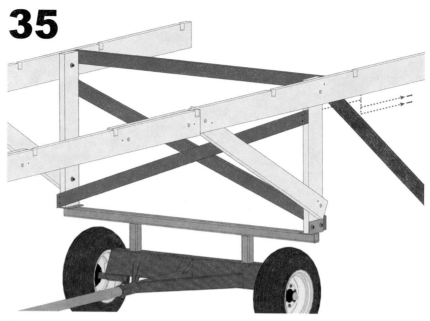

Repeat process in steps 30-34 on the rear of the structure.

36

12x 2-½" Deck Screws
2x   1x6x144"

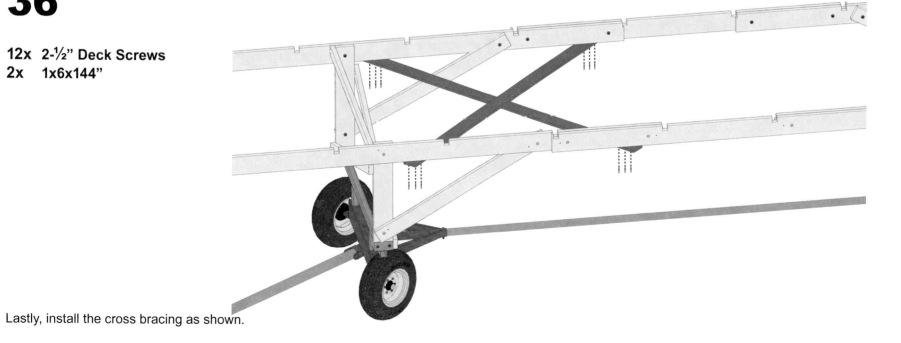

Lastly, install the cross bracing as shown.

37

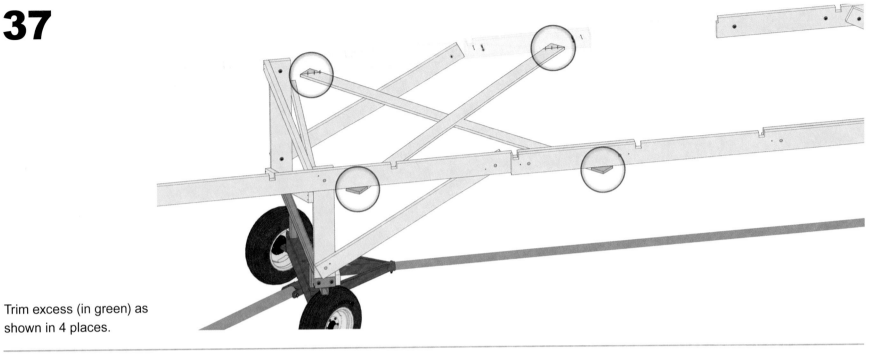

Trim excess (in green) as
shown in 4 places.

38

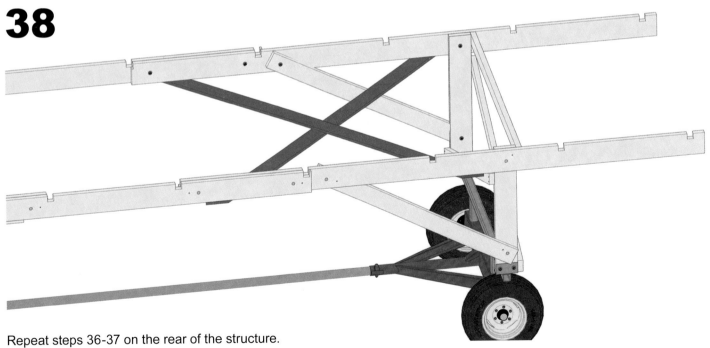

Repeat steps 36-37 on the rear of the structure.

Build The Purlins

39

1x   2x4x167"

The purlins are designed to be 20ft wide and easily removable. It is key to have precisely 94" of overlap because, as you will notice, the design locks the purlins in place from sliding side to side without having a notch in them. (We find that notching 20ft purlins creates a stress point and they always fail right at the notch.)

40

1x   2x4x167"

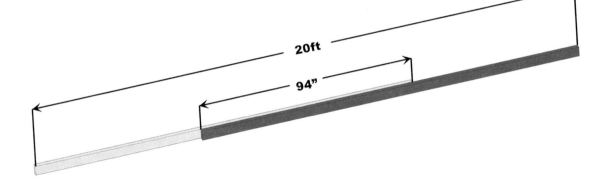

41

7x   3" Deck Screws

Fasten the overlapped joint with screws from both sides.

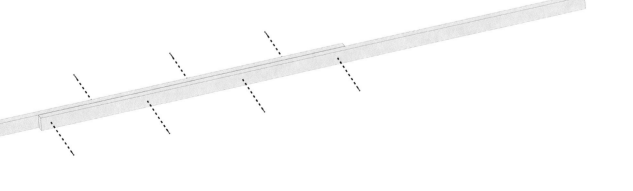

42

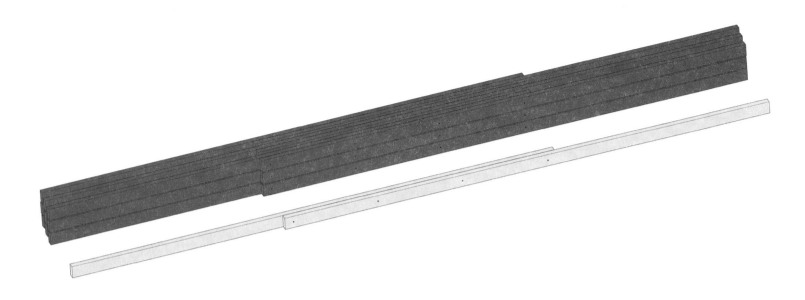

Repeat steps 39-41 to create a total of 10 purlins.

43

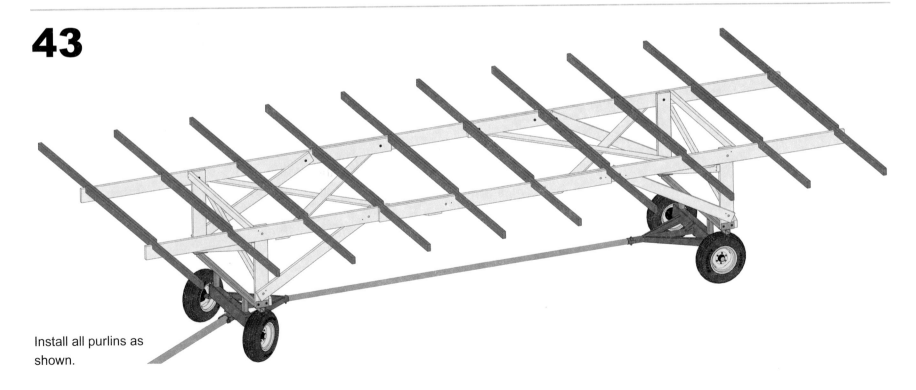

Install all purlins as shown.

Attach The Shade Cloth

44

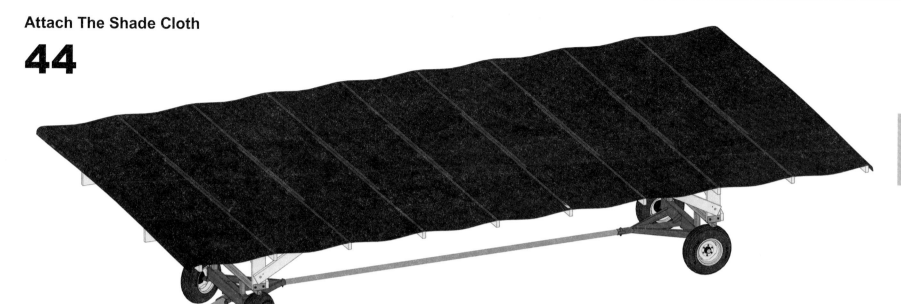

See pages 534-537 for proper folding and unfolding of shade cloth.

There are 101 ways to secure shade cloth, but we've found this to be the best. The premise of this concept is to secure the cloth using "clamping force" rather than knots and string tied through the grommets. How often have you seen tarps tear and fail right where you've tied them? That's because all of the forces are directed through the single knot location. On the contrary, by clamping the shade between two pieces of wood with a screw, you distribute the forces throughout the entire area in contact with the wood.

Once it has been initially installed, it is incredibly simple for a single individual to unfold and re-attach it. If you are methodical, you can even find and reuse the same holes in the shade cloth from year to year.

To un-attach, simply back a screw out of the hole, slide the shade cloth out from between the wood pieces, and then retighten the screw, leaving the wood strips in place. Repeat all the way around.

To re-attach, unfold the cloth as demonstrated on page 536 and then, finding the previous years hole, back each screw out, slide the shade in between the wood pieces and retighten the screw to clamp the shade cloth in place. Start at each end and then work your way down the sides last.

With diligence and care, you can get many years out of this piece of shade cloth.

45

8x 2-½" Deck Screws
2x 1x3x144"

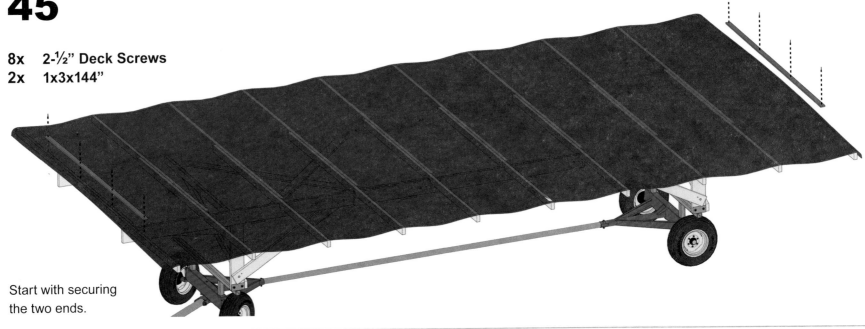

Start with securing
the two ends.

46

16x 2-½" Deck Screws
4x 1x3x144"

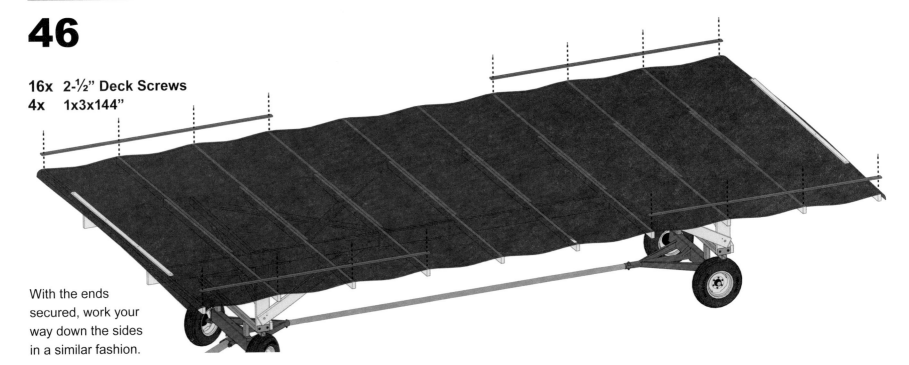

With the ends
secured, work your
way down the sides
in a similar fashion.

47

The Shademobile is complete!

COWS

CATTLE WORKING CHUTE

I remember growing up how I dreaded working cattle. When we got them in the corral we never knew if they would stay or break through the half rotten baler-twined fences. We hadn't been trained on proper handling techniques so it was always frustrating getting them to go where we wanted them to, or stop when we wanted them to stop.

Eventually everyone got short-tempered and then the animals felt the stress and they got antsy. It was not something people or animals enjoyed.

Then along came Temple Grandin and Bud Williams. I think I saw the first article published about Temple's work and it resonated so deeply with me I pulled out all of our old sorting fences

and rebuilt everything from scratch with her three overriding rules in mind:

1. Cattle like to turn, preferably to the right. They don't like long, straight chutes; they like curves.

2. They want to exit where they enter. If they think they're going out where they came in, they're far easier to work.

COWS

3. They want to go toward light but not into direct sunlight. They don't like dark corners and variegated shade.

Both Bud and Temple make a big point of how much the animal sees and how it processes information. Animals don't have calendars and schedules and licenses and dinner to fix. They are 100 percent in the moment, so things that seem incidental to us can fill their whole context, like a tiny movement, odd clothing, odd noises.

While you can certainly purchase cattle working chute kits with all the bells and whistles, we never had the money to do that and so relied on simple posts and wooden boards. We've found that this set-up is quieter than steel structures; nothing bangs and clangs. Cattle hear extremely well; anything we can do to minimize noises in the handling area is a good thing.

In this design, we incorporated all of Temple's ideas. I built this cattle working chute nearly half a century ago and we have not modified it at all. Now, instead of dreading working cattle, we look forward to it. The animals sense that attitude and work quietly as well. Probably the most important thing I learned from Bud Williams was to never get in a hurry. He had one piece of advice for everything: "slow down." It fixes a lot of problems. As long as you don't get in a hurry, this set-up works. If you get agitated, the animals won't turn around or flow and you'll get more frustrated. If you find yourself heading that direction, just stop, take some deep breaths, ask the animals to forgive you, and start over gently.

photo courtesy of Jessa Howdyshell

If you've ever seen our cattle working chute in person, you'll laugh at the level of care and precision we take in illustrating the steps for replicating our design. I can assure you that no tape measure was wielded and certainly no string lines were used when we built ours.

However, in writing this book Chris and I wanted to highlight the skill set of properly laying out and squaring a building structure, and this was the best possible place to insert that information. If you've never used "batter boards" and string lines to layout a building, you are in

for a treat. We encourage you to give it a try on something simple like this. These same principles have been used for centuries, and can be applied to laying out anything from a pole barn to the foundation of a house! You may be surprised to know that unless your contractor has a high dollar computer gizmo, he's going to lay your buildings out the same way.

So humor us if you will—get those string lines and tape measures out and strive to build the straightest truest cattle working chute there ever was.

General Design

COWS

Our setup has doors on all four sides that enable us to extend our capacity to further sort and contain our herd. To the south and west, we have additional holding pens built to contain anything from the largest bull to the smallest piglet. The entire cattle working chute, and the southern and western holding pens, are all under the shelter of a barn roof. To the north is our actual corral setup featuring 3 way sort capabilities, and to the east is our trailer loading area.

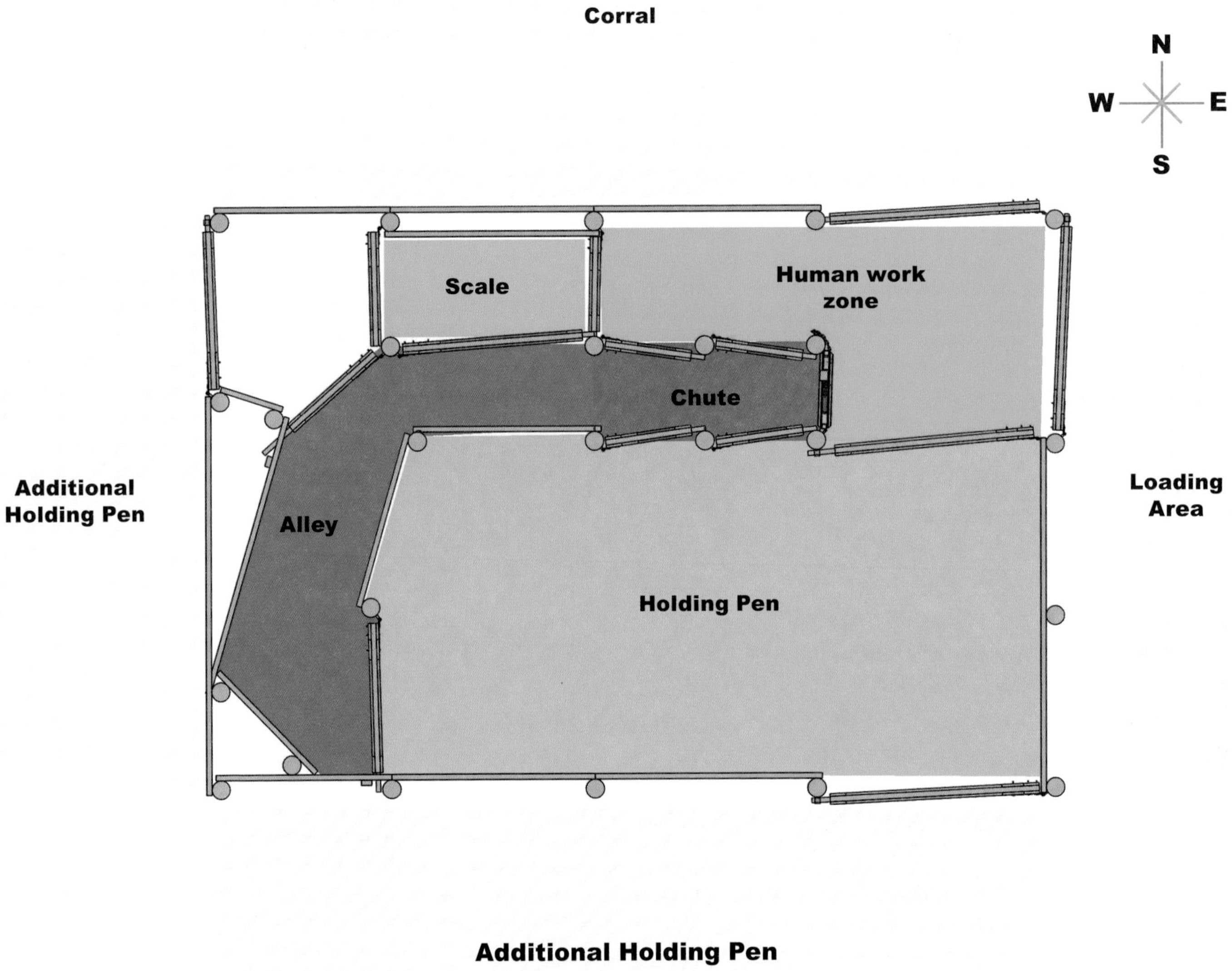

Corral

N
W — E
S

Scale

Human work zone

Chute

Additional Holding Pen

Alley

Loading Area

Holding Pen

Additional Holding Pen

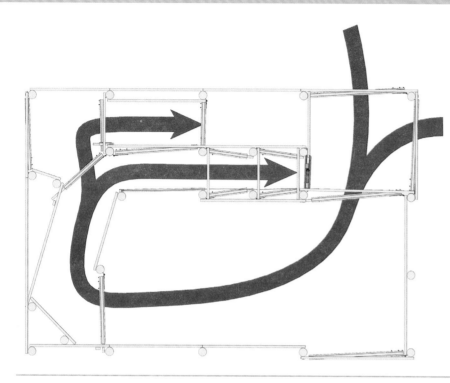

Our chute layout is rather simple. It consists of a holding pen, curved alley, chute, and scale. We also have an area that remains separated from cattle so that humans can move around safely. The holding pen can comfortably hold 8 mature cows, and has high sides to eliminate the temptation for an animal to jump over if stressed. The alley works the cattle to the right and draws the animals back towards both the daylight and where they came in. The chute consists of our homemade head gate (see page 266). With the help of a long rope, a single person can work an animal through the alley while still being able to engage the head gate from behind. The chute also provides access to the front and rear of the animal on both sides via 4 wooden gates. This allows us the flexibility to easily reach any part of the animal safely.

COWS

Animals typically cycle into the cattle working chute via the north or the east. When we are finished working them, we have the option to sort that animal in one of three different directions. This flexibility has proven invaluable time and time again.

This entire setup regularly handles 100 head with no hassles. Our livestock holding facilities aren't something you'll find on the cover of *American Cattleman*, but they've served us well and we didn't break the bank building them.

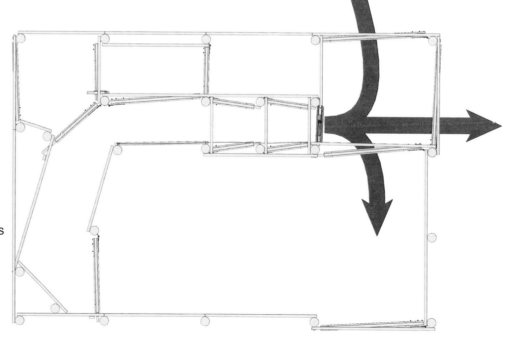

Tools

- ❏ Power Drill
- ❏ 8 or 10ft Ladder
- ❏ Circular Saw
- ❏ Socket Driver Set
- ❏ Angle Grinder
- ❏ Grinding Wheel
- ❏ Cut-Off Wheel

- ❏ 50-100ft Reel Tape Measure
- ❏ Marking Paint
- ❏ Box-End Wrench
- ❏ Chisel
- ❏ Sledgehammer
- ❏ Post Hole Digger
- ❏ Digging Bar

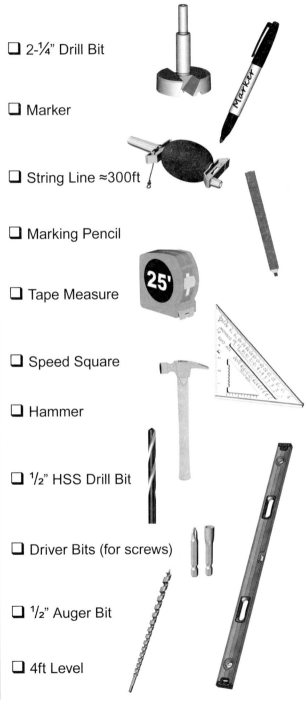

- ❏ 2-¼" Drill Bit
- ❏ Marker
- ❏ String Line ≈300ft
- ❏ Marking Pencil
- ❏ Tape Measure
- ❏ Speed Square
- ❏ Hammer
- ❏ ½" HSS Drill Bit
- ❏ Driver Bits (for screws)
- ❏ ½" Auger Bit
- ❏ 4ft Level

Scale Hardware

❑ **1x**   Scale Kit

Scale kits vary widely in price and features. Generally speaking, they consist of a set of load bars (load cells that measure weight) and a controller with a display readout. Scales require electricity to function. Some plug into an outlet while others are battery/solar operated. Keep this in mind when choosing your scale system.

You can purchase a pre-fabricated base as part of a kit, or you can build your own depending on the scale system you buy. The base we built for this pen is approximately 36" wide by 76" long. If you build your own, wait until you have built the sort pen and size it accordingly. If you are purchasing a scale base or an entire scale package, make sure it will fit within the space provided in these plans. Some modifications may need to be made to this pen layout if your scale is larger.

You want the platform and holding box big enough to accommodate the largest animal you can imagine putting in it, but nothing bigger. When an animal has extra room, it wants to jump around, turn around, even climb the walls. A tight fit encourages the animal to settle quickly – like Temple Grandin's proverbial hug – and get still enough for an accurate weight read.

Because you're dealing with electricity in a dirty place, make sure you can keep the whole area dry. Our barn is in a low place so we had to do a bit of shovel work to make sure water couldn't come in and that any accumulated moisture inside would drain away. Scales need to be stable as well, so don't skimp on getting a good foundation. A bit of concrete or heavy rot-resistant wooden block are essential to keep things sound. The load bars are sensitive and you don't want them moving around, getting off level, or sinking. After use, scrape away manure; a little maintenance goes a long way to many years of trouble-free operation.

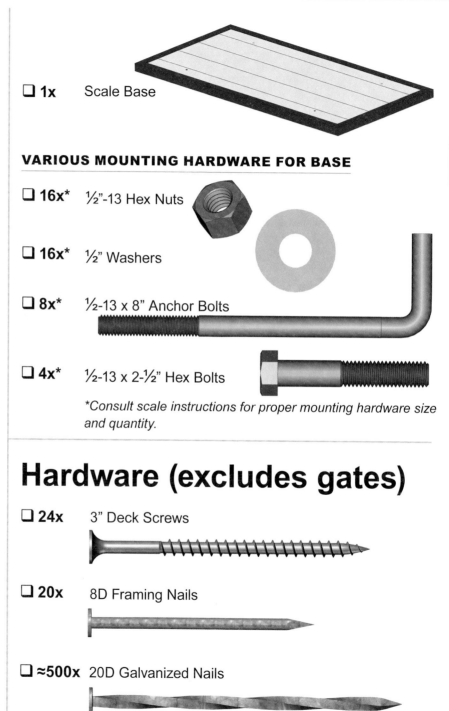

❑ **1x**   Scale Base

VARIOUS MOUNTING HARDWARE FOR BASE

❑ **16x***   ½"-13 Hex Nuts

❑ **16x***   ½" Washers

❑ **8x***   ½-13 x 8" Anchor Bolts

❑ **4x***   ½-13 x 2-½" Hex Bolts

*Consult scale instructions for proper mounting hardware size and quantity.

Hardware (excludes gates)

❑ **24x**   3" Deck Screws

❑ **20x**   8D Framing Nails

❑ **≈500x**   20D Galvanized Nails

COWS

Wood List (excludes gates)

COWS

It is impossible to provide exact lengths in a cutlist because your as-built dimensions may vary from these plans. It is easiest to measure and cut as you go and plan to have some extra lumber left over afterwards. Per the computer model calculations, this cattle working chute will require 510 lineal feet of 2x8s for the fence rails and supports. Ordering the aforementioned sizes and quantities will get you close with approximately 60 lineal feet of extra material.

Polyface uses nominal dimension lumber for this project. Meaning, a 2x4 is actually 2" x 4" as opposed to a dimensional 2x4 which is 1-$\frac{1}{2}$" x 3-$\frac{1}{2}$".

Materials for Posts/Rails

QTY	SIZE
❑ 20x	2x8x8ft
❑ 12x	2x8x10ft
❑ 24x	2x8x12ft
❑ 25x	7" x 12ft Posts*

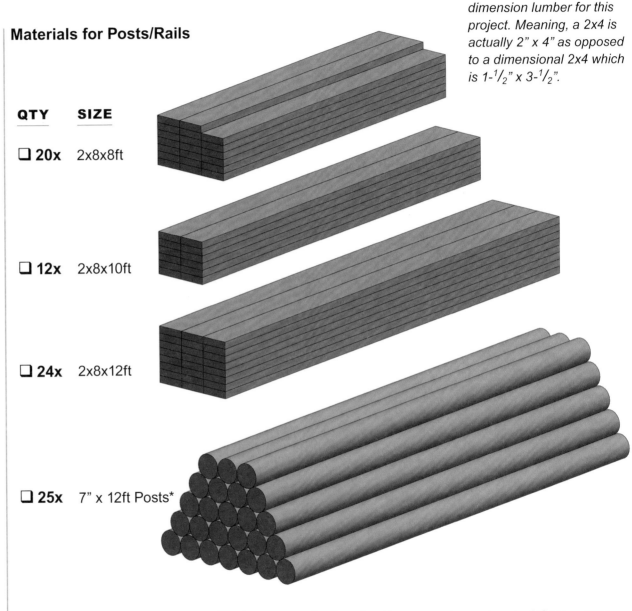

Materials for Batter Boards

QTY	SIZE
❑ 12x	2x4 Scraps

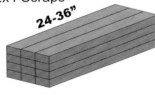

24-36"

QTY	SIZE
❑ 24x	2x2 Staves

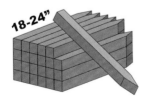

18-24"

*Posts may vary in diameter. Square posts can be used if necessary. If this pen will be part of a barn structure, the height of your posts may need to vary. Plan accordingly before procuring posts. If in doubt, make them taller because they can always be cut down shorter.

Post Layout

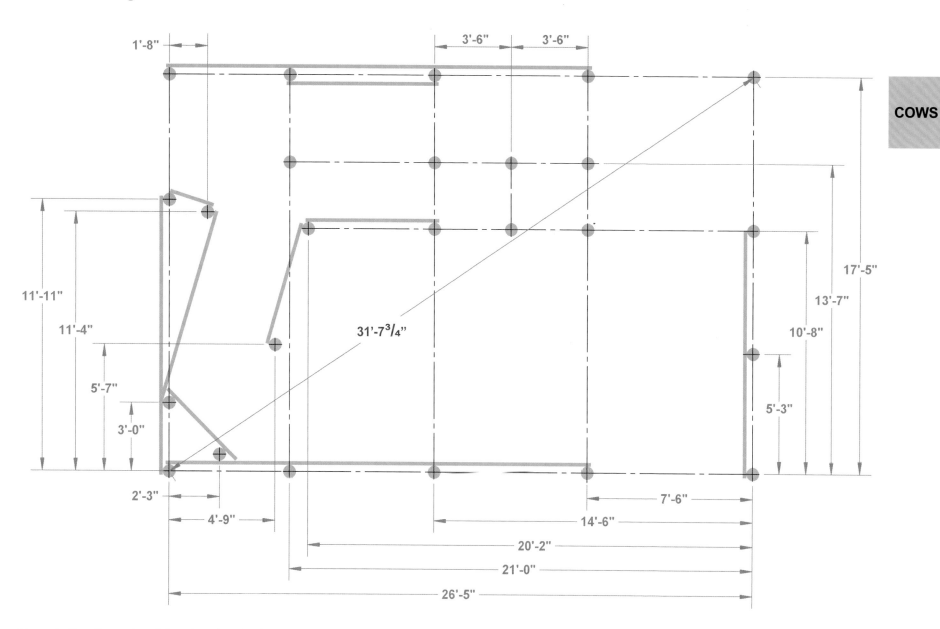

Above is the dimensional drawing that depicts the locations of each post. Refer to the instructional steps that follow to locate and square the structure prior to setting post locations.

1

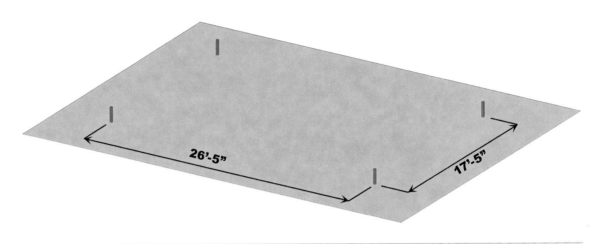

4x   Wood Stave, Flags, or Stakes

Roughly lay out the general location of the cattle working chute. It does not have to be perfectly square. The purpose is to decide on the location and orientation. We call these pins "site stakes" and they are used for reference only.

COWS

26'-5"

17'-5"

2

1x   String Line
4x   Wood Staves
2x   24-36" 2x4 Scrap Wood
4x   3" Deck Screws
2x   8D Framing Nails

If it is critically important that one particular side of the working facilities be oriented in a certain position (i.e. parallel to road or a building) then start setting the batter boards on that side.

String lines need to be very tight with secure knots. (Any slipping of the line during this process can mess up your measurements.)

Joel's Tip

A " batter board" is commonly used in construction. Place batter boards in line with the site stakes. Notice that they are not placed right next to the site stakes either. This is intentional. How far out the batter boards are placed will be determined by the access you will need around the pen while building it. If you are augering the post holes with a tractor then you will need room to maneuver without the risk of bumping a batter board. (If this were a footer for a building then you would need room for excavation equipment and potentially a concrete truck.)

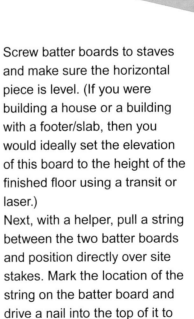

Screw batter boards to staves and make sure the horizontal piece is level. (If you were building a house or a building with a footer/slab, then you would ideally set the elevation of this board to the height of the finished floor using a transit or laser.)

Next, with a helper, pull a string between the two batter boards and position directly over site stakes. Mark the location of the string on the batter board and drive a nail into the top of it to secure the string.

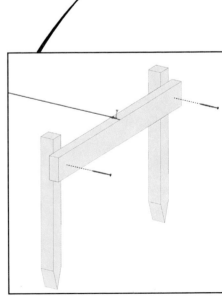

3

Place two more batter boards parallel to the existing string. Keep the batter boards in line with each other. Measure 17'-5" and mark the new batter boards. Add nails and secure the string line.

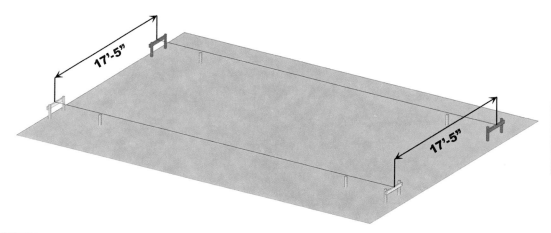

4

Set the first corner location. Choose whichever corner you deem as the most important location to be exact. Mark your string line with a permanent marker.

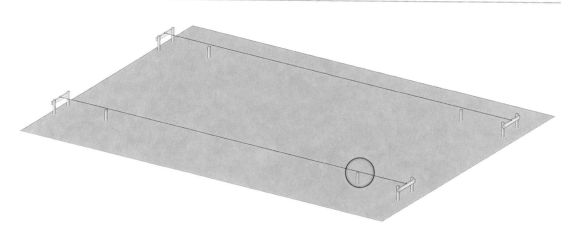

5

Measure across the string line and mark your second corner with permanent marker.

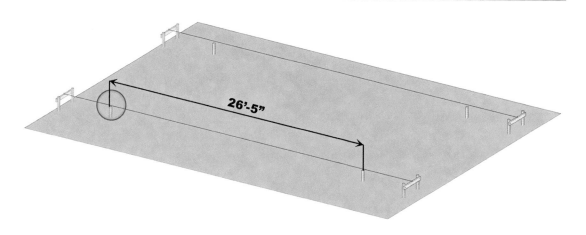

6

If you can visualize it, you are now working with a virtual right triangle with known length and width measurements. Using the Pythagorean theorem (see page 122), calculate the hypotenuse length. Measure across diagonally and mark the 3rd corner of your triangle with a permanent marker.

COWS

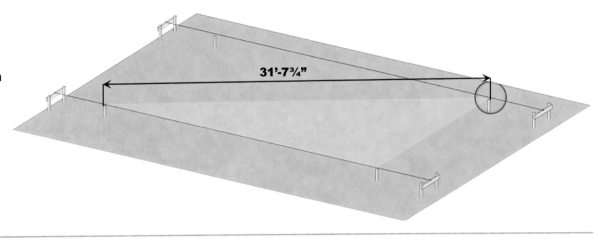

31'-7¾"

7

Set two more batter boards and, using the marks on the existing string lines, align your third string.

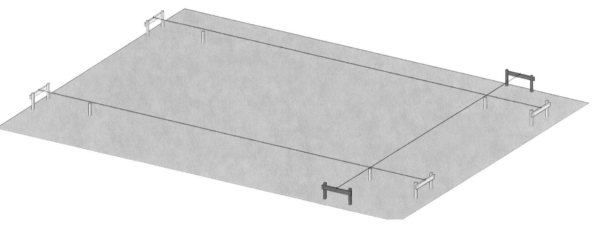

8

Place two more batter boards parallel to the existing string. Keep the batter boards in line with each other. Measure over 26'-5" and mark the new batter boards. Add nails and secure the string line.

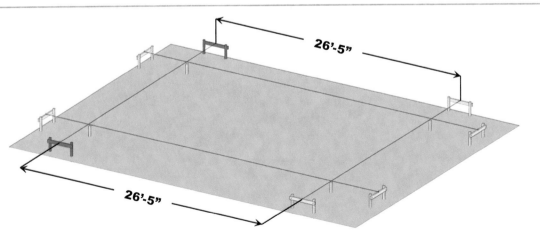

26'-5"

26'-5"

9

Check your diagonals and make adjustments accordingly. If this were a building foundation, anything within $^1/_4$" would be acceptable. Since this is a cattle working chute, not a house, it is not that critical. However, with practice, it is very doable to achieve perfection on the first try using this method!

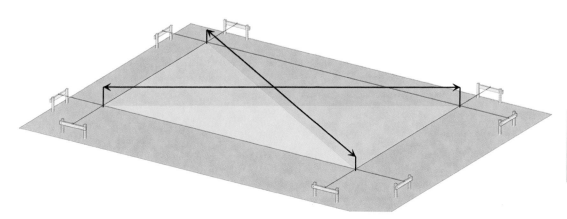

COWS

10

Measure parallel to the long side and insert 2 more string lines at the measurements shown.

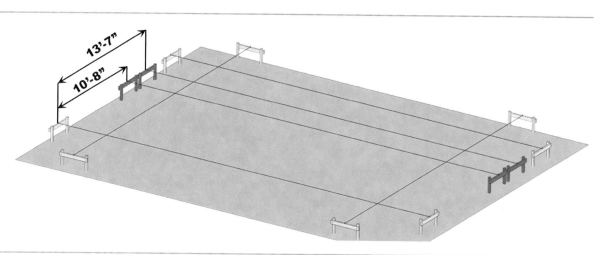

13'-7"

10'-8"

11

Using the dimensional drawing on page 237, mark all 25 post hole locations with marking paint.

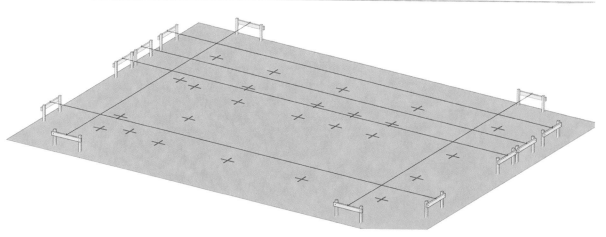

12

Measure and offset a nail on every batter board. Since all of the marks delineate the center of the posts, you will need to offset each string so that they will now line up with the outside of each post. The distance you move it will depend on the diameter of your posts. For example, if you use an 8" diameter post, then you will offset the string 4" towards the outside. Once you have measured all of your offsets, remove any old nails to avoid confusion.

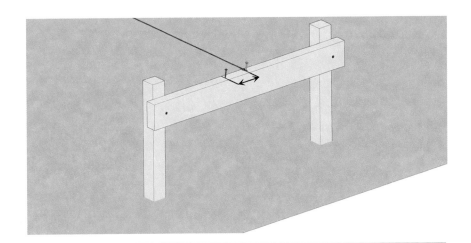

13

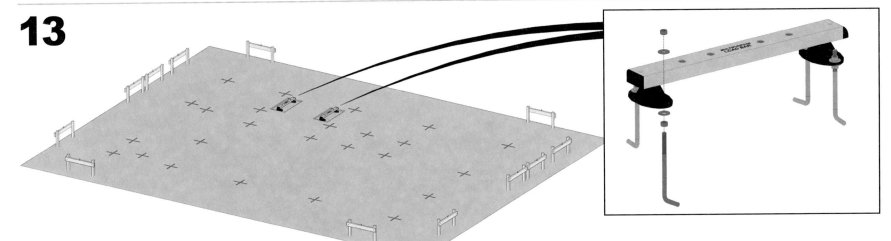

If you plan to install a scale, you will need to consult the installation manual for your product. The load bars typically require a concrete footing of some sort and some light digging work may be required. Doing this step first may prevent hassles down the road when all posts and rails are installed.

If using anchor bolts, purchase some extra nuts and washers and fasten them snuggly (not tight) to the top and bottom of the mounting foot on the load bar. After leveling the concrete, set the anchors – load bar and all – into the wet concrete. If it starts to sink, wait a little longer for the concrete to firm up slightly. Level the load bars in both directions and leave them alone until the concrete hardens for a few hours more. Then, while the concrete is still "green", carefully untighten the top nuts and remove the load bar. Do not worry about removing the bottom nut and washer if they are stuck in the concrete; leaving them there will not hurt anything. You will reinstall the load bars at a later time when the sort pen is completed.

14

Dig all of the post holes using an auger or hand digger. Posts should be at least 3ft deep where possible. If setting poles for a building, check with local ordinances for code requirements on depth.

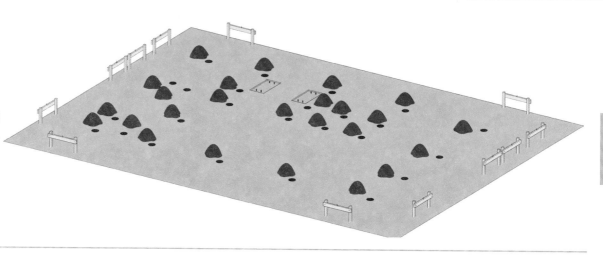

COWS

15

Put the string lines back up as a guide and set the first corner post and make sure it is level in all directions. Using a digging bar, tamp the soil firmly all of the way around each post. All posts should extend a minimum of 9ft above the ground except where otherwise noted.

Do not make contact with the string line when aligning anything to it. (If the string made contact with each post, the row could be off by a considerable amount.) String contact is hard to avoid, but prevents bigger problems.

16

Set the next corner post in a similar way.

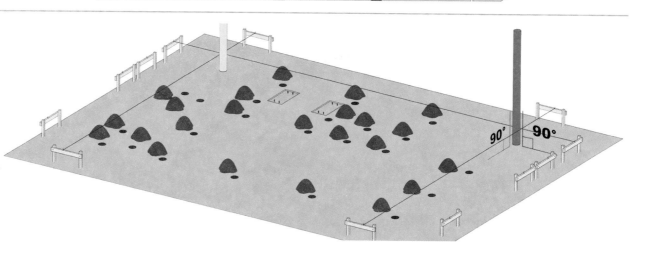

17

Set nails and place a string line towards the top of the two corner posts. This is to help keep the row of posts in alignment with each other.

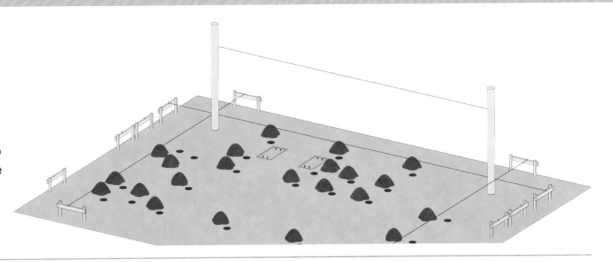

18

Set all of the middle posts, making sure each is plum in all directions. The middle post does not need to be as tall as the others. Ours is set at 7ft above ground. Any material above and beyond is not utilized. If however, you plan to use these posts to support a roof structure, you may need to make them as long as the others.

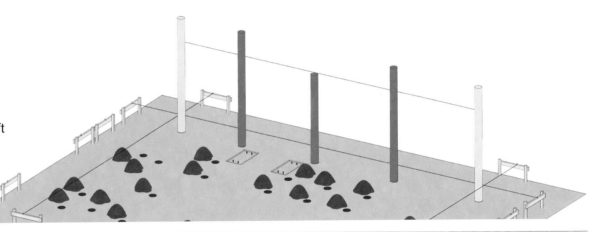

19

Continue on setting the next row of posts, starting with the first and last posts in a sequence.

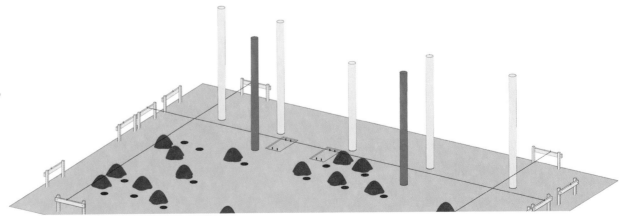

20

Place the upper string line and then set posts in between.

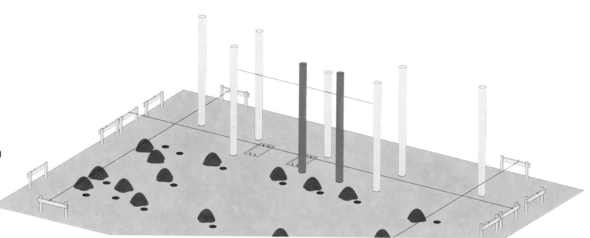

21

Set the next corner.

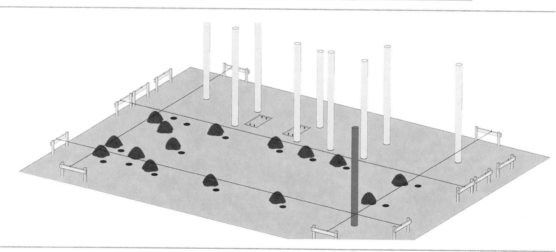

22

Set the next row of posts, starting with the first and last posts in the sequence. The far left post needs to extend a minimum of 5-$\frac{1}{2}$ft above ground.

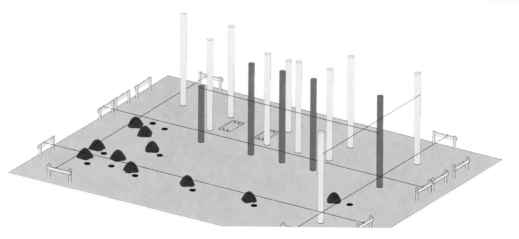

COWS

23

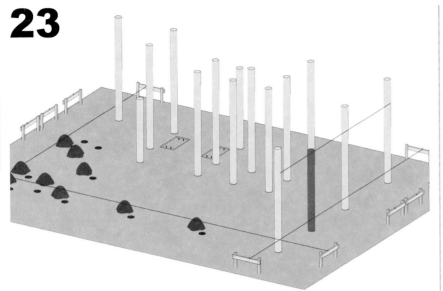

Set the remaining post in the row. This post only needs to extend a minimum of 7ft above ground.

25

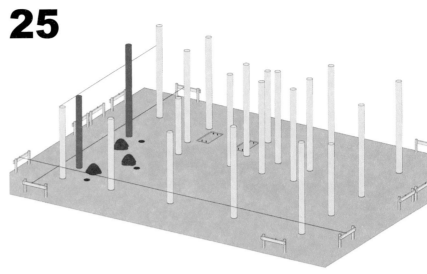

Set the remaining posts in the row. The left post only needs to extend a minimum of 7ft above ground.

24

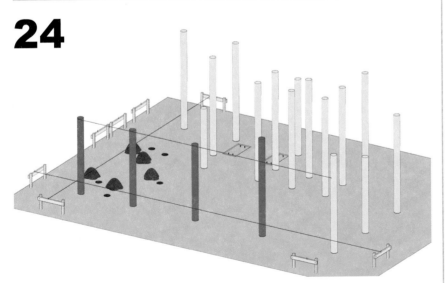

Set your next corner, and then install the posts in between. The three left posts only need to extend a minimum of 7ft above ground. If you plan to use these outer posts to support a roof, make them whatever length you need them to be.

26

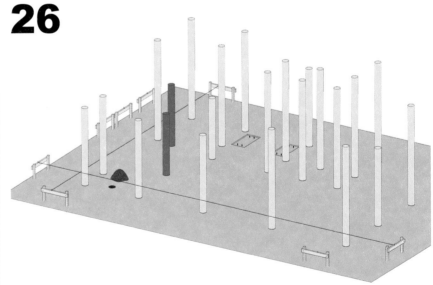

Position the two lone inner posts measuring off of the perimeter string lines. Leave the final post out at this time. These two remaining posts only need to extend a minimum of 5-$\frac{1}{2}$ft above ground.

27

Batter boards can be removed!

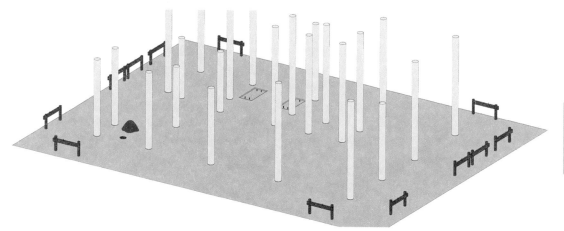

COWS

28

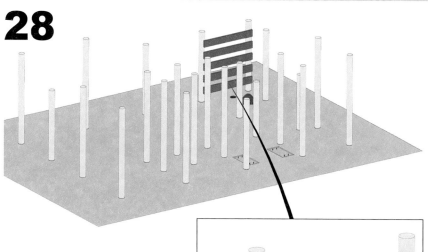

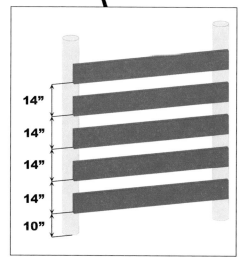

Nail 2x8s as shown with roughly a 6" space between each board. End the board centered on each post so that the next board can meet against it and be nailed on.

14"

14"

14"

14"

10"

29

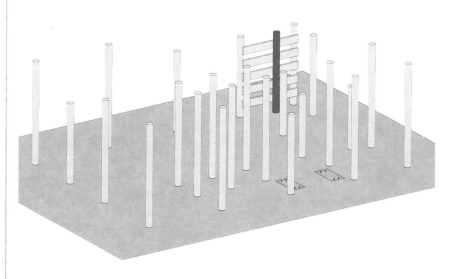

Set the final post tight against the fence rails.

30

This is the trickiest section. Cut and piece as needed to accomplish the configuration shown.

 If you are struggling to visualize where the rails are supposed to be, the Post Layout on page 237 gives a top down view which may help.

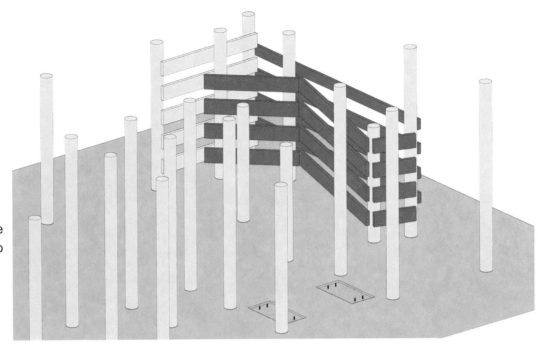

31

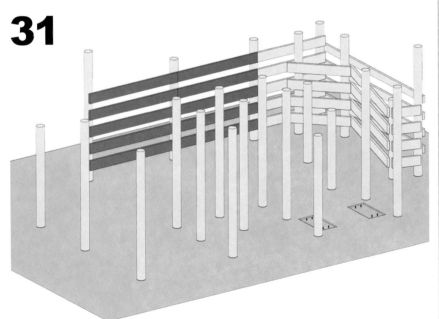

Install rails where shown.

32

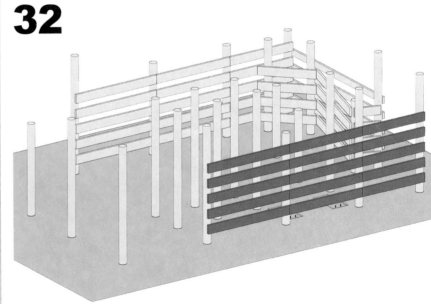

Install rails where shown.

33

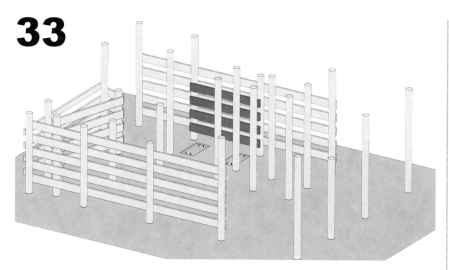

Depending on the scale you choose, you may or may not need to place rails on the inside as shown. The goal is to keep the animals from stepping off the scale platform. Eliminate gaps between the scale platform and the sides where an animal could catch its foot.

34

Fasten load bars per manufacturer recommendations.

35

Install the scale base per manufacturer recommendations.

36

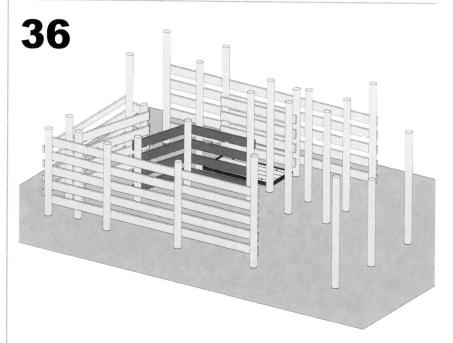

Install rails where shown.

37

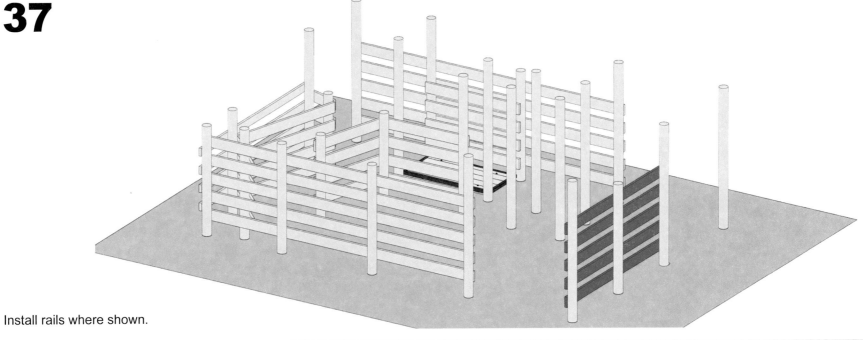

Install rails where shown.

38

Cut and nail these headers over every door opening. They will add rigidity to the posts and prevent sagging and spreading over time from the weight of the gates and the wear and tear of the cattle.

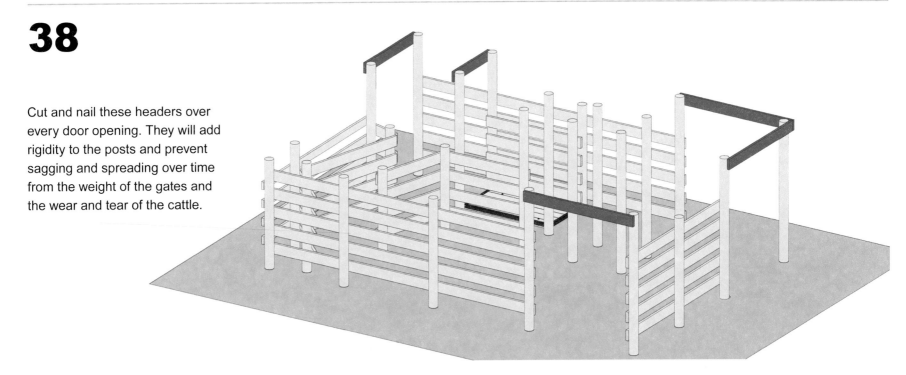

39

Continue cutting to fit and nailing headers between all posts in the chute area for added support.

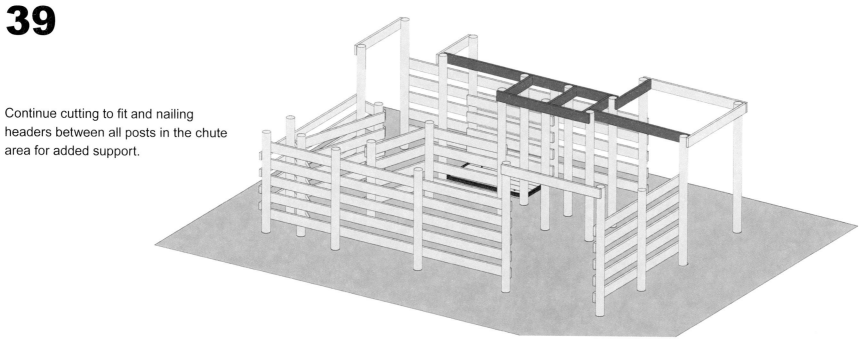

40

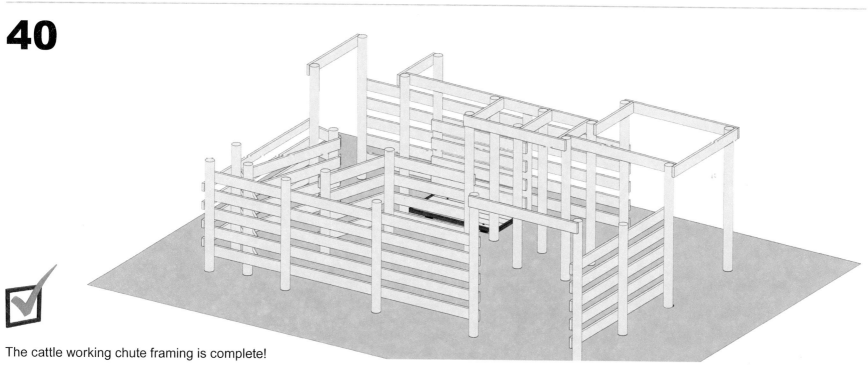

The cattle working chute framing is complete!

Plan Your Gates

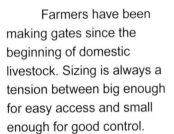

COWS

Farmers have been making gates since the beginning of domestic livestock. Sizing is always a tension between big enough for easy access and small enough for good control. The bigger the gate the harder it is to swing. And the harder it is to keep from sagging. Like us, you'll want to use different sizes in different spots. In general, anything bigger than 80" becomes unwieldy for one person to easily swing. Anything less than 36" limits flow to one animal at a time. As a general rule, I always build as small a gate as necessary; they're lighter and more manageable. No reason exists for making a gate bigger than necessary.

A functional gate requires two things. First, it must have a diagonal brace, best placed from the top of the hinged end to the bottom of the swinging end. You can invert it, but this direction is standard. Secondly, use big enough hinges. Normally you'll be buying straps and pins for the hinges; the tendency is to get smaller and cheaper ones but this is a mistake. Especially in a working chute arrangement, these gates will get a lot of abuse no matter how gentle your animals. One 800 pound push can bend a light hinge. Go ahead and get beefy hinges; you won't regret it.

The following are instructions on how to make a typical wooden gate. This is our favorite gate design because it is robust and (if made correctly) will outlast many of us. We also prefer wood over metal because it is much quieter. Metal banging and clanging elevates the stress of the animal. We also highly recommend spending the extra money and using white oak timber for these.

Gate Hardware

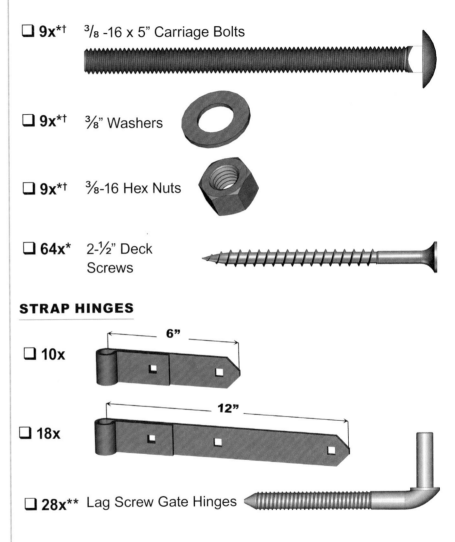

☐ **9x*†**　⅜ -16 x 5" Carriage Bolts

☐ **9x*†**　⅜" Washers

☐ **9x*†**　⅜-16 Hex Nuts

☐ **64x***　2-½" Deck Screws

STRAP HINGES

☐ **10x**　6"

☐ **18x**　12"

☐ **28x****　Lag Screw Gate Hinges

*　Quantity listed is per gate. Multiply this by the number of gates (14 per our diagram).

**　The size of the lag screw gate hinges will depend on the straps used. Be sure to pair them at the time of purchase to ensure proper fit.

†　The size of the bolts, washers and nuts may also vary with the size of straps used. Verify compatibility prior to purchase.

Gate & Hinge Locations

This diagram shows the perspective range of motion – that is, which direction and how far each gate will swing. It also labels the location of each gate listed on the following pages.

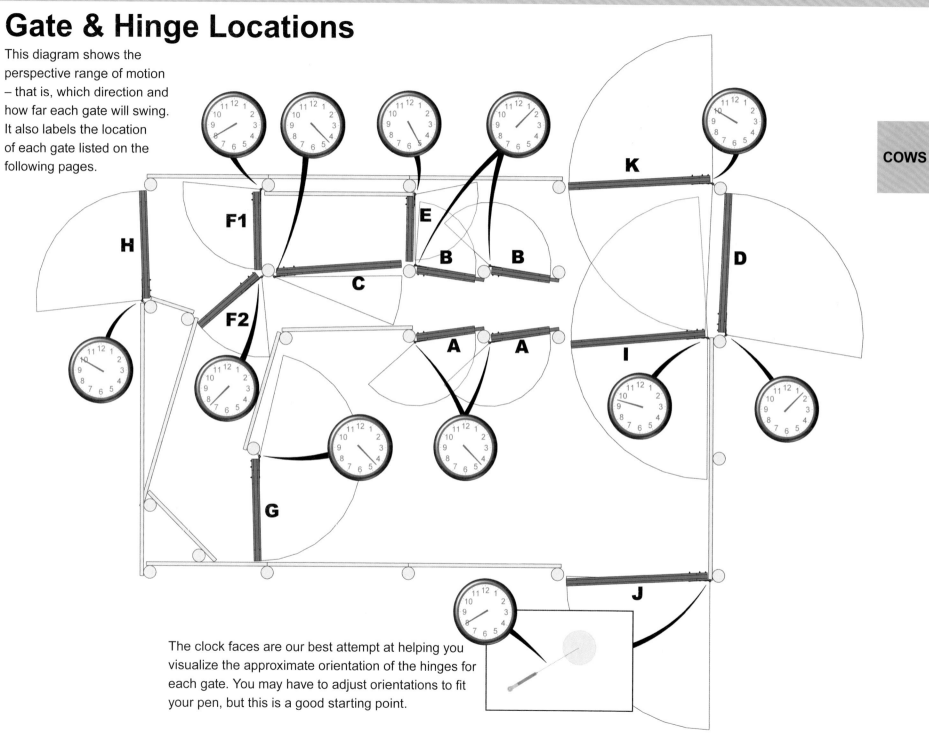

The clock faces are our best attempt at helping you visualize the approximate orientation of the hinges for each gate. You may have to adjust orientations to fit your pen, but this is a good starting point.

List of Gates & Cutlists

Pay attention to which side of the gate the hinges are located. The widths of these gates may change based on the as-built dimensions of the posts you set. By observing the range of motion diagram shown on the previous page, you may need to adjust the sizes accordingly so that the gates operate as intended. The letters on the gates correspond to the locations on the previous page.

A

2x

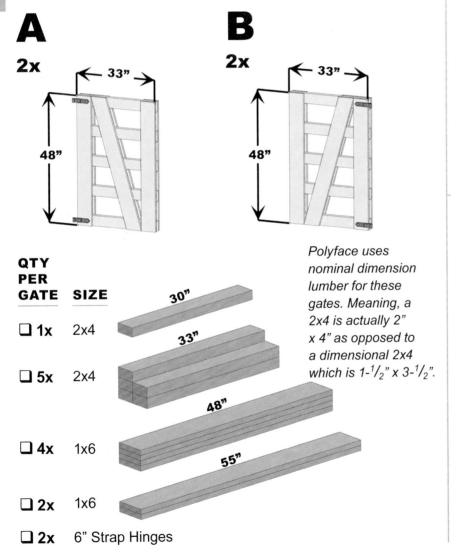

QTY PER GATE	SIZE
☐ 1x	2x4
☐ 5x	2x4
☐ 4x	1x6
☐ 2x	1x6
☐ 2x	6" Strap Hinges

B

2x

Polyface uses nominal dimension lumber for these gates. Meaning, a 2x4 is actually 2" x 4" as opposed to a dimensional 2x4 which is 1-$\frac{1}{2}$" x 3-$\frac{1}{2}$".

C

1x

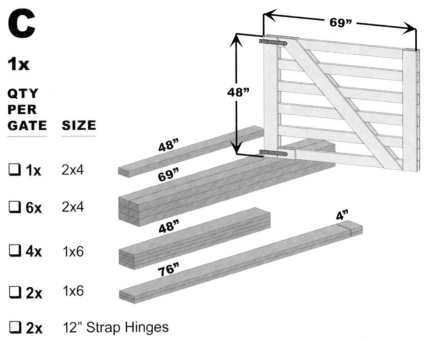

QTY PER GATE	SIZE
☐ 1x	2x4
☐ 6x	2x4
☐ 4x	1x6
☐ 2x	1x6
☐ 2x	12" Strap Hinges

D

1x

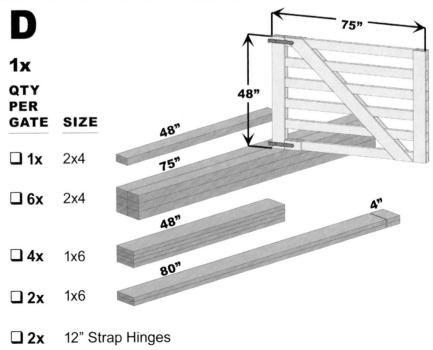

QTY PER GATE	SIZE
☐ 1x	2x4
☐ 6x	2x4
☐ 4x	1x6
☐ 2x	1x6
☐ 2x	12" Strap Hinges

E

1x

QTY PER GATE	SIZE
☐ 1x	2x4
☐ 6x	2x4
☐ 4x	1x6
☐ 2x	1x6
☐ 2x	6" Strap Hinges

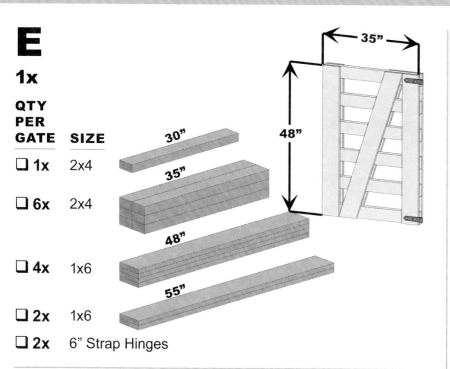

G

1x

QTY PER GATE	SIZE
☐ 1x	2x4
☐ 6x	2x4
☐ 4x	1x6
☐ 2x	1x6
☐ 2x	12" Strap Hinges

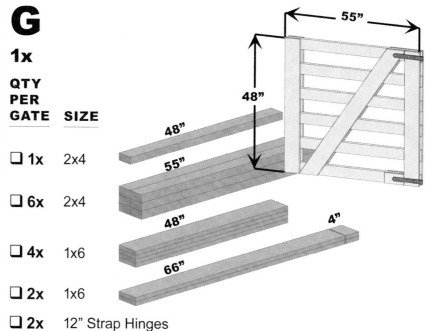

COWS

F

2x (1 for F1, 1 for F2)

QTY PER GATE	SIZE
☐ 1x	2x4
☐ 6x	2x4
☐ 4x	1x6
☐ 2x	1x6
☐ 2x	12" Strap Hinges

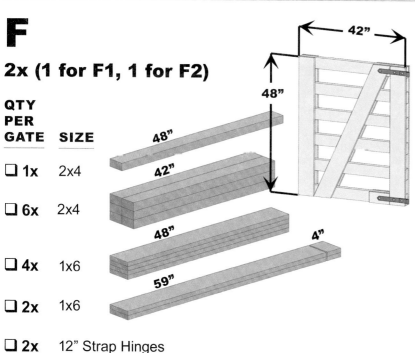

H

1x

QTY PER GATE	SIZE
☐ 1x	2x4
☐ 6x	2x4
☐ 4x	1x6
☐ 2x	1x6
☐ 2x	12" Strap Hinges

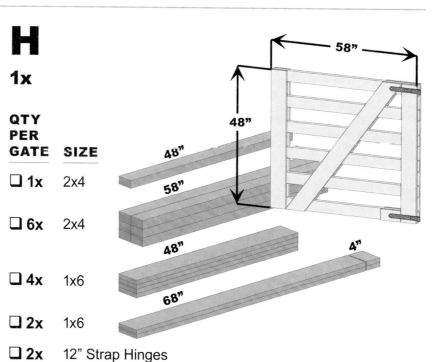

COWS

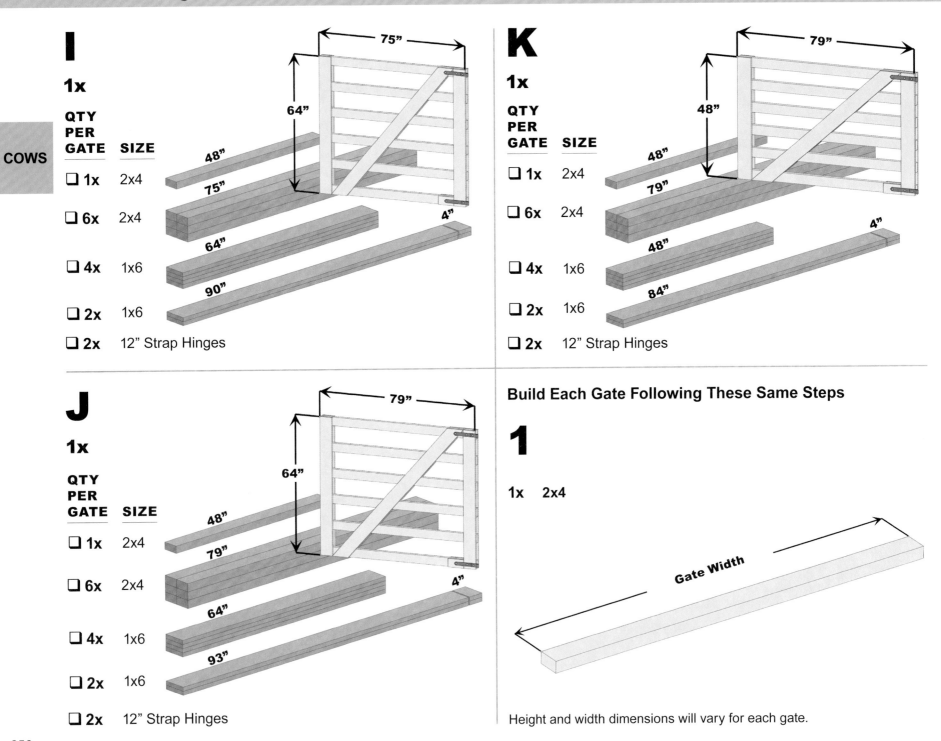

I

1x

QTY PER GATE	SIZE
❑ 1x	2x4
❑ 6x	2x4
❑ 4x	1x6
❑ 2x	1x6
❑ 2x	12" Strap Hinges

75"
64"
48"
75"
64"
90"
4"

J

1x

QTY PER GATE	SIZE
❑ 1x	2x4
❑ 6x	2x4
❑ 4x	1x6
❑ 2x	1x6
❑ 2x	12" Strap Hinges

79"
64"
48"
79"
64"
93"
4"

K

1x

QTY PER GATE	SIZE
❑ 1x	2x4
❑ 6x	2x4
❑ 4x	1x6
❑ 2x	1x6
❑ 2x	12" Strap Hinges

79"
48"
48"
79"
48"
84"
4"

Build Each Gate Following These Same Steps

1

1x   2x4

Gate Width

Height and width dimensions will vary for each gate.

2

1x **1x6**
1x **2-½" Wood Screw**

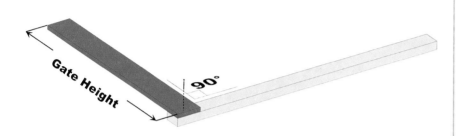

Place only one screw at each joint until everything is squared. (This makes it easier to make adjustments as needed.)

4

1x **2x4**
2x **2-½" Wood Screws**

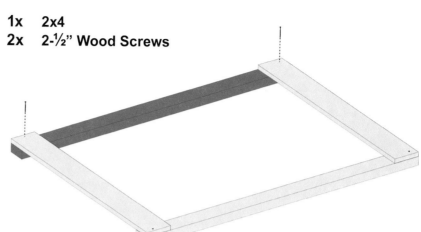

3

1x **1x6**
1x **2-½" Wood Screw**

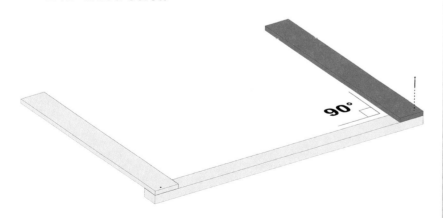

5

4x **2x4**
8x **2-½" Wood Screws**

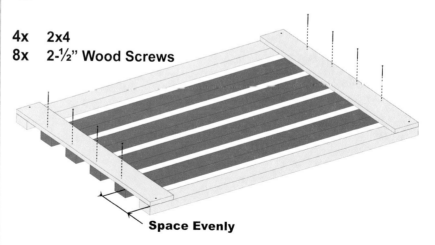

Space Evenly

Space the middle horizontal members evenly apart. On the 48" tall gates the spacing shown will be approximately 8-¾" and on the 64" tall gates the spacing will be 12".

COWS

6

1x 1x6
2x 2-½" Wood Screws

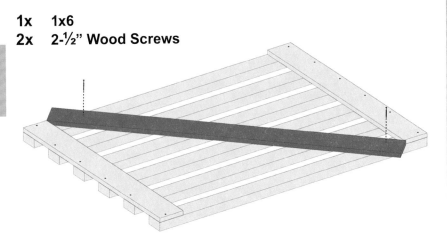

Attach the diagonal brace.

7

16x 2-½" Wood Screws

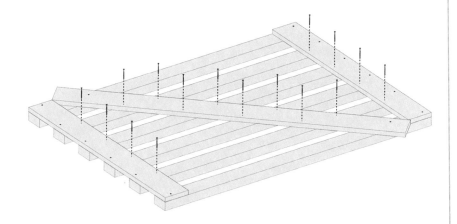

Secure with additional screws as shown.

8

1x 1x6x4
2x 2-½" Wood Screws

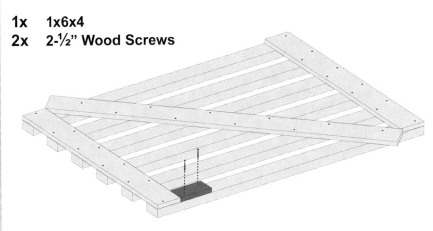

Fasten the 1x6x4 block to the gate as shown.

9

4x 1x6 (Varying Lengths)
32x 2-½" Wood Screws

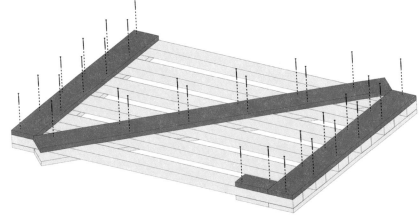

Flip the gate over and add 1x6s and fasten as shown.

10

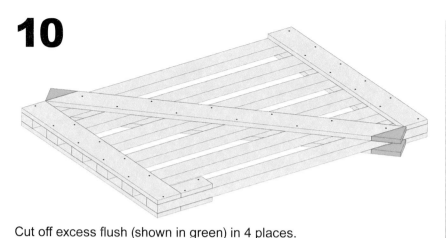

Cut off excess flush (shown in green) in 4 places.

11

2x   **Strap Hinges**
9x   **⅜-16 x 5" Carriage Bolts**
9x   **⅜" Washers**
9x   **⅜-16 Hex Nuts**

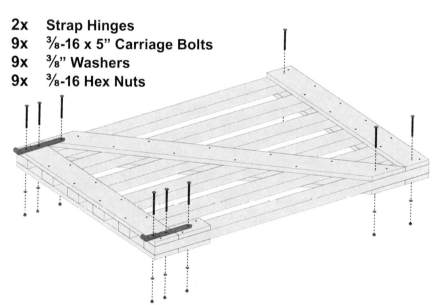

Pre-drill appropriate sized holes. Strap hinges should extend approximately 1-¼ to 1-½" past the edge of the gate. Bolting the corner joints adds to the durability of the gate. Grind off any excess threads that are sticking out to prevent accidental lacerations when brushing up against gates.

12

1x   **2x4**

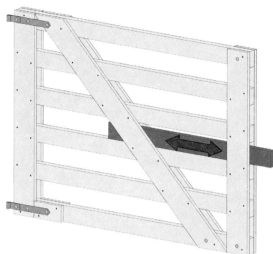

This piece of 2x4 is referred to as a "board slide", and is the mechanism for latching the gate shut. Sand the sides of it down until it slides in between the 1x6s without binding. This gate slide will be approximately 30" long on the shorter gates, and 48" long on the larger gates.

Skip this step for Gate F1, and any other gate that you plan on using a gravity latch on. See the latch sections on the following pages for further clarification.

13

One of the gates is complete!

259

Mount The Gates

14

2x   Lag Screw Gate Hinges

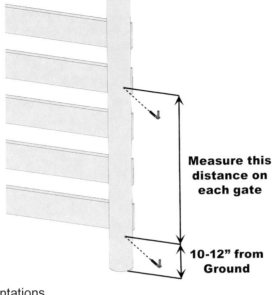

Measure this distance on each gate

10-12" from Ground

Pre-drill holes in the orientations denoted by the clock faces found on page 253.

16

1x   Gate

Hang gate and check range of motion. You may need to adjust the depths of the hinges to get the gate to hang level and swing freely.

15

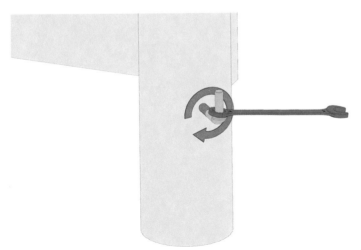

Tighten the lag hinges to the desired depth. A box-end wrench works very well for this step.

17

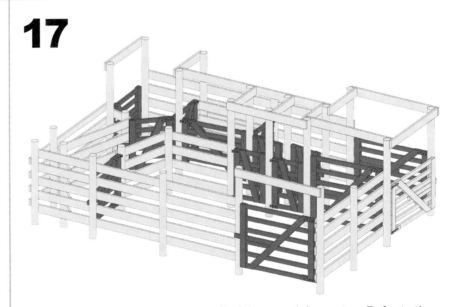

Repeat the mounting process on all of the remaining gates. Refer to the gate and hinge location diagram on page 253 to better understand how each gate is intended to function.

Gate Latches

There are countless ways to create latching mechanisms for wooden gates in cattle handling facilities. It is impossible for us to instruct you as to what kind of latching mechanism to use in each situation you may encounter. So, we will share 5 of our favorites and explain the pros/cons of each in hopes that you will be able to take that knowledge and apply it to your situation. This is not an exhaustive list, so use your imagination and be creative!

Fence Rail Pocket

Use on Gates G and F2. The fence rail pocket is one of our favorites because it is extremely simple and uses existing structure to create a secure latch.

Pros:
- Quick and easy
- Strong

Cons:
- Only works when gate lines up with the gaps in the fence rails.

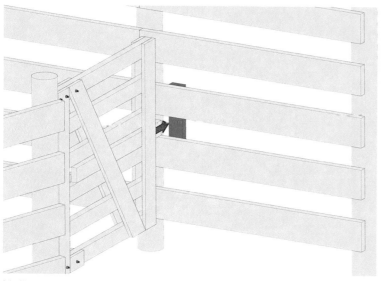

Nail or screw scrap lumber to fence rails in the desired location to create a pocket for the board slide to nest.

Mechanism-less

Use on Gates A and B. The mechanism-less latch is our favorite gate because it requires nothing! We use this one on the small access gates on either side of the chute because the animals are only ever pushing on it from one direction. There are no metal straps or hooks to worry about injuring the animal and it is easy to open/close when needed.

Pros:
- No work necessary
- Nothing protruding to potentially injure livestock/handlers

Cons:
- Only works in one direction.

The overhead view of the gate illustrates that this latch only works in one direction.

COWS

Simple Metal Strap

Use on Gates D, I, J, H and K. The metal strap is another favorite because of its simplicity and security.

Pros:
- Relatively simple to make
- Strong
- Flexible mounting options

Cons:
- Because it protrudes outwards, it has the potential to injure livestock/handlers, if it contains any sharp edges.

ADDITIONAL GATE HARDWARE

QTY PER LATCH

☐ **2x** ⅜ x 1-¾" Hex Head Lag Screws

☐ **1x** ³⁄₁₆ x 2" Bar

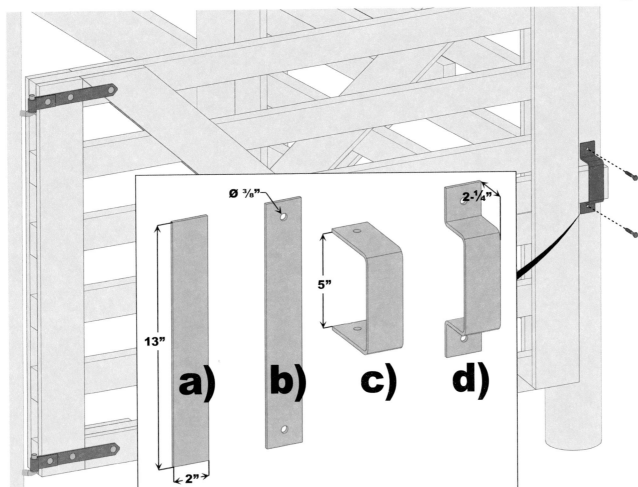

Ø ⅜"

2-¼"

5"

13"

2"

a)   b)   c)   d)

It is preferred to drill holes prior to bending. This metal is thin enough to bend with a sturdy vice, and some heat. Any local machine shop can quickly bend these for you if you lack the proper equipment.

Bored Notch

Use on Gates E and C. The bored notch latch is created by boring holes into a post to create a notch that will allow the sliding board to nest into it.

Pros:
- Unobtrusive
- With the correct tools, it is easy to do

Cons:
- By the nature of the design, there is not a lot of contact area between the board and the notch.
- Excessive rattling of a gate can cause the board to slip out of the notch.

COWS

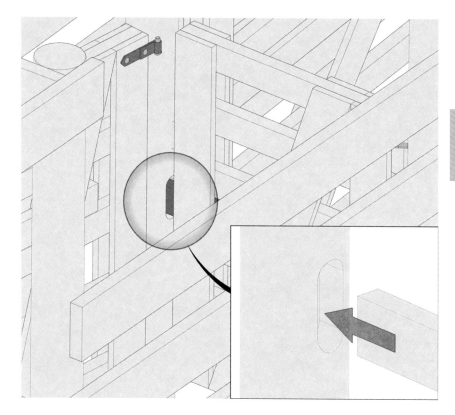

a)

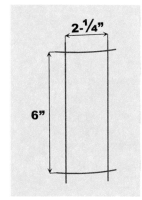

Measure and mark notch in desired location.

b)

Drill the first hole, 2-3" deep.

c)

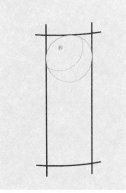

Drill the second hole the same depth.

d)

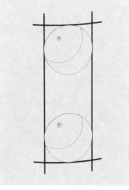

Drill the third hole in the middle.

e)

Using a chisel and a hammer, square up and clean the notch.

f)

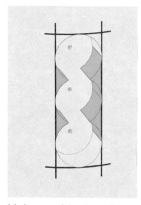
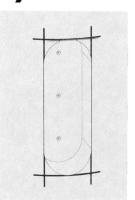

The notch is finished.

Gravity Latch

Use on Gate F1. The gravity latch is by far the most complex of the four designs, but has its place. This type of latch is great in situations where it is difficult to reach the sliding board to open and close the gate. Areas such as in front of, or behind the scale are perfect for this setup because you can reach over the fence and lift or close the latch while safely remaining on the other side of the fence.

Pros:
- Unobtrusive
- Accessible from multiple angles
- Gravity works with you

Cons:
- Difficult to make and install correctly
- Cost of hardware

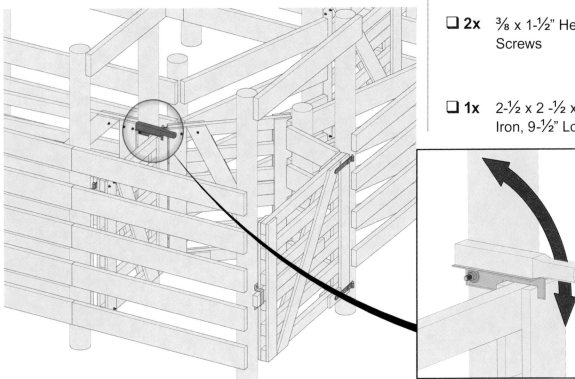

ADDITIONAL GATE HARDWARE

QTY PER LATCH

☐ **1x**   ½ -13 x 10" Carriage Bolts

Length of carriage bolt depends on the size of the post. Choose bolt that is at least 3" longer than post diameter.

☐ **1x**   ½"-13 Hex Nut

☐ **3x**   ½" Washers

☐ **2x**   ½-13 Nylock Nuts

☐ **2x**   ⅜ x 1-½" Hex Head Lag Screws

☐ **1x**   2-½ x 2 -½ x ³/₁₆ Angle Iron, 9-½" Long

☐ **1x**   2x2x15" Wood

Build A Gravity Latch

a)

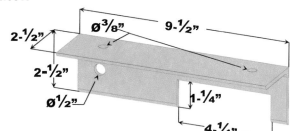

**1x 2-½ x 2-½ x ³⁄₁₆ x 9-½"
Angle Iron**

b)

1x 2x2x15

Round and shape a handle on one end as shown. Cutting the end off of an old wheel barrow handle works great too!

c)

2x ³⁄₈ x 1-½" Hex Head Lag Screws

d)

**1x ½" Carriage Bolt
1x ½" Washer
1x ½" Nut**

Length of the carriage bolt depends on the size of the post. Plan on a bolt that is at least 3" longer than the diameter of post.

e)

**1x ½" Washer
1x ½" Nylock Nut**

f)

Insert latch onto bolt assembly.

g)

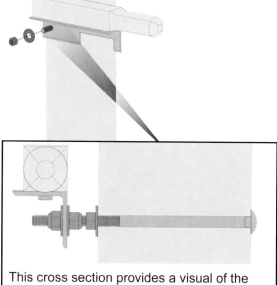

**1x ½" Washer
1x ½" Nylock Nut**

Do not over tighten the nut. The latch should be able to swing freely.

This cross section provides a visual of the spacing.

HEAD GATE

photo courtesy of Chris Slattery

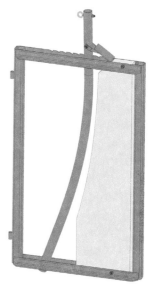

When our family bought this farm, we inherited the cattle chute and head gate arrangement, which was pretty crude. We didn't know any better and for years wrestled with that thing. It was a simple V with a rope on top of the sliding catch board. You had to pull it at the right moment and then wrap the rope around a post quick to hold tension against the animal's neck.

Of course, while you were trying to hold tension and then wrap the rope around the post it would often give a little, often enough for the cow to escape. The whole procedure took two people: one to put the cow in the chute and the other one to do the wrestling match with the rope and catch gate. It was horrible.

Then I met Teresa; I was 16 and she was 15. Yes, we were high school sweethearts and have now been sweet-hearting for nearly 40 years. But I digress. I got invited to her maternal grandparents' Christmas gathering and headed out to the barn to see what I could see. There I spied it: a one-man wooden head gate with a pawl and notches. I couldn't believe how simple it was, and how functional.

Her Granddaddy didn't weld so his was all of wood. I knew how to weld so I decided to make mine out of steel (a guy always has to tinker, right?). I built that thing in an afternoon, mounted it in the chute, and boy howdy, we were in business. Since then we've run thousands of animals through this head gate and it's still going strong.

The pull rope going back along the access chute allows one person to operate this efficiently and safely. It makes almost no noise. The squeeze pipe is pinned at the bottom so if you have an animal get stuck, you can pull that bottom pin and let the whole sliding pipe go to the side. Because a lot of light can come through it, animals don't balk like they do at self-catch gates. The whole gate probably weighs a quarter what a self-catch outfit weighs.

I welded a bolt on the end of one mounting hinge pin so no animal can pick it up off the hinge pins. The head gate is the main working end of a chute, but if the chute can wiggle around, the head gate isn't very secure. When I built this chute, I made all the poles tall so I could brace them with triangles. They've been tested, for sure, but they've never failed.

I've built half a dozen of these gates over the years and they always work well. Make sure you get the notches deep enough and the pawl frictionless enough so it drops down good and solid into the notch. If the pawl doesn't swing easily it might not drop all the way into the notch. If you don't weld, you can certainly make this whole head gate out of wood like Teresa's grandfather. His worked like a charm, but I did notice that one of his notches had busted out. That's the risk of using wood. But I know he used his for many, many years. Mine isn't snazzy, but it's simple, cheap, and it works well.

Tools

☐ Circular Saw

☐ Socket Driver Set

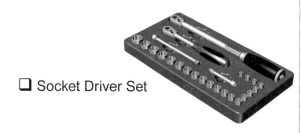

☐ Power Drill

☐ ²⁵/₆₄" HSS Drill Bit

☐ Driver Bits (for screws)

☐ Welder

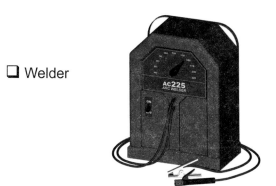

☐ Jigsaw

☐ Grinding Wheel

☐ Cut-Off Wheel

☐ Angle Grinder

☐ Speed Square

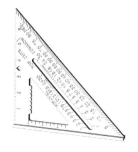

☐ Marking Pencil

☐ Tape Measure

☐ Hammer

☐ 4ft Level

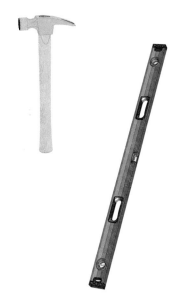

COWS

Hardware

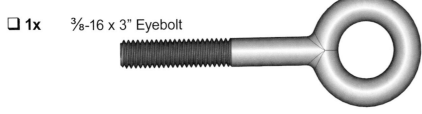

COWS

- ❏ **1x**   ⅜-16 x 3" Eyebolt
- ❏ **≈4x**   ⅜ x 2-½" Lag Screw Eyebolts
- ❏ **3x**   ⅜-16 x 3" Hex Head Bolts
- ❏ **2x**   ½-13 x 2" Hex Head Bolts
- ❏ **2x**   ⅜ x 2-½" Hex Head Lag Screws
- ❏ **1x**   Retainer Clip
- ❏ **1x**   ⅜ x 3" Hitchpin

- ❏ **4x**   ⅜" Washers
- ❏ **5x**   ⅜-16 Hex Nuts
- ❏ **2x**   ½-13 Nylock Nuts
- ❏ **2x**   ⅜-16 Nylock Nuts
- ❏ **≈16"**   ¼" High Test Chain
- ❏ **≈25ft**   Braided Rope
- ❏ **2x**   ½ x 6" Lag Screw Gate Hinges

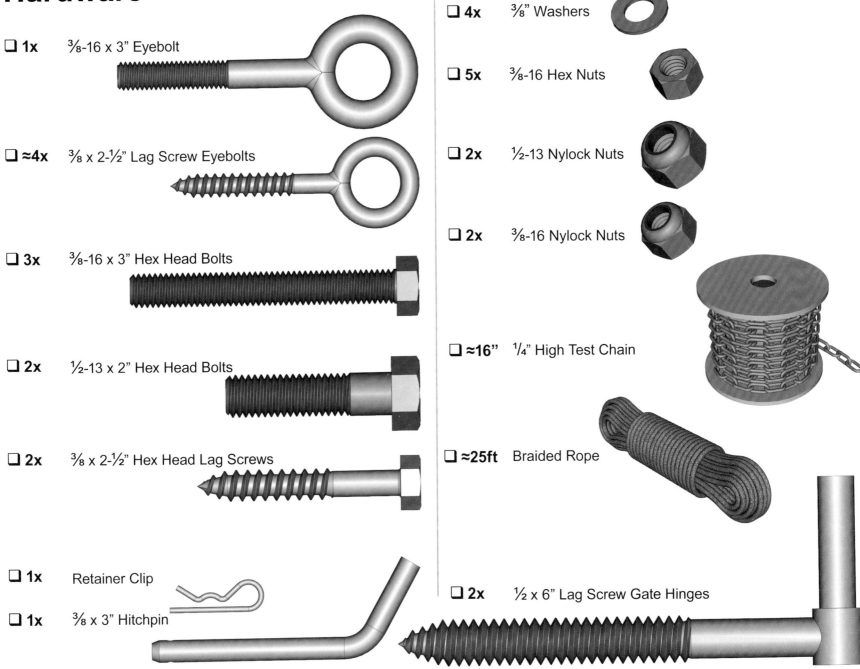

Metal & Wood Cutlist

QTY	SIZE
☐ ≈3"	NPS Ø½" SCH80 Pipe
☐ ≈10"	³⁄₁₆ x ¾" Bar
☐ ≈5"	³⁄₁₆ x 3" Bar
☐ ≈6"	2 x 2 x ¼" Angle
☐ ≈8ft	2 x 2 x ³⁄₁₆" Square Tube
☐ ≈10ft	1-½ x 1-½ x ¼" Angle
☐ ≈6ft	NPS Ø1-½" SCH40 Pipe*
☐ 4ft	2x12x48" Oak Plank

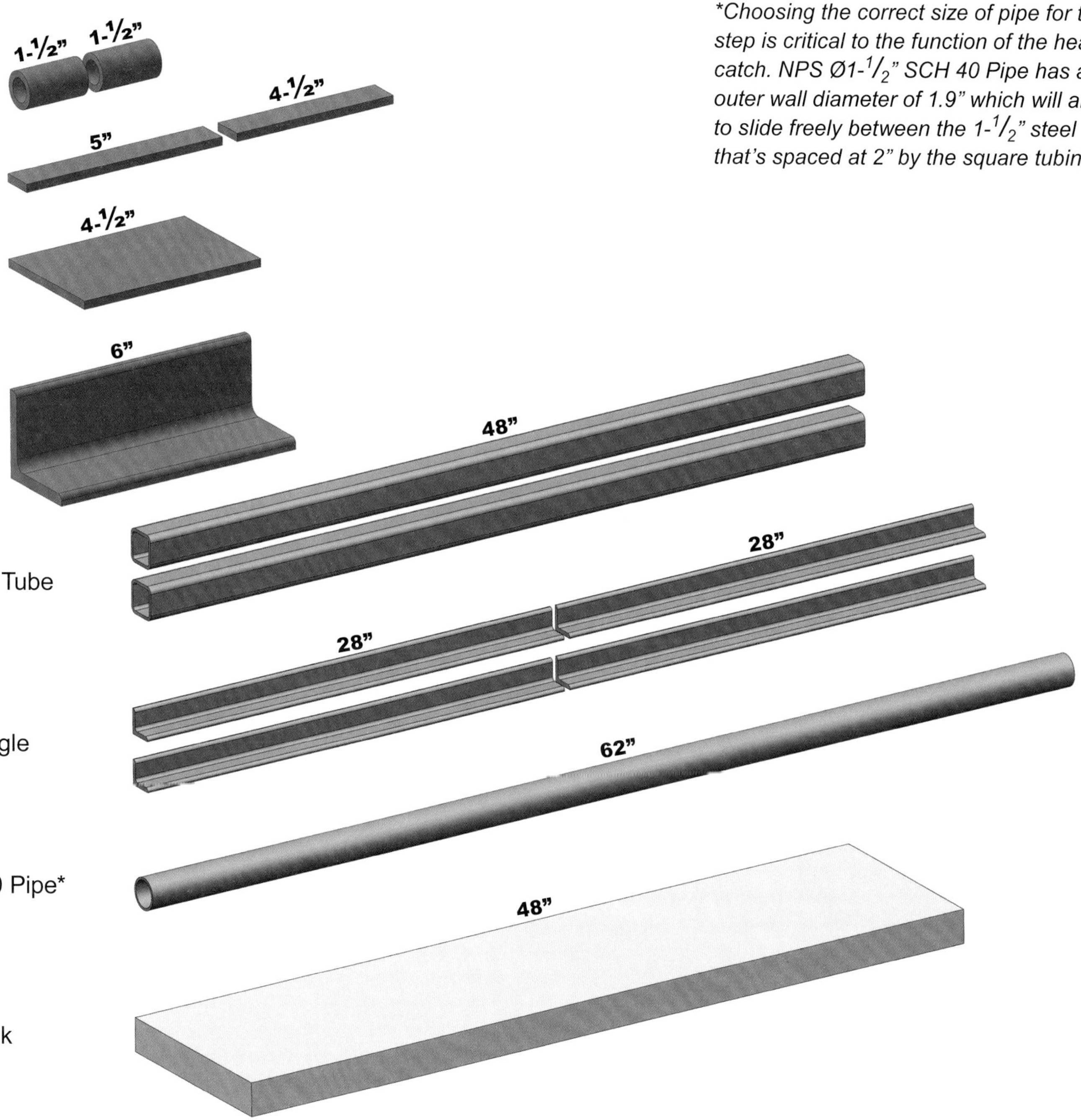

1-½" 1-½"
4-½"
5"
4-½"
6"
48"
28"
28"
62"
48"

Choosing the correct size of pipe for this step is critical to the function of the head catch. NPS Ø1-¹⁄₂" SCH 40 Pipe has an outer wall diameter of 1.9" which will allow it to slide freely between the 1-¹⁄₂" steel angle that's spaced at 2" by the square tubing.

COWS

COWS

1

1x   1-½ x 1-½ x ¼" Angle 28" Long

28"

a)

1"

b)

c)

Remove corner tips (in green) as shown.

2

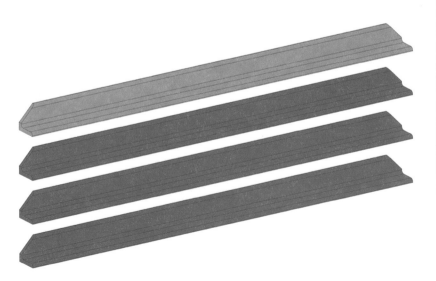

Repeat step 1 to create a total of 4 pieces.

3

2x   28" Angle From Step 2

Note that the two pieces in this step are oriented opposite of each other.

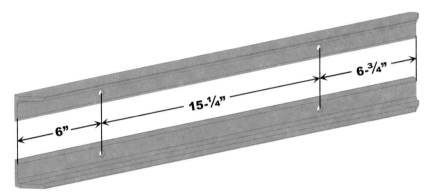

6-¾"

15-¼"

6"

All holes must be large enough to accommodate ⅜" hardware. Choose the next drill bit size up in your drill bit set. In a fractional bit set it will most likely be a $^{25}/_{64}$" bit. An X or Y size bit will also suffice.

4

2x   28" Angle From Step 2

Note that the two pieces in this step are oriented opposite of each other.

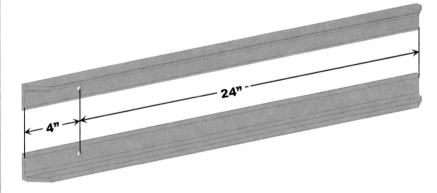

24"

4"

All holes must be large enough to accommodate ⅜" hardware. Choose the next drill bit size up in your drill bit set. In a fractional bit set it will most likely be a $^{25}/_{64}$" bit. An X or Y size bit will also suffice.

5

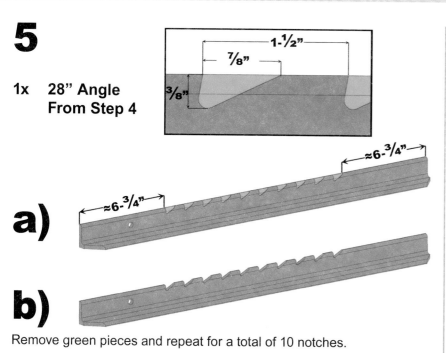

1x 28" Angle
    From Step 4

a)

b)

Remove green pieces and repeat for a total of 10 notches.

6

1x 2x12x48" Oak Plank

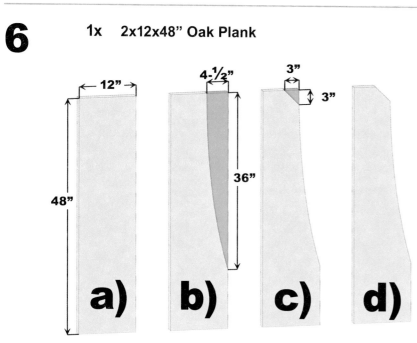

a) **b)** **c)** **d)**

Shape plank by removing the parts shown in green.

7

1x 2x2x³/₁₆" Square Tube, 48" Long

8

2x NPS Ø½" SCH80 Pipe, 1-½" Long

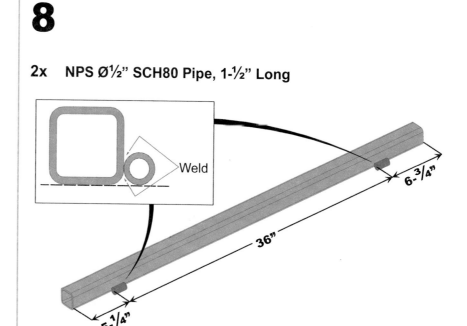

Weld

9

Observe the orientation of the parts in the illustration carefully to ensure proper alignment of all pieces.

1x   2x2x³/₁₆" Square Tube, 48" Long
1x   Angle From Step 3
1x   Angle From Step 5

COWS

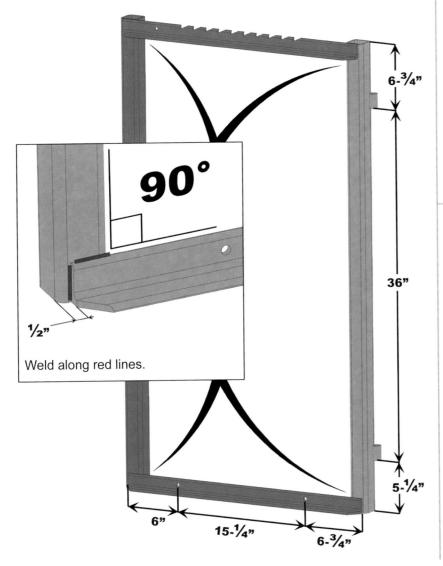

90°

½"

Weld along red lines.

6-¾"

36"

5-¼"

6"

15-¼"

6-¾"

10

1x   2x12x48" Oak Plank Cut From Step 6

Ensure the thickness of the wood is less than 2". If not, you will need to sand down the areas that are in contact with the metal to ensure a proper fit.

11

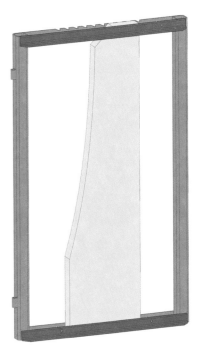

1x   Angle From Step 3
1x   Angle From Step 4

With the wood sandwiched in between, weld the two remaining pieces of steel angle in place. Be careful that they are oriented correctly. The holes should all align with the previously welded pieces. Leave the wood centered for now to prevent it from catching on fire from excessive heat during welding.

12

Position the plank about ½" from square tube.

13

2x ⅜-16 x 3" Hex Head Bolts
2x ⅜" Washers
2x ⅜-16 Hex Nuts

With the plank in position, drill through the wood using the existing holes as pilots. It is a good idea to drill and fasten the first hole before drilling the second to be sure nothing moves around.

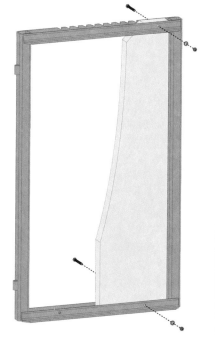

14

1x NPS Ø1-½" SCH40 Pipe, 62" Long

Choosing the correct size of pipe for this step is critical to the function of the head catch. NPS Ø1-½" SCH40 Pipe has an outer wall diameter of 1.9" which will allow it to slide freely between the angle iron which is spaced at 2".

15

14"

≈10"

The bend profile is not critical, but do not over bend it or else the catch will not close tightly enough.

COWS

16

50-1/2"

1"

2"

All holes must be large enough to accommodate ⅜" hardware. Choose the next drill bit size up in your drill bit set. In a fractional bit set it will most likely be a ²⁵⁄₆₄" bit. An X or Y size bit will also suffice.

18

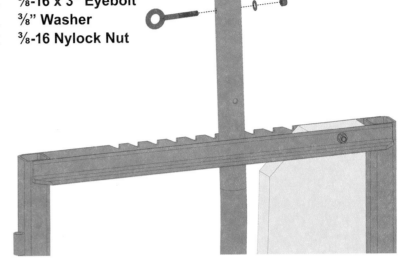

1x   ⅜-16 x 3" Eyebolt
1x   ⅜" Washer
1x   ⅜-16 Nylock Nut

17

1x   ⅜ x 3"
Hitchpin w/
Retainer Clip

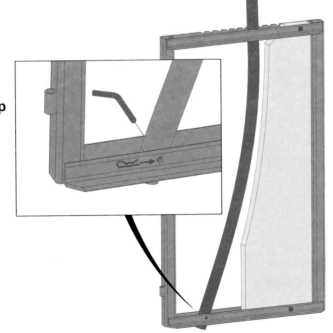

19

1x   ³⁄₁₆ x ¾" Bar, 5" Long

a)   5"   ¾"

b)   ½"   Drill hole the same size as in step 16.

c)   Remove corner piece shown in green.   ¾"

d)

20

1x   ³⁄₁₆ x ¾"
Bar, 4-½" Long

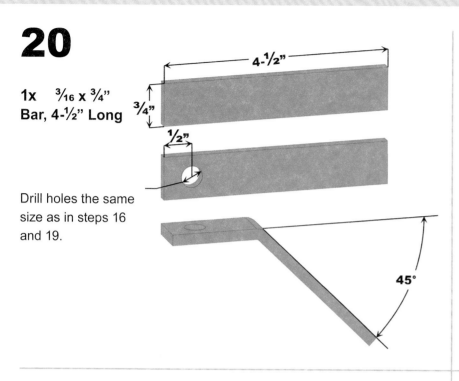

Drill holes the same size as in steps 16 and 19.

22

1x   ³⁄₁₆ x 3" Bar, 4-½" Long

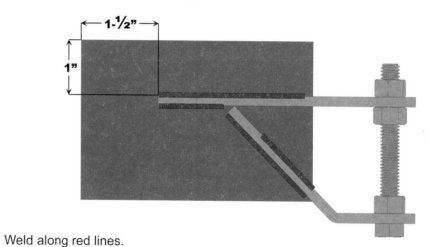

Weld along red lines.

21

1x   ⅜-16 x 3" Hex Head Bolt
3x   ⅜-16 Hex Nuts

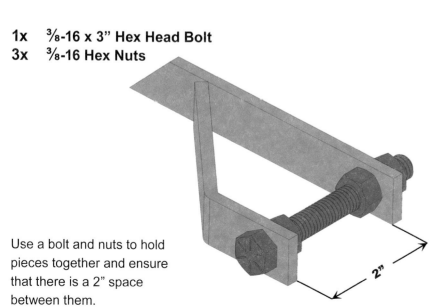

Use a bolt and nuts to hold pieces together and ensure that there is a 2" space between them.

23

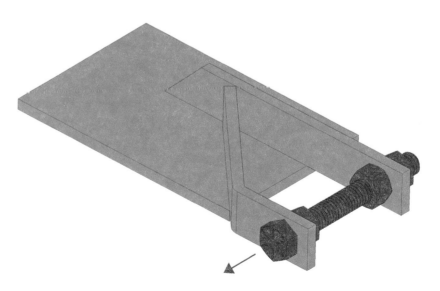

After the welds have cooled, remove nuts and bolt from the latch.

24

1x ⅜-16 x 3" Hex Head Bolt
1x ⅜" Washer
1x ⅜-16 Nylock Nut

COWS

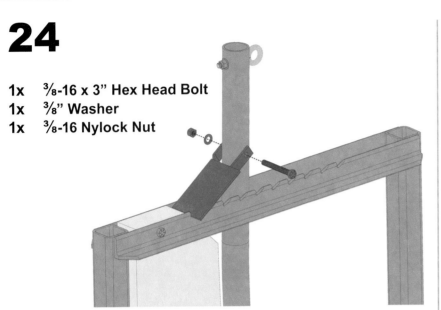

Be careful not to over tighten. The latch should swing freely but with minimal play.

25

≈16" ¼" High Test Chain

Weld along red lines to attach chain link approximately halfway up the square tube as shown.

26

The head gate is complete!

27 Modify A Standard "Off-The-Shelf" Lag Hinge By Welding A Bolt To It

2x ½ x 6" Lag Screw Gate Hinges
2x ½-13 x 2" Hex Head Bolts

a)

b)

Cut the top of the hinge off (remove the part shown in green).

c)

Weld the bolt on the top.

d)

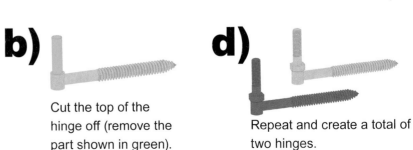

Repeat and create a total of two hinges.

28

1x   2x2x¼" Angle, 6" Long

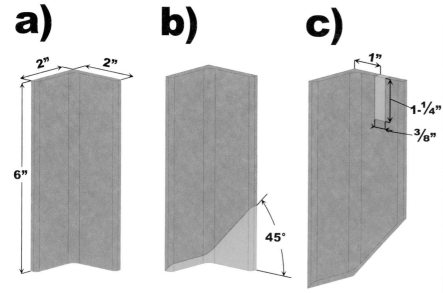

a)
2"   2"

6"

b)
45°

c)
1"

1-¼"

⅜"

d)
Round corners to eliminate sharp edges.

e)
1"

3-¾"

Drill same size holes as in step 16.

29

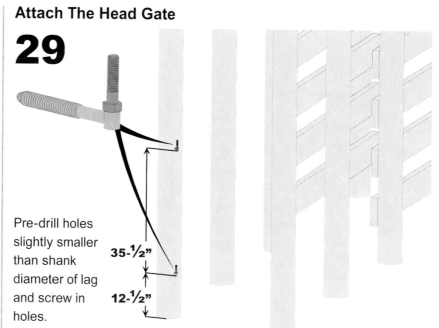

Pre-drill holes slightly smaller than shank diameter of lag and screw in holes.

35-½"

12-½"

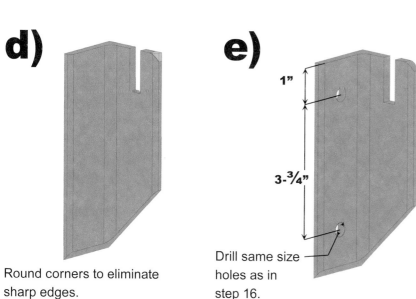

This overhead view of the head gate illustrates the swing radius.

Hang the head gate on the lag screws. Orientation of the lag screw hinges may vary from installation to installation. Notice how the head gate rests against the right hand post when closed, but can still swing wide open without binding at the hinge. Experiment with your hinge placement to ensure proper functionality of the head catch.

30

2x   ½-13 Nylock Nuts

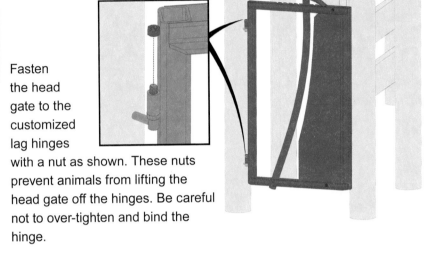

Fasten the head gate to the customized lag hinges with a nut as shown. These nuts prevent animals from lifting the head gate off the hinges. Be careful not to over-tighten and bind the hinge.

32

Check that the chain will reach and fasten securely.

31

2x   ⅜ x 2-½" Hex Head Lag Screws
1x   Angle From Step 28

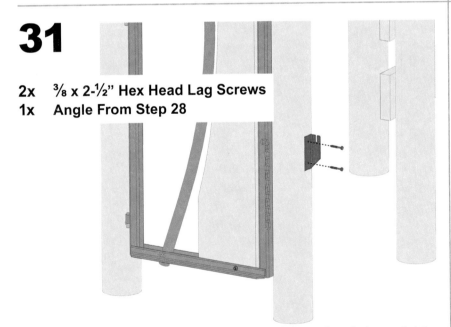

Mount the chain catch bracket at the same height as the chain or slightly higher.

33

≈4x   ⅜ x 2-½" Lag Screw Eyebolt

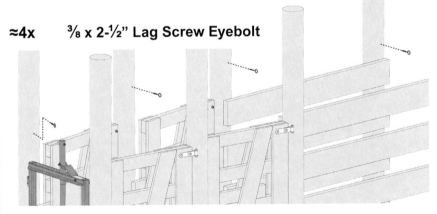

Pre-drill holes and install eyebolts (to hold the rope) inside of alleyway as needed. The eyebolts should be high enough to not be in the way but low enough to comfortably reach as you walk cattle into the chute. Rope should be accessible to handler at least 2-3 cow lengths back. Install as many eyebolts as necessary.

COWS

34

≈25ft   Braided Rope

Tie one end of rope onto head
catch and the other end onto
the last eyebolt in the alleyway.

35

Check proper operation of head catch. Make sure it opens and closes
without binding onto rope or anything else. Make adjustments as
necessary.

36

The homemade head gate is installed!

Pic 1

photo courtesy of Jessa Howdyshell

WINTER HAY FEEDING

Back in the 1960s Dad conceived of a hay feeding plan where we'd move a feeder gate through the hay pile as the cows ate it. Rather than bring the hay out to the cows, we'd simply meter the hay pile and let the cows eat through it. It was genius but it had one flaw: what to do with the manure build up behind.

At the time we only had a dozen or so cows so each week we'd shovel out the manure into a little manure spreader and spread it on the pasture. We noticed that what we spread early in the winter didn't seem to make any difference but what we spread late in the winter made green stripes.

We surmised that the vaporization and leaching robbed the nutrients before the dormant soil warmed up enough to metabolize them. We had to figure out a way to let that bedding stockpile so we could clean it all out in the spring when the soil awakened and could eat all those goodies. That moved us into the "carbonaceous diaper" idea. Since we had lots of forest and cut a lot of firewood, both to burn for our own use and to sell, wood chips seemed the obvious best carbon source.

We purchased a used industrial chipper and began turning branches left over from firewood operations into chips for bedding. It worked beautifully. Since we were moving this V-slotted feeder gate (see picture 2) through the

Pic 2

Pic 3

both photos courtesy of Jessa Howdyshell

hay, the buildup of bedding was not a problem because as we moved the gate through the hay pile it moved onto fresh ground and away from the constantly building material behind. By the end of the winter, the earliest floor area might be a couple of feet deep in manure and chips and the end was a few inches.

As this change and others brought on increased fertility, we began expanding the cows and making way more hay. We built another barn with the hay stored in the center square and feeder gates all around. Since we fed on all four sides

of the hay pile, we couldn't walk the feeder gates through the hay like we'd done in the past. We made the gates vertically movable and lifted them with the tractor front end loader every couple or three weeks as the bedding built up.

The V-slots worked okay, but they were rigid. We made them big enough for cows, which wasted space if we used them for calves. I began wondering if there was a way to make a vertically mobile arrangement that allowed any size animal to seek its own space. Why not make hanging boxes?

The Birth Of The Hay Feeding Bunks

With some pulleys and a hand winch, we made hanging boxes that work perfectly. Animals can get as close or as far apart as they want; nobody ever gets stuck (that happened a few times in the V-gate) and we don't need a front end loader to elevate it. All we do is give the hand winch a few cranks and the whole contraption comes up. (See picture 3).

This way we can have a 4ft. bedding depth in the feeding area, which gives us lots of manure storage. We add corn to it as we build. The cows

Pic 4

photo courtesy of Jessa Howdyshell

tromp out the oxygen and the whole pack turns into anaerobic fermentation, which ferments the corn as well. When the cows exit to spring grass, we put in pigs and they seek the fermented corn, stirring the whole pack, injecting oxygen, and turning it in aerobic compost. That is the heart and soul of Polyface.

The hay feeding boxes, then, are part and parcel of an efficient feeding system that enables us to capture all the winter nutrients, compost them, and use them as our fertilizer. We don't have to carry hay anywhere; we just throw it in the boxes from the stack. The boxes are on a cantilever so the cows can't back up and poop

into them. This keeps things clean so the animals aren't ingesting soiled feed. The boxes, then, are also part of our sanitation program.

The first ones are finally beginning to rot after 20 years. We're rebuilding those this year (2020), but they've lasted a long time. Again, because we have our own band saw mill, wood is cheap and our favorite building material. You could build these out of metal, of course, but they'd be far heavier and for us, much more costly.

The boxes are segmented like a caterpillar. The fact that they dangle and move doesn't seem to affect their functionality. The vertically mobile feed bunk enables us to accumulate the entire

winter's manure and urine under the protection of the roof and get maximum benefit from these nutrients. Although we don't buy chemical fertilizer, we constantly think about fertility and this design is part of that whole closed-loop thinking.

Visitors to our farm notice that our barns look like a hodge podge (See picture 4). That's because they've expanded over time to accommodate the meteoric rise in fertility. Back when we could scarcely feed a dozen cows we never imagined that we'd now be feeding 200. The barns at the front of the farm were actually designed to handle a 20 ft long dump trailer with loose hay (that idea didn't work) and today are typically filled with small square bales: approximately 10,000 in all. We typically house 180 head in this barn in 4-6 different groups.

Our newest barn is farther back on the property—we call it "far barn" – and it houses 120 head in two groups, and has a storage capacity of 800 3x3x7ft large square bales of hay (see picture 1). The barn is long with hay storage running down the middle. On either side of the haymow are awnings running the length of the building. These awnings house the livestock. This long straight barn has easy access for hay stacking and maintenance of the bedding areas.

The small square bales are easier to handle for feeding, but of course much more difficult

Joel's Tip

Notice how high our haymow is in picture 5. Generally speaking, raising the height of the building is an inexpensive way to gain more storage capacity because the roof is the same cost regardless of how high it is.

Pic 5

photo courtesy of Jessa Howdyshell

COWS

to make and stack (see picture 5). Because large squares pack tighter in the baler, much more hay can be stored in the same space compared to small squares. Square bales of any dimension are easier to stack, transport, and handle as compared to round bales. In this type of feeding set-up, the squares are easy because the individual pats flake off. Round bales require some sort of unrolling technique.

When setting out to write this chapter we asked ourselves the question: <u>What would we do differently if we could start from scratch and design the perfect winter feeding barn?</u> We agreed that "far barn" would definitely be a good

starting point, but we've always wished that we made the livestock areas a little wider and taller for better access with tractors, especially with 4ft of manure pack by the end of the winter.

So, this chapter is a culmination of what works in our current setup with potential improvements based on our lessons-learned over the years. This first section provides an overview of our ideal winter hay feeding barn layout. This is not a step-by-step tutorial; there are plenty of barn building books and how-to's out there already. Our intent here is simply to guide you through the thought process when designing your winter feeding arrangements.

Immediately following this section is a step-by step tutorial that does guide you through the build process of the hay feeding bunks.

Lastly we'd like to reiterate that our fear through this entire book has been the risk of stifling individual ingenuity. Please do not surmise that you need a set-up just like this to get into cattle. We encourage you to be resourceful, use what you have, and improvise. As your herd and business grow, you'll upgrade your infrastructure. By then too, you'll have more experience about customized edits to your building plans. Glean what you can, learn from our mistakes, and use your imagination.

Design Your Own Feeding Barn

COWS

If you have the luxury of building a barn from scratch, then the length, width, and height of your hay feeding barn will ultimately be determined by several questions you need to answer for yourself. If you are retrofitting an existing structure, then use this section as a set of guidelines/best practices when implementing your own setup.

Questions to consider before building:

1) **Livestock Capacity:** How many animals will you be housing? We plan for a minimum of 30ft² per animal and/or 2ft of feed bunk space per animal.

2) **Storage Capacity:**

a. What is the length of your hay feeding season? How many bales will you need to store? A mature cow of 1,000 pounds (that's a small cow on most modern American farms) will eat about 28 pounds per head per day. A normal big square bale weighs 600 pounds, which will handle 20 cows per day (they never eat every morsel). If you need to feed for 100 days, you'll need 5 of these bales per cow. A herd of 100 cows needs 500 of these bales; each bale (assuming 3ft X 3ft X 7ft) requires 63 cubic feet, so 500 would need 31,500 cubic feet. If you can store them 4 across and 5 high, that's 28ft X 15ft X 3ft per rank, meaning you'll need 25 ranks of 3ft or a total of 75ft. This is simple math, but it's how we figure storage volume requirements for the winter supply.

photo courtesy of Jessa Howdyshell

b. If you're making small squares simply adjust your dimensions to arrive at the volume. Ditto for round bales, although they don't fit the neat cubic volume formulas we've used. Normally, stacking round bales on end is the easiest way to get the maximum capacity in any given structure.

3) **Site location:** How big can my building realistically be? Will it necessitate earthwork to be done?

4) **Proximity to water sources:** While we frown upon unmetered access to ponds and creeks, when managed correctly these can be reliable sources of winter water for livestock. Otherwise you need to consider running a frost-free water line to your barn. Proximity to water sources also needs to be considered when dealing with the issues of run-off. While this system of deep bedding under shelter eliminates excessive runoff, there

is still a loafing lot that can build up with manure. We encourage you to do your best to be a good steward and protect your water sources to the best of your ability.

5) **Proximity to electricity:** We discourage putting electricity in hay barns because of the risk of fire, but when properly encased in conduit and installed professionally it can serve as an added luxury for lighting, hay elevators, water de-icers, etc. If you are someone who has a day job and feeds before/after work, then some form of lighting may be a worthwhile investment. Statistically, more than 95 percent of all barn fires are caused by electrical wiring problems. Rats chew off insulation; they get bumped with machinery, etc. Modern LED head lamps provide a low risk alternative to electrical cords running around the barn.

Feeding Area

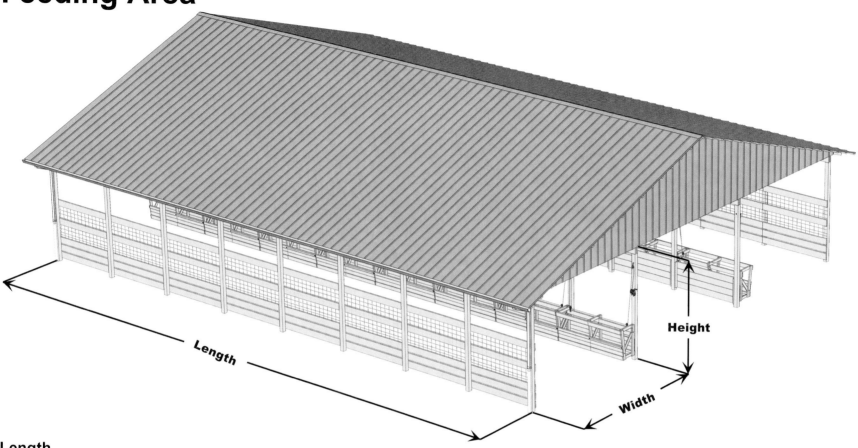

COWS

Length

The length of your building is ultimately determined by the number of animals that you need to house and feed, but it can also be limited by the lot size that you have to build on. As mentioned previously, our general rules of thumb are 30ft^2 per animal and/or 2ft of feed bunk space per animal.

Width

Our feeding areas are narrower than 20ft (16ft to be exact), and we have always wished them to be bigger. Cows seem to get congested easily. Not only that, but it can get

tight when backing manure spreader loads of chips in to spread. Narrow is not a deal breaker, but we would not recommend less than 16ft. Bigger is better in this case. We have one section that is 19ft wide and that extra room is downright commodious.

Height

The height of the feeding area is determined by two factors:

1) The length of your feeding season (longer feeding season means deeper bedding pack).

2) The height of your tractor or other equipment used for spreading bedding material.

Our typical bedding packs for a 100 day feeding season can accumulate to 3-4ft thick! The barn needs to be tall enough to allow your equipment to access it even with 4 or more feet of bedding. The barn ceiling clearance shown in this book is 14ft high, which is an ideal place to start.

Hay Storage

COWS

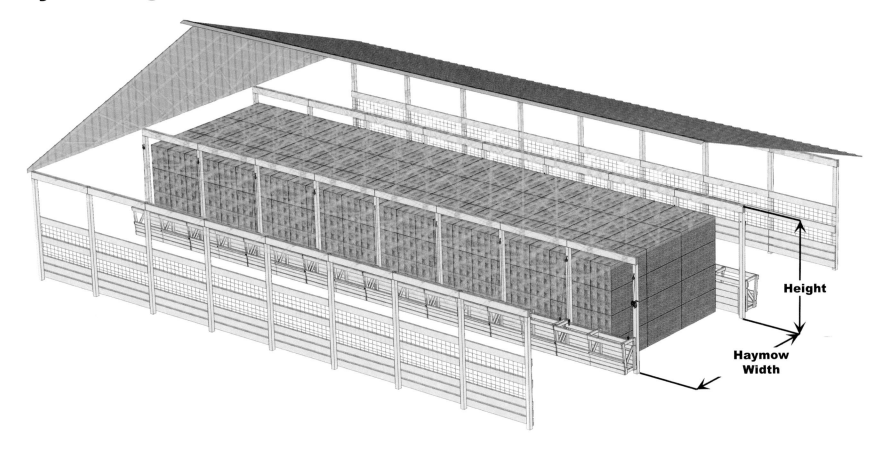

Height

Haymow
Width

Haymow Height

The size of your hay storage area largely depends on the amount of hay and the size of your bales. Height also depends on equipment limitations and how high you are able to stack. Generally speaking, raising the height of the building is an inexpensive way to gain more storage capacity because the roof is the same cost regardless of how high it is.

Haymow Width

The width is determined entirely by the amount of feed you need to store, the size of the bales, and the available footprint you have to work with. Ultimately, we want the hay to nest in place with very little dead space between the hay stack and the bunks. The reason is that we feed by walking on the haymow and drop flakes directly into the bunks below. A large gap creates the risk of a person falling while feeding.

After calculating your feed requirements, find out the size of your bales and determine how many you need to place end to end across the width. That will be your working measurement. Keep in mind that the wider your stack is, the farther you will have to walk from the center of the pile to the bunks to drop it in. With that in mind we would not recommend a 50ft wide haymow.

Example Scenario

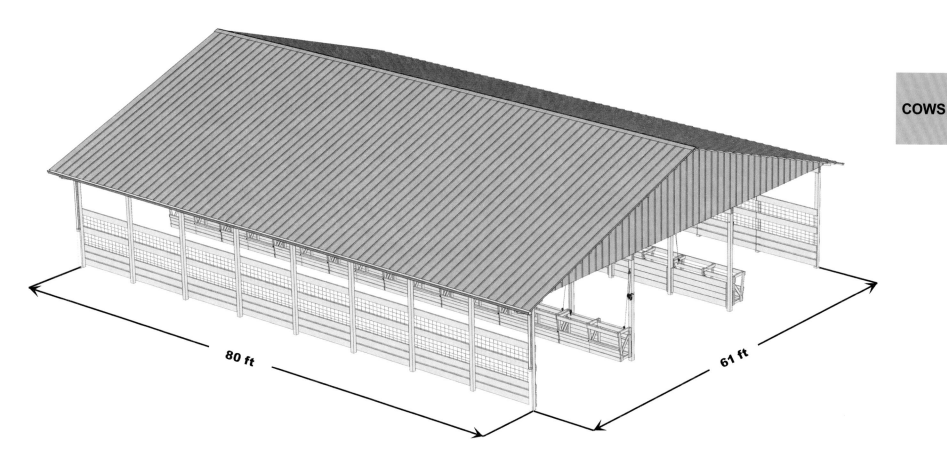

COWS

For illustrative purposes, we have drawn a hay barn that is 80ft long by 61ft wide, with 14ft of head clearance. This barn can hold forty 1000lb cows on either side and boasts 40ft² of space per animal. With a feeding period of approximately 100 days, this barn needs to hold 373 large square bales of hay. If our length is 80ft, that means we are limited to 26 three foot wide ranks of hay. Since this barn is only wide enough to accommodate 3 bales width-wise, they will need to be stacked 5

high. As you'll notice with the wooden trusses we've illustrated, we actually only have enough space for 4 bales high. This is something to keep in mind when selecting your roof design. Check out the metal truss alternatives later in this section for additional vertical clearance if needed.

Loafing Lot

COWS

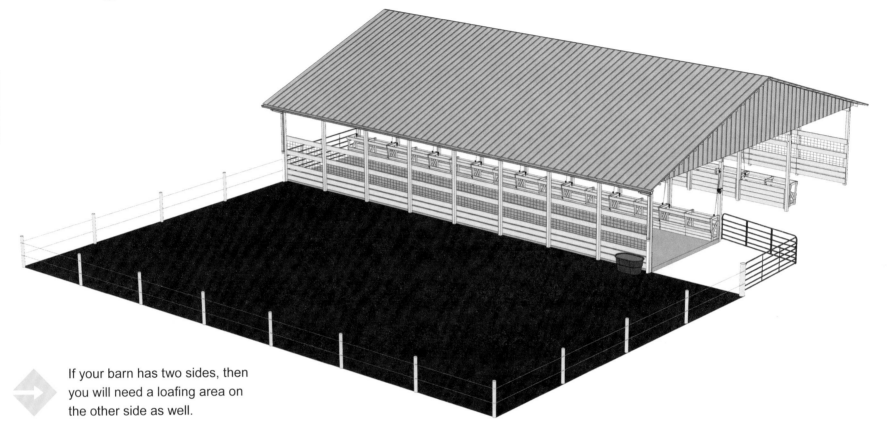

If your barn has two sides, then you will need a loafing area on the other side as well.

Here is a basic illustration of a loafing lot. These lots serve several purposes:

1) They are a sacrificial parcel of land that let cattle have more room to spread out without damaging pasture fields.

2) These lots provide a way to contain cattle outside of the feeding area for re-bedding and doing other maintenance.

Some things to consider when creating a loafing lot:

1) Ideally, select an area that is relatively flat and on high ground for maximum drainage and minimal runoff.

2) The loafing area should be a minimum of 400ft² per cow.

3) Sturdy perimeter fence should be at least 2 electric wires to contain cattle. (It will be quite a temptation for cows staring at green pasture across the fence all winter long.)

4) It should have a way to block cattle off from feeding area (i.e. metal gate or electric fence wire.)

5) The churned up and heavily fertilized loafing lot can offer great gardening opportunities during the summer. We've grown potatoes and pumpkins in these lots.

If you are watering your cows with drinkers or water tubs, position them somewhere you won't have to constantly move them every time you add bedding. A fixed drinker inside the barn would eventually get buried. Keep all of this in mind when you install your water system. Long garden hoses are a pain because you need to remember to drain them every night to keep them from freezing.

You may wonder why we don't install permanent stationary frost-free drinkers. Everything about Polyface revolves around mobility. The fewer stationary installations we have, of any kind and for any purpose, the more flexibility we have to make adjustments and upgrades in the future. Remember the bedding inside builds to 4 ft deep, so any stationary watering system is problematic when the floor elevates that much. A permanent installation outside is okay, but again, what if we want to put a hügelkultur bed under the eaves in the future? A few minutes of extra work on the coldest days of the year in order to protect future flexibility are well worth it.

If your loafing lot is oriented next to your barn as shown, then we highly recommend installing a gutter system to prevent all of the excess rainwater from mucking the lot up any more than necessary. As a bonus, it can be captured and fed into a cistern or pond for future use. Be sure to stop aluminum downspouts well above cow height. Install 4" schedule 40 PVC the remainder of the way down the posts and fasten it securely. (Cows tend to rub on posts and will easily crush an aluminum downspout.)

This is an example of how we orient our gates in and out of the hay barn. This gate set up allows us to push the cows out of the barn and lock them in the loafing lot while we add more bedding, etc. With creativity and forethought you can save yourself a lot of work with routine tasks.

Roof Designs

COWS

Ultimately, your roof design will be determined by your needs, builder preference and availability, and last but not least, cost. After a quick walk around our farm you will notice a wide range of roof structures utilized. We've used whole round timbers, rough cut dimensional beams, manufactured trusses, as well as hoop-style structures for roofs. We cannot and will not give you a one size fits all approach for your building's design. However, we will say that manufactured trusses, whether they be metal or wood, have come a long way in both design and price since we built our first barns. We would think seriously about the clear-span style metal trusses because they are less obstructive and offer a lot more vertical clearance than a standard wooden truss design.

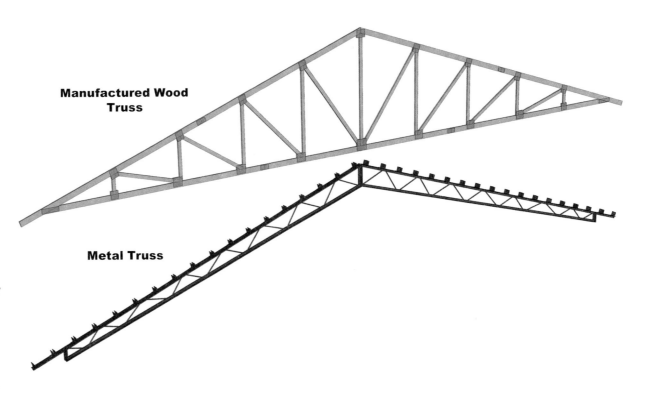

Manufactured Wood Truss

Metal Truss

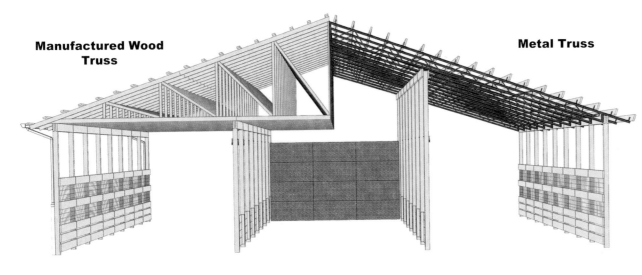

Manufactured Wood Truss

Metal Truss

Notice the increased clearance with the metal truss design, especially in the hay storage area. The added clearance aids the person who will be on top of the haymow, feeding the hay out. In one of our barns we have to crawl in and out of rafters when it is completely full. It is an added inconvenience that we never anticipated when we built it. (Now we know there's a better option, so we are sharing it with you.)

Deep Bedding

COWS

photo courtesy of Jessa Howdyshell

Because the bedding builds up so high, you need some sort of retaining wall to hold it in. While our illustrations show dimensional rough-sawn lumber, our preferred material is junk pine pole timbers from the forest. These native Virginia pines have no real commercial value, grow straight, and begin dying at 50 years old. With a chainsaw we flatten one end so we can fasten them to the poles with long screws. They rot in 10 or 12 years, but that's fine. They don't cost anything but a bit of time. When they rot, we simply remove them and install a new set. Cheap and strong.

Above this heavy wall we run cattle panel and/or a couple of wires so the cows don't fall out of the shed when the bedding gets deep.

We apply the chips with a manure spreader after running the cows out of the shed. Normally two and at the most three beddings per week are adequate. We put in about 80 pounds of corn per cubic yard of carbon material. We place more corn in the bottom than in the top because the pigs have to work harder to get to the bottom. Whole shelled corn will only last about 120 days before it ferments into nothing but skin. We've

used un-hulled barley or rye to extend this window of palatability, but the smaller grains are not as efficient for the pigs to find and eat.

Because the wood chips have such a high Carbon : Nitrogen (C:N) ratio, every fourth bedding we use old junk hay. It's a great salvage for rained on or moldy hay. We throw the hay on the floor, twice as much as they should eat. They eat half and incorporate the other half into the bedding. This keeps the C:N ratio close to 30:1 which is ideal for good compost. In general, if the top of the bedding gets mucky or if you start smelling ammonia, it's time to re-bed. We put in about 3 inches in depth per bedding.

As the bedding deepens, the ferment-

ation gradually shrinks the mass and definitely uses up the moisture (urine) in the pack. At first it seems like you can't keep up with them but once things get deeper than 18 inches the pack develops a life of its own and it takes much less carbon to keep up with the urine. We stack the wood chips in a separate nearby shed so they go in dry; that maximizes the amount of urine each piece can absorb. The chips dry down in that storage stack due to decomposition. Sometimes it looks like the shed is on fire with all the steam coming off the drying piles.

When the cows exit in the spring, we put in the pigs and they seek the imbedded fermented corn, slinging the material around

291

COWS

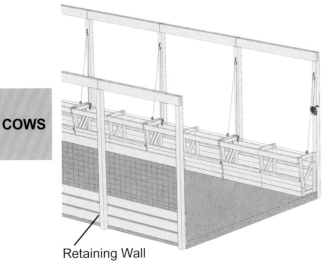

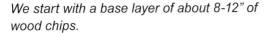
Retaining Wall

We start with a base layer of about 8-12" of wood chips.

We add more chips 2-3 times a week in 3" increments. We sprinkle corn every second or third bedding.

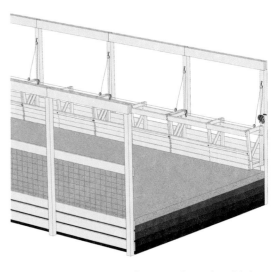

As the bedding gets deeper, the microbial action becomes more evident. Before long, we've got a high octane microbial engine working beneath the animals feet.

and aerating it. The bigger the pig, the deeper it can dig. Pigs under 100 pounds can only dig about 18 inches deep. A 300 pound hog can go 3ft deep. We have not found one that can go 4ft deep. If the bedding is that deep, we let the pigs go down a couple of feet, clean that top layer off with the front end loader, and then put the pigs back in to go down to the bottom.

The amount of chips needed varies based on type. Chips made during the green season from deciduous trees requires more volume than chips made in the winter when the leaves are off. The leaves move the C:N ratio from a pure chip of about 250:1 to a 75:1. With the cow excrement average 18:1, obviously maintaining a 30:1 or 25:1 C:N ratio takes more chips if it's 75:1 versus 250:1. The junk hay will

run less than 40:1, which is why a bit of that helps us keep our C:N ratio more balanced.

The cows drop about 50 pounds of material a day between manure and urine. Soaking all of that up requires a hefty amount of carbon, but remember, this is our fertility program. Rather than buying chemicals, we invest in weeding (upgrading) the woodlot, maintain cow comfort, and produce a soil-building elixir. What's not to love?

In general, a 200 pound pig will turn about 8 cubic feet of material a day. We don't use groups of more than 15 hogs because bigger groups incentivize loafers. Then the workers get tired and frustrated. We use simple pipe gates to create sections so the pig groupings are smaller. We try to put in enough

pigs to get through the material in less than a month. Remember, the corn deteriorates a little bit each day. The faster you can get through it, the less corn you risk losing. The pigs never eat every kernel so once they get irritated (listen to them--they'll tell you) it's time to quit.

We immediately spread the compost and then run the Eggmobile nearby so the chickens can pick out the last of the uneaten kernels of corn. We feed the pigs nothing else while they are pigaerating. Their favorite treat is when they come upon a fermented, urine-soaked glob of junky hay. It smells like silage and they eat it like cotton candy. This whole system is elegantly simple and certainly has been indispensable in building soil fertility here at Polyface.

Hay Feeding Bunks

❑ Socket Driver Set

❑ Power Drill

❑ ½" Auger Bit

❑ ½" HSS Drill Bit

❑ Driver Bits (for screws)

❑ Speed Square

❑ Marking Pencil

❑ Hand Saw

❑ Hammer

❑ 4ft Level

Tools

❑ 8 and/or 10ft ladder

❑ Tape Measure

❑ Table Saw

Hardware (Bunk Installation)

Refer to step 22 on page 303 to determine how many bunks you will string together in each "battery." Quantities in this bunk installation section are on a "per battery" basis.

This quantity equals the total number of bunks plus one.

❑ __x ½-13 x 7" Eyebolts (see note A on next page)

These quantities match the total number of eyebolts.

❑ __x ½" Washers

❑ __x ½-13 Hex Nuts

❑ 1x Cable Thimble (see note C on page 294)

Hardware (Bunk Installation)

COWS

☐ **3x**   Cable Clamps
(see note C)

☐ **1x**   Hand Crank Winch
(see note B)

☐ **__x**   ⁵/₁₆ Quick Links
(see note D)
This quantity equals the total number of bunks.

☐ **__x**   Cable Pulleys (see note C)
This quantity equals the total number of bunks.

☐ **__ft**   Steel Cable (see note B)
See page 296 to calculate length

Notes:

A). Length of eyebolt will depend on how thick the post is in your barn (that you will be hanging the bunks from).

B). Each bunk weighs approximately 500 lbs empty. Choose a cable and winch strong enough to support whatever number of bunks you plan to string together. If in doubt, consult a professional.

C). Choose pulleys, thimbles, and cable clamps based on size of cable used.

D). Quick links are a simple way to connect the pulleys to eyebolts, but they are not the only way. Be creative but be safe.

Hardware (Per Bunk)

☐ **≈100x**   20D Galvanized Nails

☐ **≈100x**   8D Ribbed Nails

☐ **6x**   ½-13 x 4" Carriage Bolts

☐ **3x**   ½ x 4" Hex Head Lag Screws

☐ **13x**   ½" Washers

☐ **8x**   ½-13 Hex Nuts

☐ **2x**   ½-13 x 3" Eyebolts

☐ **2x**   ⁵/₁₆ Quick Links
(see note D)

☐ **2x**   Cable Pulleys (see note C)

Wood Cutlist (Per Bunk)

This cutlist assumes a bunk length of 10ft. The length of your bunks may vary depending on the dimensions of your barn. See the page 297 for additional details prior to procuring lumber.

Wood scraps are highlighted in red.
Do not discard until project is completed.

COWS

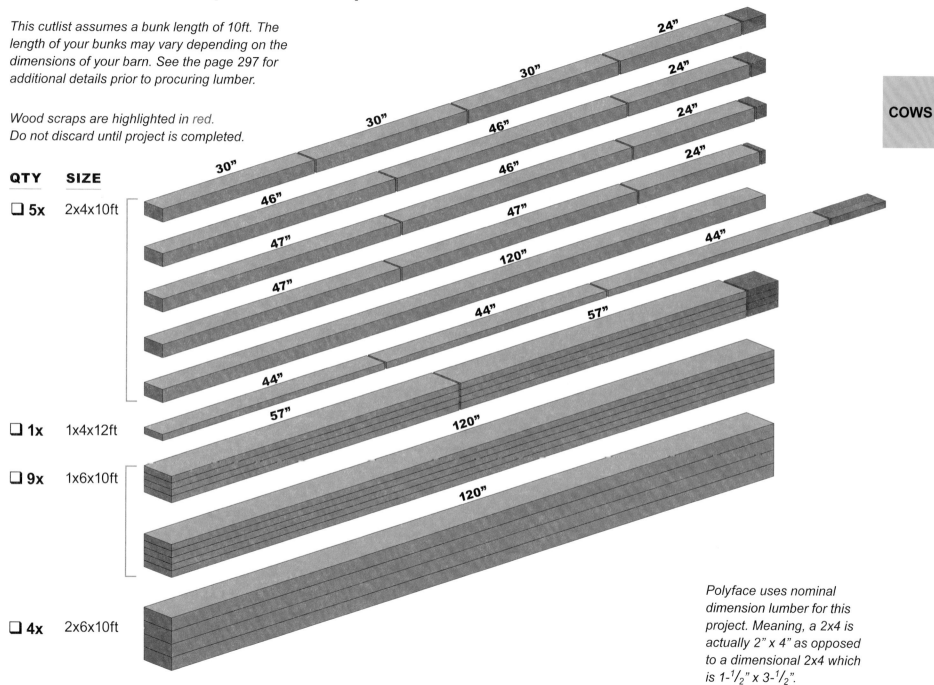

QTY	SIZE
☐ **5x**	2x4x10ft
☐ **1x**	1x4x12ft
☐ **9x**	1x6x10ft
☐ **4x**	2x6x10ft

Polyface uses nominal dimension lumber for this project. Meaning, a 2x4 is actually 2" x 4" as opposed to a dimensional 2x4 which is 1-1/2" x 3-1/2".

Determining Cable Length

It is difficult for us to tell you how much cable to buy because it depends on several factors:

1.) How wide are your bunks?

2.) What is the distance between the bunk and the upper pulleys at the bunk's lowest position?

3.) What is the number of bunks on a single winch?

COWS

Estimate the amount of cable needed for your bunks:

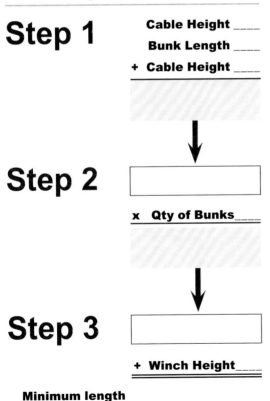

Step 1

Cable Height _____
Bunk Length _____
+ Cable Height _____

⬇

Step 2

x   Qty of Bunks_____

⬇

Step 3

+ Winch Height_____

Minimum length of cable required

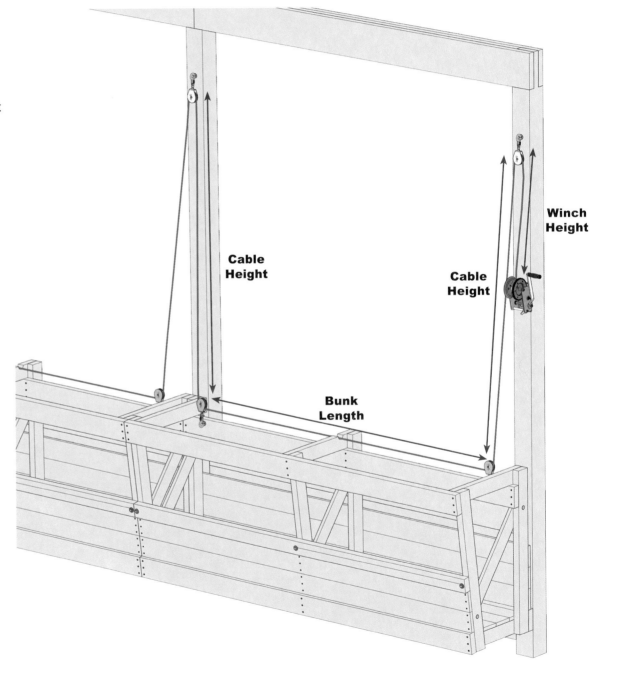

Determining Bunk Length

Build The Bunk

1

1x   2x4x46

COWS

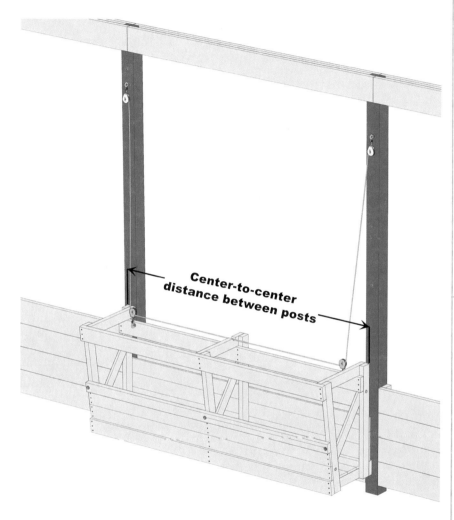

Center-to-center distance between posts

Bunks need to be sized to fit between the barn posts as shown in this illustration. If the posts are centered at 10ft apart, then the bunk should be 10ft long. The goal is to keep the edges of the bunks centered on the posts. Twelve feet is the maximum bunk length that we recommend, but 10ft is ideal.

2

1x   2x4x30
4x   20D Galvanized Nails

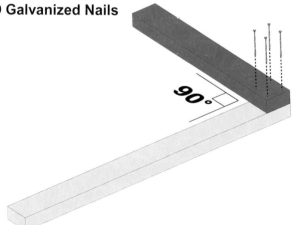

90°

3

COWS

1x   **2x4x24**
4x   **20D Galvanized Nails**

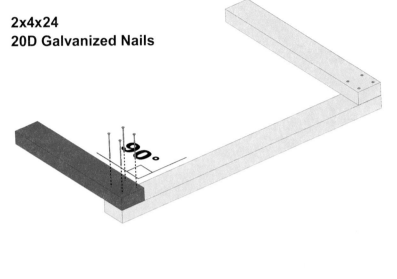

4

1x   **2x4x47**
8x   **20D Galvanized Nails**

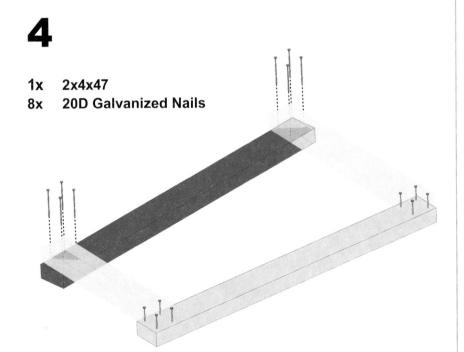

5

1x   **1x4x44**

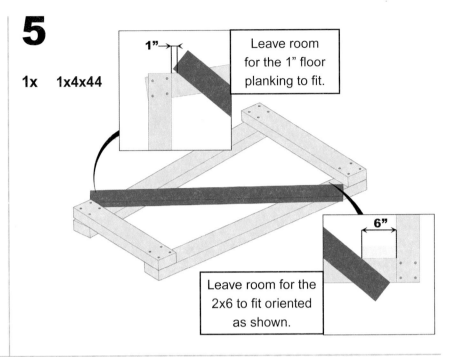

Leave room for the 1" floor planking to fit.

Leave room for the 2x6 to fit oriented as shown.

6"

6

2x   **½-13 x 4" Carriage Bolts**
2x   **½" Washers**
2x   **½-13 Hex Nuts**

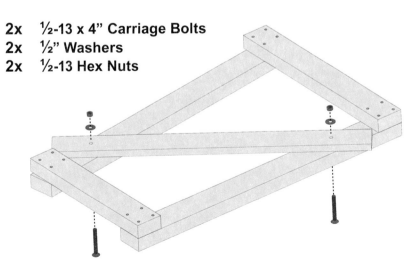

Drill ½" holes and fasten bolts.

7

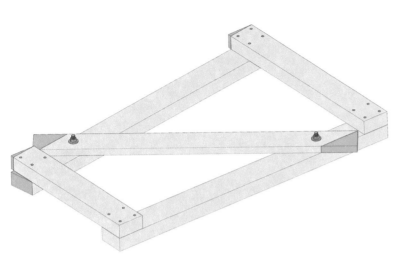

Trim excess (shown in green) from all corners.

9

1x   **2x4x24**
8x   **20D Galvanized Nails**

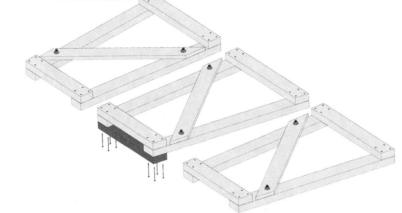

Attach to middle section and trim excess per step 7.

8

The two additional sides are mirrors of the first.

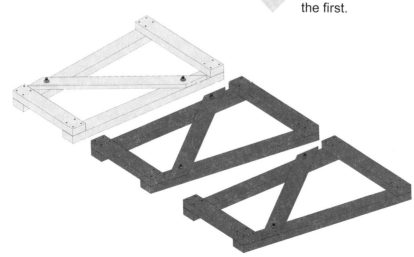

Repeat steps 1-7 to create 2 more sides.

10

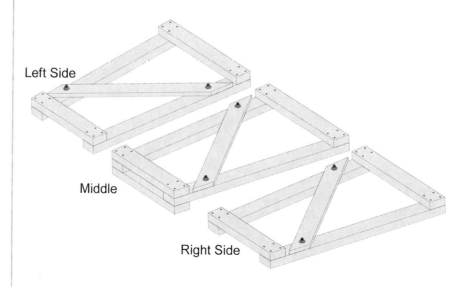

Left Side

Middle

Right Side

COWS

11

1x   2x6x120
9x   20D Galvanized Nails

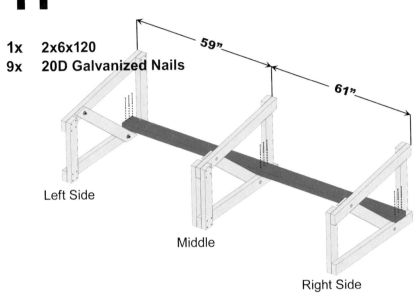

Left Side

Middle

Right Side

13

1x   2x6x120
9x   20D Galvanized Nails

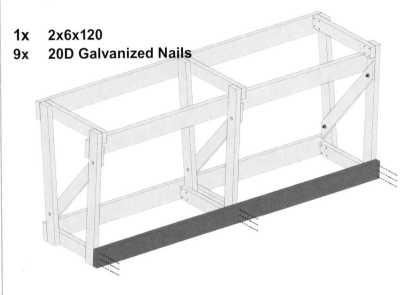

12

2x   2x6x120
18x   20D Galvanized Nails

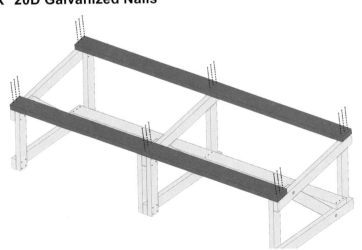

14

1x   1x6x57
6x   8D Ribbed Nails

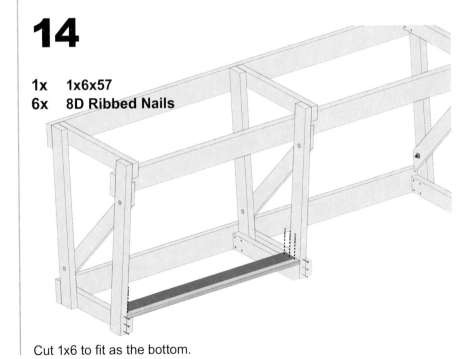

Cut 1x6 to fit as the bottom.

15

7x   1x6x57
42x   8D Ribbed Nails

Repeat step 14 for the rest of the flooring. Trim pieces to fit as needed.

17

We specify heavy duty lag screws to secure this board because it take a large amount of abuse. Cows continuously rub on this board so it needs to be fastened securely!

1x   2x4x120
3x   ½ x 4" Hex Head Lag Screws
3x   ½" Washers

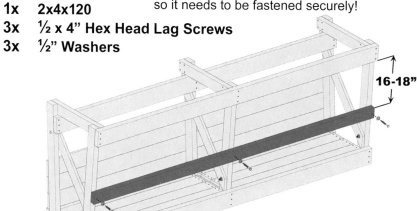

16-18"

COWS

Drill holes and fasten lag bolts. Hole size may vary and should be slightly smaller than the thread diameter of the lag.

16

3x   1x6x120
27x   8D Ribbed Nails

18

2x   1x6x120
18x   8D Ribbed Nails

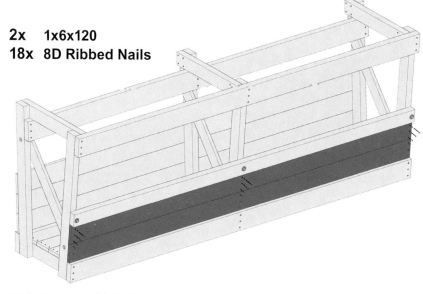

Fill in the gap with 1x6's. You may need to trim a board to fit.

19

COWS

2x   ½-13 x 3" Eyebolts
4x   ½" Washers
2x   ½-13 Hex Nuts
2x   Cable Pulleys
2x   ⁵⁄₁₆" Quick Links

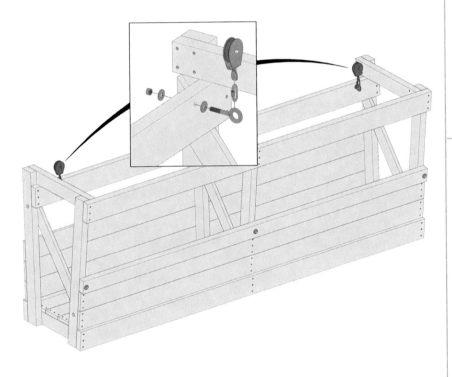

This is just one of many ways to connect the pulleys to the bunks. Keep in mind that this bunk alone weighs about **500lbs**. It will also be filled with hay and have several cows pushing on it. When in doubt, go "beefier".

20

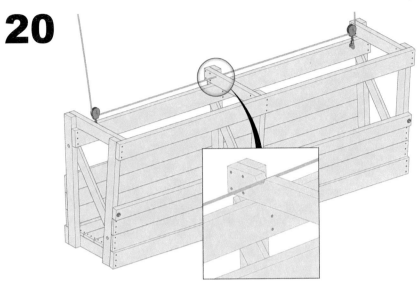

Depending how you mount your pulleys, the cable may interfere with the 2x4x30" section on the middle support. Notch or cut as needed to ensure the cable doesn't rub on anything.

21

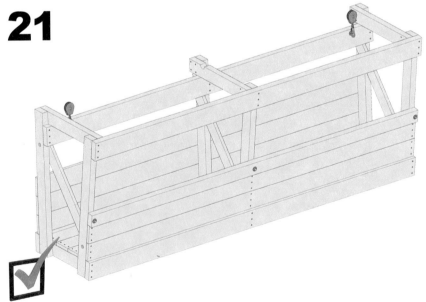

The hay bunk is complete! Repeat the steps to build as many bunks as needed.

Install The Bunks

22

5x ½-13 x 7" Eyebolts
5x ½-13 Hex Nuts
10x ½" Washers

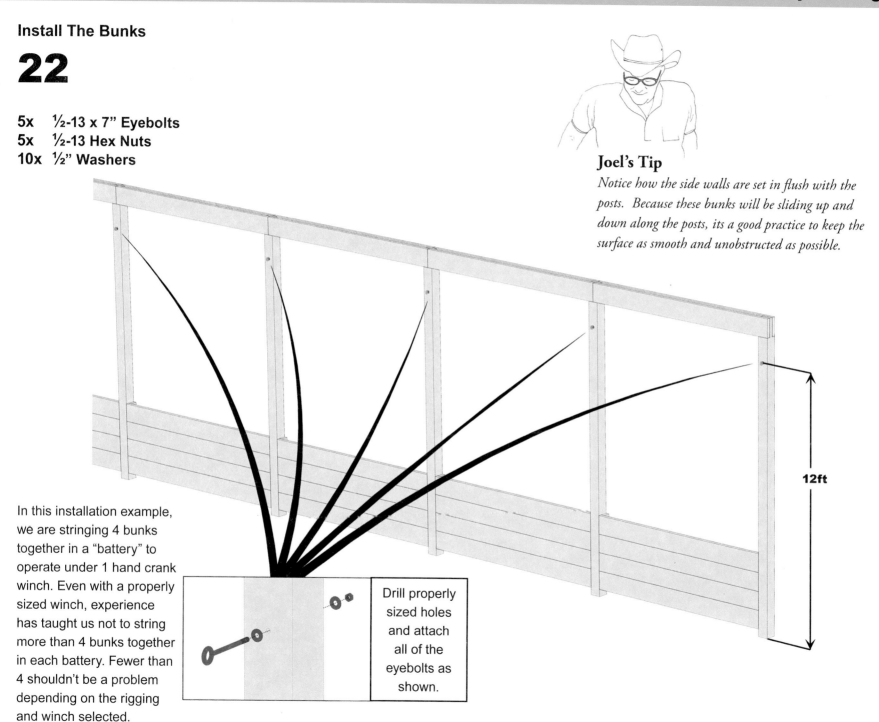

Joel's Tip

Notice how the side walls are set in flush with the posts. Because these bunks will be sliding up and down along the posts, its a good practice to keep the surface as smooth and unobstructed as possible.

COWS

12ft

In this installation example, we are stringing 4 bunks together in a "battery" to operate under 1 hand crank winch. Even with a properly sized winch, experience has taught us not to string more than 4 bunks together in each battery. Fewer than 4 shouldn't be a problem depending on the rigging and winch selected.

Drill properly sized holes and attach all of the eyebolts as shown.

23

You will need one less pulley than the number of eyebolts because the end of the cable will be tied to the far left eyebolt.

4x   Cable Pulleys
4x   ⁵⁄₁₆" Quick Links

COWS

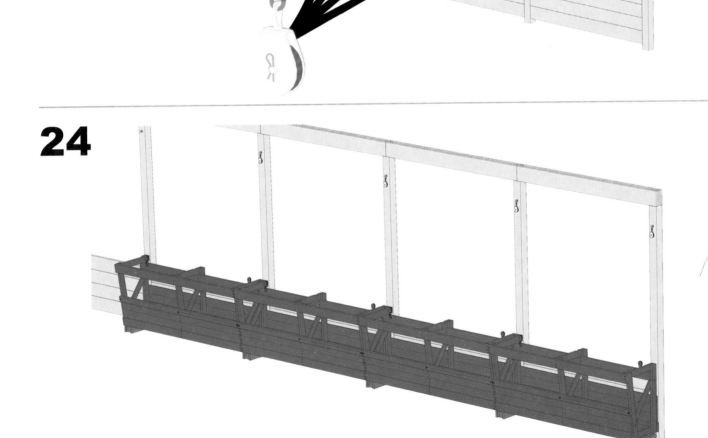

24

Joel's Tip

It is helpful to block the bunks up 12" or so off of the ground. Cinder blocks and scrap lumber work well for this task.

Orient bunks between posts as shown.

25

1x   Hand Crank Winch

COWS

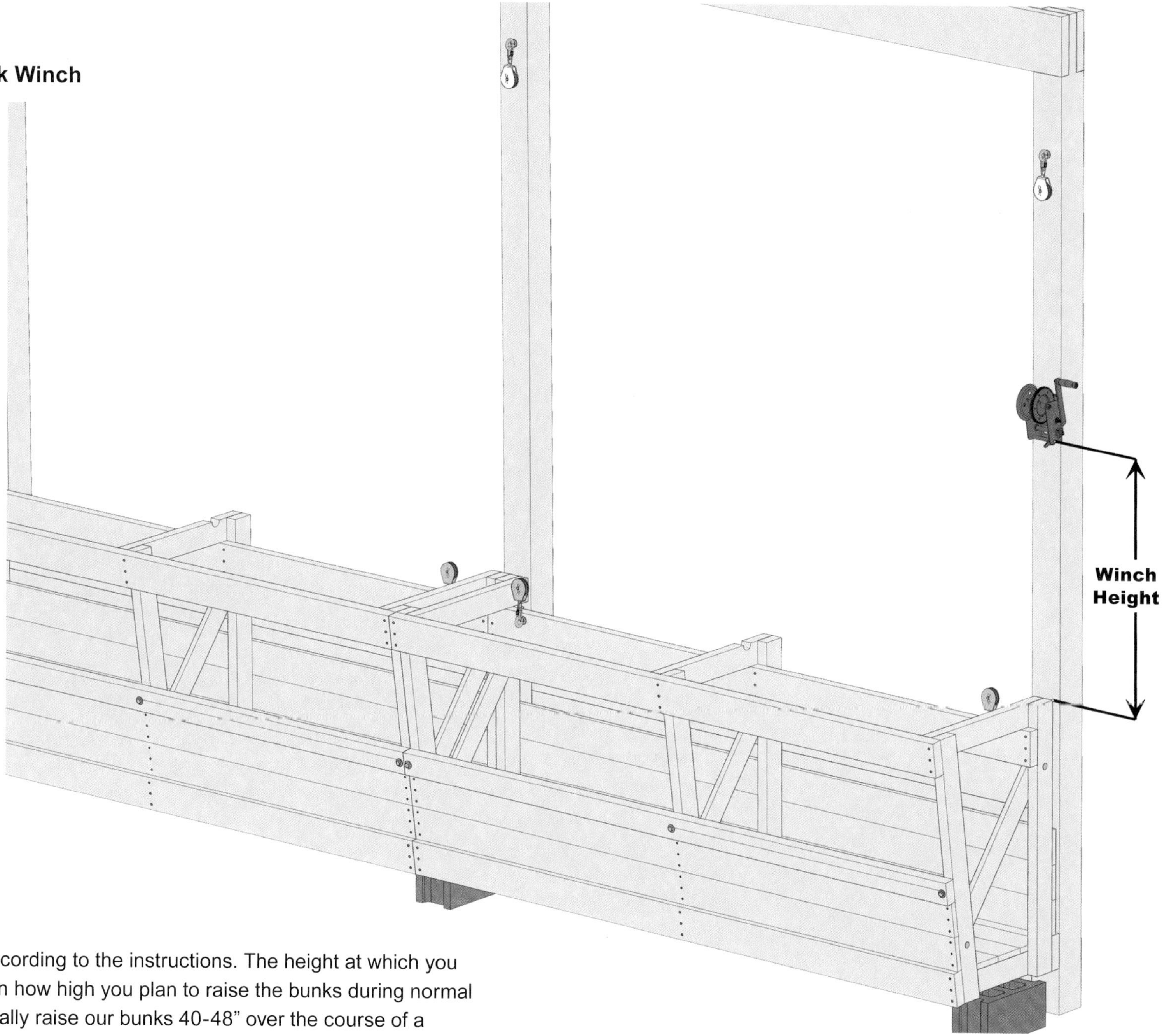

Winch Height

Mount the winch according to the instructions. The height at which you mount it depends on how high you plan to raise the bunks during normal operation. We typically raise our bunks 40-48" over the course of a season so plan accordingly.

26

Steel Cable
1x Cable Thimble
3x Cable Clamps

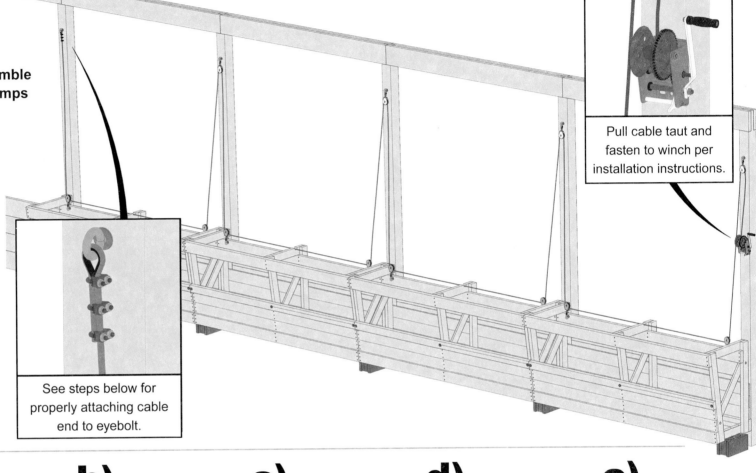

Pull cable taut and fasten to winch per installation instructions.

See steps below for properly attaching cable end to eyebolt.

a)

Stretch thimble over eyebolt. (It may be easier to remove eyebolt and do this on the ground).

b)

Thread cable through eyebolt and around thimble.

c)

With 4-6" of cable sticking past the thimble, also referred to as "turnback", fasten the first cable clamp as shown.

d)

Fasten the 2nd clamp as close to the thimble as possible.

e)

Secure the third and final clamp.

27

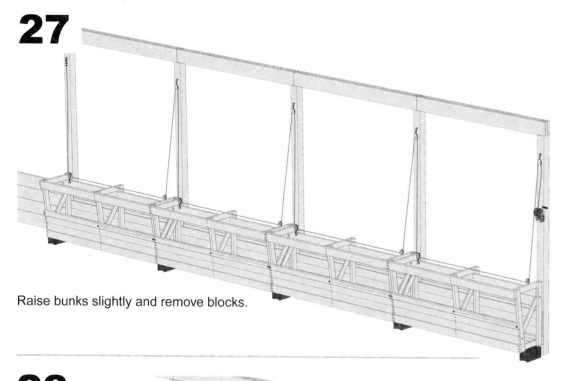

Raise bunks slightly and remove blocks.

Joel's Tip

When raising and lowering the bunks, you may encounter one bunk that is higher or lower than the others. However, we've never had this be a problem for us. A little jiggling and yanking normally frees everything up and equalizes the bunks.

Now, if you have a bunk that is an odd length – say it is shorter to fit on the end – realize that it will always tend to want to be higher than the other bunks because it weighs less. For this reason we try to keep all of our bunks uniform in length.

COWS

28

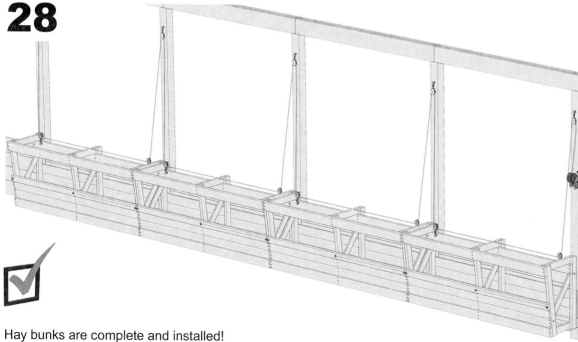

✓

Hay bunks are complete and installed!

photo courtesy of Jessa Howdyshell

photo courtesy of Chris Slattery

PIG BASICS

Here at Polyface we've never been extremely breed specific; rather, we've been phenotype specific. In the modern pig industry and show ring, the pigs are over bred to be lean, long, extra muscular on the rear end. They actually walk with a stilted gait; we call them "Arnold Schwartsapiggies." In a pastured situation, we need a pig that will put on fat, has what we call a "torpedo" phenotype: slightly sloped at front and rear but flush from shoulder to belly to ham. We call these "state of the art 1950s genetics."

We prefer piggies from farrowing operations that practice good pasture management outdoors; those are few and far between. Sometimes you have to take what you can get; the main thing is to watch the phenotype so you don't get stuck with these poorly-formed muscle-bound pigs that don't want to walk or move. If you decide to farrow yourself, all of the above applies to breeding stock. If you're planning to raise them on pasture, you want as compatible a source as you can find. The closer your provider outfit looks to yours, the higher your chances of success.

Exotic breeds certainly have their place, but remember that weird animals come with their own liabilities. For example, if a full-grown pig is only 220 pounds, processing will be extremely expensive because the kill fee at the abattoir is the same whether it's a 220 pound pig or a 350 pound pig. Spreading out that overhead over more pounds is cost effective. Extra hairy pigs are much harder to control because hair is an insulator. Cute has its place, but if you're going to operate a profitable farm business, you want function over artsy. Standard bred hogs with the proper phenotype are normally easier to find and more economical to process.

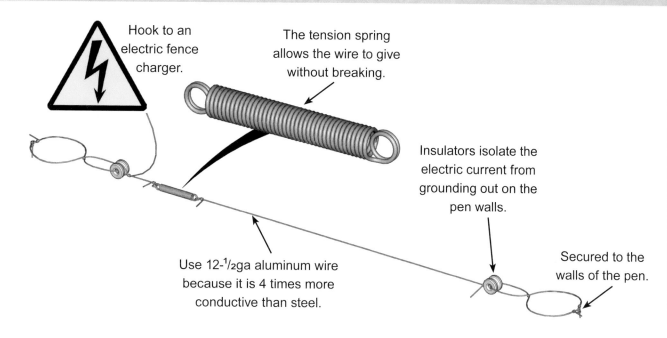

Hook to an electric fence charger.

The tension spring allows the wire to give without breaking.

Insulators isolate the electric current from grounding out on the pen walls.

Use 12-¹/₂ga aluminum wire because it is 4 times more conductive than steel.

Secured to the walls of the pen.

PIGS

photo courtesy of Jessa Howdyshell

Key Components of a Pastured Pig Operation

Training Fence

The training fence is the most crucial part of the entire pastured pig model. Without successfully breaking a pig to electric fence you will have a perpetual headache on your hands until the day when your pig finally turns into bacon. Following these simple tips will ensure a memorable first experience that is certain to leave a lasting impression with your pigs.

It will be necessary to either have some sort of secured receiving facility for new piglets arriving at your farm, OR be certain that the piglets have been broken to electric prior to arriving. A great example of a "pig-tight" pen facility can be found in the hoop house section of this book on pages 446-451.

Position the training fence in a corner of the pen opposite of the gate you let them in/out of. Use aluminum wire because it has 4x the conductivity compared to steel, and situate it 8" from the ground. Tie a lightweight tension spring from the hardware store to one end of the training wire and use insulators to isolate the current from the pen structure that it is connected to. For how-to's on knot tying, insulators, etc. see pages 538-549.

It is a good idea to have your electric charger hooked up and ready to shock prior to installing the fence inside the pen. If someone can lend a hand in keeping the piglets away while you work it'll be a great help. If your experience is anything like ours, the curious piglets will be munching on the fence before

you finish setting it up!

And now for the fun part: turning the power on. It is rather amusing to watch each pig find out for itself what the wire is and does. Pigs are dramatic and will run back and forth, squeal, and generally sound like mayhem, but the shock causes no physical harm. However, in order to be effective, the initial experience must be memorable enough to create a lasting psychological barrier.

We want to reiterate, don't forget the spring – it is very important. Unlike cattle, which tend to jump backward when shocked, pigs tend to jump forward. The spring permits enough give so that the wire does not get broken when the piglets run through it during the training process.

309

Pig Waterers

PIGS

After trying many different options, we've had the best luck with an industry standard 85-gallon drinker. This drinker consists of an 85-gallon tank (white) attached on top of a drinking trough (blue). There are two float valves: the first float is attached to the black screw on lid, and it regulates the height of the water in the tank. The second float is attached underneath the white tank and regulates the height of the water in the trough.

Flushing the system is easy. Pulling the rubber stoppers in both the bottom of the trough and the tank allows a surge that is typically enough to clean the trough. These waterers are equally as effective in a barn as they are on pasture. Again, dual purpose. It's this versatility that helps us stomach the $300+ sticker price.

While we do not personally own any, manufacturers do make deicing systems that pair seamlessly with these drinking units. Due to the number of drinkers we own, and the fact that we do not have electricity in most of our animal facilities, this is not an option for us. When it gets so cold that these drinkers freeze up, we typically resort to shallow watering troughs with a board bolted across the top to prevent pigs from climbing in. (See image above). We fill these "emergency troughs" every morning and dump them every night. It's a cheap insurance policy for those handful of times you may need it.

We also make a habit of draining hoses completely after every use in the dead of winter, which is a bit of a hassle, but it's a lot

Industry standard 85-gallon drinker.

Emergency water trough: Bolt a 2x8 or 2x10 across the top of the trough to prevent pigs from climbing in.

cheaper than running 1000-watt deicers all of the time and the risk of damaged electrical wires in buildings full of combustible material.

A downside to these drinkers is that they need to be fairly level to operate efficiently, and must remain full, unless they are securely attached to something underneath. The only thing that keeps the pigs from turning them into soccer balls is the weight of the water in the tank. When used as medication dispensers (the original design plan), the molded base has knock-outs to fasten them securely to the ground. But in our portable applications, we simply keep them full of water and that's plenty of weight to keep 300-lb. pigs from pushing them around. Another downside is that we waste 80+ gallons of water every time we move the drinker to a new location.

We've used nipples, nose drinkers, and open troughs. I'm convinced pigs take a "waste water" schedule. "Porky, you take the 8-10 shift; Hammie, you take the 10-12 shift, and Chops, you take the 12-2 shift." They

literally stand there with their nose on the valve dispenser letting water run. Then they root up the area for a wallow, dig out the T-post or platform holding the waterer, upend the whole deal, bite through the hose, and make a pond before supper.

During a pasture move, the drinker is often the last thing we move – especially when the weather is hot and dry. If you drain the water before you move the pigs, they may become more interested in the newly acquired mud, than a move to fresh pasture. Use your discretion here – this is a rule of thumb meant to be broken.

The downside to moving the water last, is that you really need to wait around until it has ample water in the tank to prevent it from being tipped over after you leave. You can imagine what would ensue if a drinker were left tipped over and running nonstop for 24-48 hours. The pigs would have a wallowing good time and you'll be left with a mess of a paddock and a pricey water bill.

Water System

Routing water is extremely simple. Since our pastured pig operation is strictly seasonal, we run most of our water lines above ground. We typically like to lay our pipe next to the road, outside of the paddocks and reach of the pigs. If given the choice, lay it on the high side of the road so that you can trace leaks easier. Seeing a trickle of water across the lane on a dry day is a sure sign that you've got a leak. If you lay the pipe on the down side of the lane, leaks can go out in the paddock and will be harder to detect.

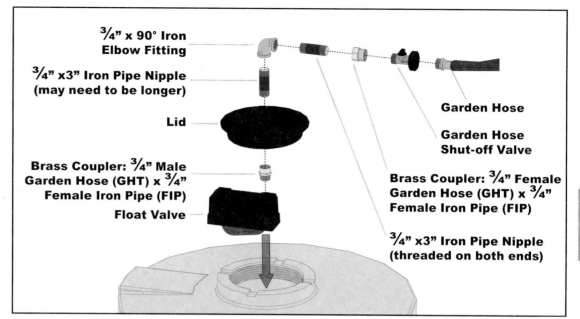

¾" x 90° Iron Elbow Fitting

¾" x3" Iron Pipe Nipple (may need to be longer)

Lid

Brass Coupler: ¾" Male Garden Hose (GHT) x ¾" Female Iron Pipe (FIP)

Float Valve

Garden Hose

Garden Hose Shut-off Valve

Brass Coupler: ¾" Female Garden Hose (GHT) x ¾" Female Iron Pipe (FIP)

¾" x3" Iron Pipe Nipple (threaded on both ends)

Your tank should come with instructions on installing the top float, but in case it doesn't, we've illustrated how we typically set ours up. Lengths and sizes may vary based on your tank.

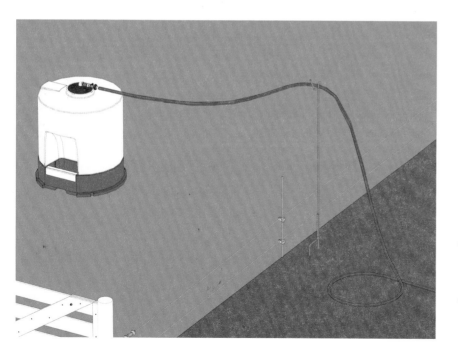

When our pig pastures do not have any trees on which to hang our water lines, we use simple homemade posts to do the job like we've shown here. A favorite material is ½" rebar. We bend the top around something round to make a pigtail and we weld a foot on the bottom by which to tread the post into the ground. Simple and cheap.

Locate the pigtail hose holder just outside of the paddock or else the pigs will knock it over.

Lastly, it is imperative to keep the supply hose out of reach of the pigs. In a hoop house, we string the hose through the purlins, in a barn we hang it from the rafters or posts, and in pastures, we suspend the hoses from tree limbs or the pigtail hose holders. Anything to keep the hose up off the ground.

Pigtail Hose Holder

Pig Feeders

We've tried numerous feeders as well over the years. Many work very well, while some excel in a specific scenario more than others. For small groups of pigs in sheltered areas (say someone drops off a dozen piglets) we like using the Osborne RN1 Wheel feeders. These feeders are small and light enough for one person to handle when empty. We've also had success with the Brower SF74 feeders for piglets.

For larger groups of pigs under shelter, we also use the flip-up lid style feeders by Brower and other manufacturers. These work well, but they are being phased out of use by the newer wheel-type feeders. That does mean they are becoming readily available at auctions, and for a reasonable price. If noise is at all a concern, then these feeders may not be for you since they can be annoying to listen to. Every time a pig wants to eat, it lifts up a lid and then it slams shut when they are done.

Our favorite feeder for use in both outdoor and indoor feeding scenarios is the Osborne R045 big wheel feeder (shown above). Other companies have their own renditions of this feeder as well. You might wonder why there aren't any lids to protect the feed trough on the bottom. This is simply because it doesn't need them. There is a water tight lid to keep the stored feed dry, but any feed in the trough below that is moistened turns to slop and the pigs do not discriminate.

To prevent any blockages there is a

PIGS

The Brower® SF74 feeder works great for piglets.

Osborne® RN1 Nursery Feeder

Osborne® R045 Big Wheel Feeder

paddled wheel that rotates around the bottom of the entire trough. When the pigs push on the wheel, it turns an agitator inside of the feed bin and it dispenses more feed. The amount can be adjusted by the individual. These pig powered paddles push any leftover feed towards the animals as it releases the new. The result is a system that is self-cleaning and does not readily clog even when wet.

The only downside we see to this feeder is that it requires strength to rotate that wheel. The specs say that pigs need to be at least 50 lbs. to turn it. That means you will need something else for the smaller end of your feeding operation. Other than that, these feeders are a winner!

See the following pages for details on an important modification that we make to every feeder we intend to use out on pasture.

photo courtesy of Jessa Howdyshell

Old style flip-up lid feeders.

Pig Feeder Skid Modification

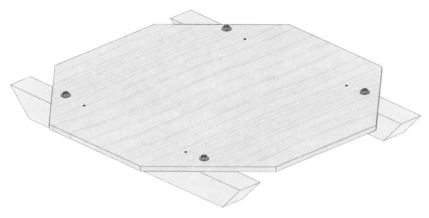

One notable modification we make to any of these feeders is a plywood base with skids. On metal feeders this is especially important because it keeps them up out of the moisture. On the Osborne feeders, the base is fiberglass so rot is not an issue. However, we still install a plywood base with skids because we often use tractor mounted pallet forks to transport the feeders. The skids make it easier to get under the feeder while the plywood protects the fiberglass bottom from being damaged by the forks scraping against it. The following are steps illustrating this simple, yet important modification.

Tools

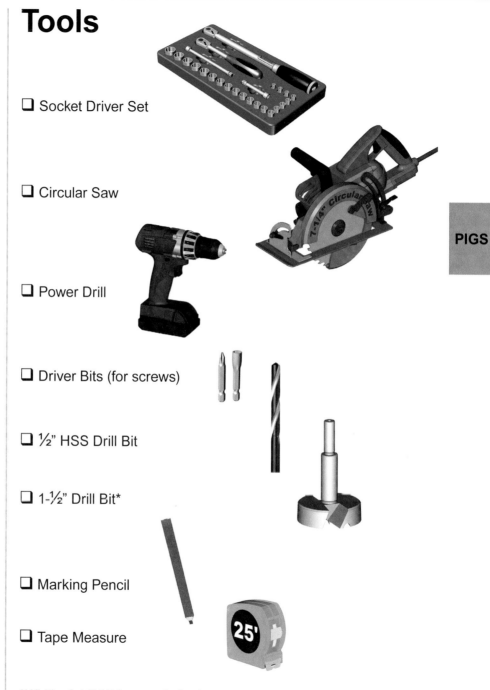

❑ Socket Driver Set

❑ Circular Saw

PIGS

❑ Power Drill

❑ Driver Bits (for screws)

❑ ½" HSS Drill Bit

❑ 1-½" Drill Bit*

❑ Marking Pencil

❑ Tape Measure

Width of drill bit for counterboring may vary. It ultimately needs to be slightly larger than the outer diameter of the washers you are using.

Wood Cutlist

White oak holds up extremely well for these skids and is worth the extra money spent upfront.

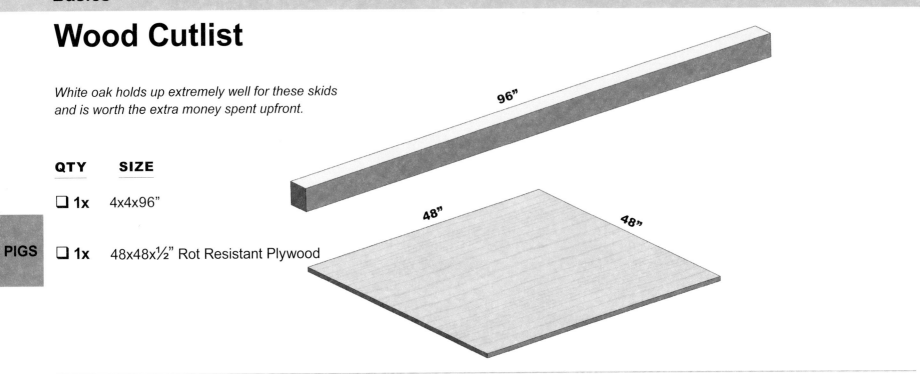

QTY	SIZE
☐ **1x**	4x4x96"
☐ **1x**	48x48x½" Rot Resistant Plywood

PIGS

Hardware

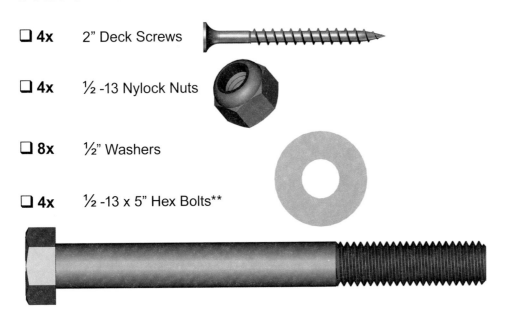

☐ **4x**	2" Deck Screws
☐ **4x**	½ -13 Nylock Nuts
☐ **8x**	½" Washers
☐ **4x**	½ -13 x 5" Hex Bolts**

**The length of hex bolt may vary depending on factors. Choose a size that works for you according to step 5.

1

1x   48" x 48" Plywood Sheet

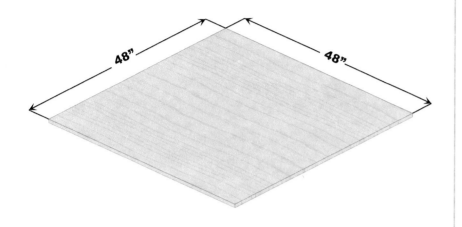

2

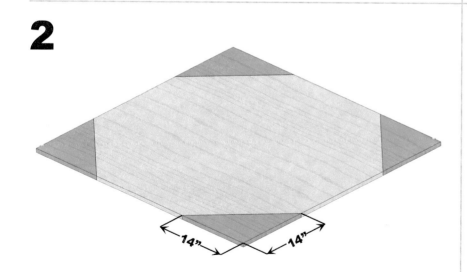

Trim off corners as shown in green.

3

1x   4x4x96" Rot Resistant Skid

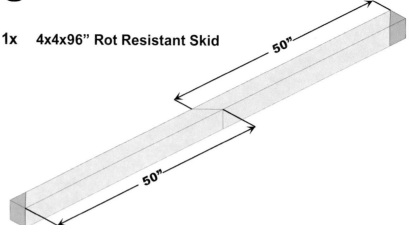

Assuming all cuts are 45° and when oriented as shown, you can get two skids approximately 50" long out of one 8ft 4x4.

4

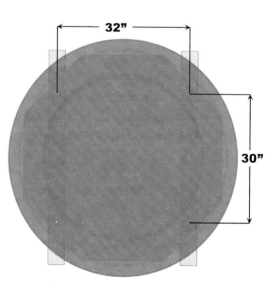

To illustrate where the holes should go look at this transparent top view. (Your dimensions may vary from ours depending on the geometry of your feeder.) Position the skids as wide as possible while making sure that the bolt holes catch the skids, plywood, and a flat surface on the bottom of the feeder.

PIGS

315

5

30"

≈4"

Choose a bolt length that will not protrude through the bottom of the skid. First, drill a counter sunk hole with a diameter larger than the washer. After that has been done, drill the rest of the way through with a drill bit slightly larger than the bolt diameter.

PIGS

6

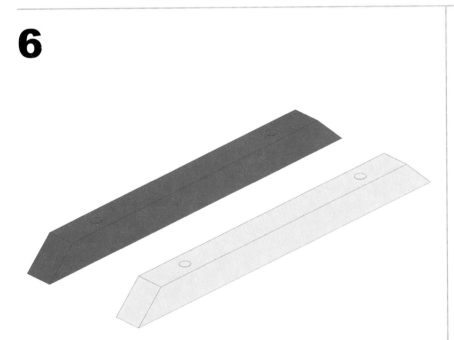

Repeat the process to create a second predrilled skid.

7

1x   Plywood From Step 2
4x   2" Deck Screws

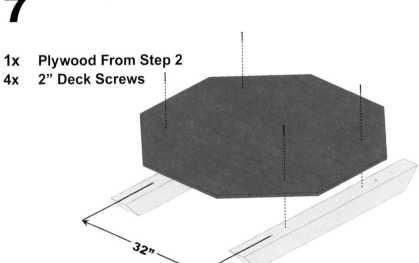

32"

The distance between your skids may differ. Consult step 4 to determine your distance. Be sure that the skids are parallel to each other and centered on the plywood. Screw them in place.

8

4x ½-13 x 5" Hex Bolts
8x ½" Washers
4x ½-13 Nylock Nuts

Align the base with the bottom of the feeder and drill holes the rest of the way through the plywood and feeder bottom. It is best to drill one at a time. Insert and fasten that bolt before drilling the next hole to ensure that everything lines up.

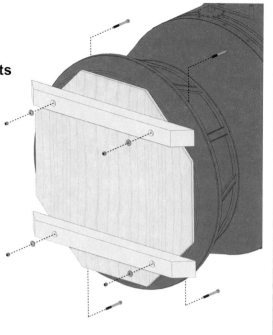

9

Your pig feeder is complete! Don't forget to adjust your tractor's pallet forks to fit between the skids.

Shade

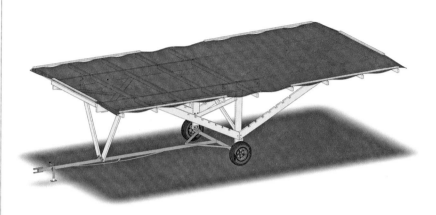

PIGS

Most of our pig pastures do have naturally occurring elements to provide shade. We do, however, have a few pastures that do not, and that's where the need for some sort of shade structure comes in handy. We've built a variety of different pig specific structures over the years. Our philosophy is multi-purpose infrastructure, and with that mantra we've found the Gobbledygo shines as a multi-use shade structure. It is stable and highly maneuverable in our paddocks. Since pigs don't fly –yet – they don't need roost bars. We typically remove and store them in a safe place. We use the same 80% shade material as we do with the turkeys and cattle. See page 205 for tips on shade cloth considerations.

PASTURE INFRASTRUCTURE

PIGS

Although pasture infrastructure is important, the most critical step occurs in training to electric fence prior to going out to pasture. Unless you're an Olympic marathoner who enjoys chasing pigs, train them before taking them out to pasture. In fact, we go so far as to say if you aren't willing to put in the meticulous procedure of training them, don't even try to take them to the field. It'll be a waste of time and loss of your religion. If you short cut the training step you'll reap a whirlwind of disappointment in what should be the more enjoyable part of the pastured pig experience: introducing them to lush forage for the first time. Pigs frolicking and dancing is certainly a delightful experience unless it occurs across and then outside your electric fence. That turns joy into a nightmare. Trust us on this; you do not want to short change the training procedure.

While a physical fence rather than electric (psychological) fence can work, the cost prohibits multi-paddock layouts and results in the inevitable moonscape. These barren, muddy, dusty pig yards brought hog cholera in the early 1900s. They stink and incubate every pathogen and parasite known to our porcine friends. In

photo courtesy of Chris Slattery

general, using all physical fencing for pigs will run about $2 per foot while the electric systems we'll show you in these pages runs only about 15 cents a foot. That's worth investing some time and patience to have a well-trained hog. We've broken this topic into the following subsections, and provide as much detail as we can on each facet of our pasture infrastructure:

- Road Access & Corral Placement
- Paddock Layout
- Corner Posts
- Orienting Gate Openings
- Line Posts
- Fence Wire
- Electric/Wooden Gates
- Routing Electric

Road Access & Corral Placement

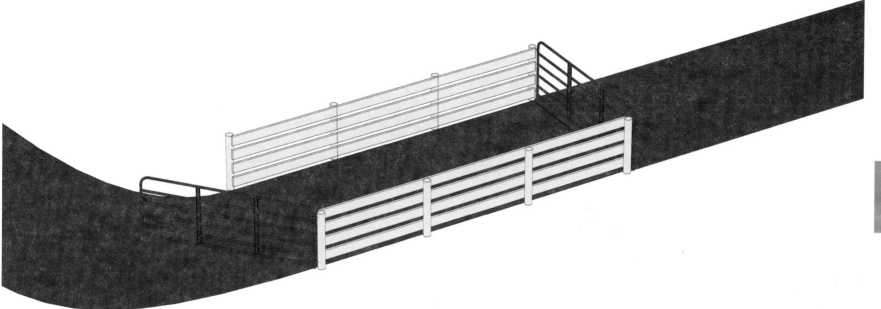

The first step in creating a pig pasture system is to determine access. If a road exists, don't change it; use it in the layout. If not, then road placement is based primarily on the topography of your land. In permaculture, we like to follow contours and use roads as pseudo-swales to slow or direct water where needed. This is a fine concept but not always practical. If a road on contour means it'll take 10 minutes to get where you need to go instead of 2, then the layout is inherently inefficient and will cost you in the long run. Generally speaking, adhere to a few basic principles when laying out a road.

1.) **Stay out of low spots.** A road on a ridge is always more stable than a road in a low valley. Keeping a road high generally means it'll dry out quicker.

2.) **Avoid steep grades.** The steeper the grade, the harder it is to keep the road from washing and eroding. Some excavation to gentle the climb is far better than a straight and steep ascent. A steep road is also hard to traverse in wet or slippery conditions. Traverse a hillside whenever possible.

We can't possibly do the whole road building issue justice in this book; plenty of manuals exist on principles for good road construction. An excellent resource for access planning and development is Darren Doherty's *Regrarians Handbook*.

The ideal orientation creates a central road with pastures on either side; think of an airplane with the fuselage as the road and the wings as paddocks. The centrally located

road offers efficient movement capabilities for both animals and machinery. The road can be as narrow as 12 ft., with both edges defined by pasture electric fences. Remember that minimizing road width ultimately reduces the amount of non-productive land in your pasture setup.

After locating the road, determine a good corral and loading spot. It needs to be an area easily accessible in all weather conditions and logistically capable of accommodating whatever hauling trailer you use. If you decide to build an earthen ramp, now is the time to do that as well. Since the corral is the most permanent infrastructure of the entire system, it is easiest to construct it first and then work around/build off of it. See pages 343-349 for details on corrals.

Paddock Layout

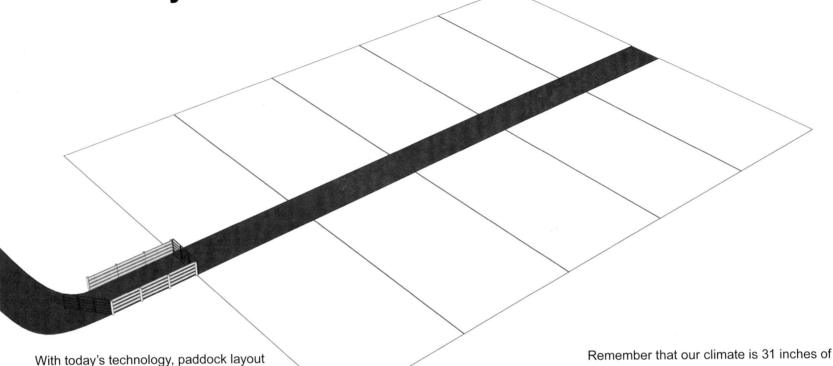

PIGS

With today's technology, paddock layout has become easier. With a Gmail account, you can log into Google "My Maps" and zoom into your field of interest. You can then draw out your proposed paddocks and Google will spit out exact acreages and perimeter distances. Armed with this information, you can head out to the field with precisely the right amount of posts and wire, and can quickly and accurately lay out your paddocks using your exact location on your GPS-enabled smartphone. However, for those of you like me who still use flip phones and an analog watch, this isn't an option. I still measure in one-yard (3ft) paces and have developed a sixth sense for measuring distances over the years. With a little math and footwork, you can still create precise paddock areas with relative ease. A

half-acre is approximately 2400 square yards, so a paddock 30 yards wide by 80 yards deep will get you pretty close.

Some things to consider

Size and number of paddocks is an inexact science and here at Polyface we've done every extreme imaginable, from too intensive to not intensive enough. In general, disturbance moves the ecology; too little makes it revert toward forest (brambles, brush, then re-forestation) and too much makes it regress toward weeds. The Goldilocks "just right" spot is different in different areas and that is why we hesitate to give hard and fast rules.

Remember that our climate is 31 inches of rainfall with 6 frost-free months per year. I think our experiences probably work in a broad band of 18-45 inches of rainfall per year in anything more than 5 months of frost-free growth.

With that caveat in mind, here are our numbers. We offer 48 square yards per hog for 5-12 days (never more than 12 days), up to 3 visits per year with an average 80 rest days in between. Pigs can do a lot of damage; the single biggest temptation is to raise too many pigs on too little acreage and turn it into a moonscape. This is not pastured pigs. With this formula in mind, our total production is about 10 pigs per acre per year plus another 10 half way. In other words, since it takes 6 months to grow the weaner pig (already 2 months old) to market weight, it'll be inside and being trained for one month, then outside for 5 months (150 days). That means in one rotation of 8 paddocks at an average of 9 days apiece, you'll take one batch all the way to market weight and another batch half way before winter.

1) **Paddock Size:** Our typical grass-based paddock is $1/2$ acre in size with pig groups ranging from 35-50 head. We've found this to be a sweet spot for maximum capitalization on our time and labor. As the groups get smaller, the money we tie up in infrastructure (waterers, feeders, paddocks) and labor eats our profits. That is also the number that can access one waterer and one feeder; if you go over 50, you need to add more of this infrastructure, in which case you'd want to go to 100 to maximize use. You can adjust paddock sizes accordingly to fit your herd numbers. The goal is to not over impact an area, and provide ample rest in-between grazing. We normally run half acre paddocks, which offers 48 square yards per hog during their 5-12 day stay. Bigger pigs, shorter stays. At a minimum 48 square yards per pig, if you have only 2 pigs you'd want paddocks of about 96 square yards apiece for 5-12 days. If you want to move them faster, you can drop the square yardage down. If you want to move them every 3 days, for example, you would drop the paddock to 16

square yards per pig (48 divided by 3). Our first prototype was a "Tenderloin Taxi" that we moved every day with 4 pigs in it.

2) **Number Of Paddocks:** The number of paddocks will depend primarily on the amount of rest/recovery in your region. In our area of the world, 2-3 months of recovery is typical, so we can get 3 grazings on a paddock per season. We use roughly 8 or 10 paddocks in rotation for our average 9-day stay. But if you want to move them faster, you'll need more (smaller) paddocks. The formula works fine going faster and smaller; it does not work if you go bigger paddocks and longer stays. Impaction time is the great determinant of recovery time. If you stay 14 days, for example, you'll need to add another month on the recovery. In a perfect world, you'd move every day, but that quickly creates a labor issue unless you're moving 100 or more. Like many things, we try to strike a balance between perfect and sensible. If you live in a more arid/brittle climate, you will have to make adjustments to paddock sizes, grazing durations, and rest periods to suite your region. In general, harsher environments need longer recovery periods. In our area, wet conditions generally mean the pigs tear things up more and dry conditions offer more forgiveness.

Joel's Tip

We offer 48 square yards per hog for 5-12 days (never more than 12 days), up to 3 visits per year with an average 80 rest days in between.

PIGS

3) **Stocking rates and rotation:** We run groups of 35-50 hogs in $1/2$-acre paddocks and move them every time they finish 2 tons of feed. The bigger they grow, the faster they eat the feed. This formula maintains a constant impaction and disturbance factor on the land. In other words, 50 pigs at 100 pounds for 12 days are the same impact as 50 pigs at 200 pounds for 6 days. It's the same weight per stay. That's the key to not over-disturb or under-disturb. Lastly, in our wooded forest glens, mainly used during the fall acorn crop drop, our paddocks range from 2-5 acres in size. Touched only once a year, these paddocks are far less sophisticated than the intensive pastures. We use a piece of nylon rope looped around trees as poor-boy insulators with a single strand of electric fence. We never put pigs in these expansive paddocks until they're 200 pounds and exceptionally well trained.

Corner Posts

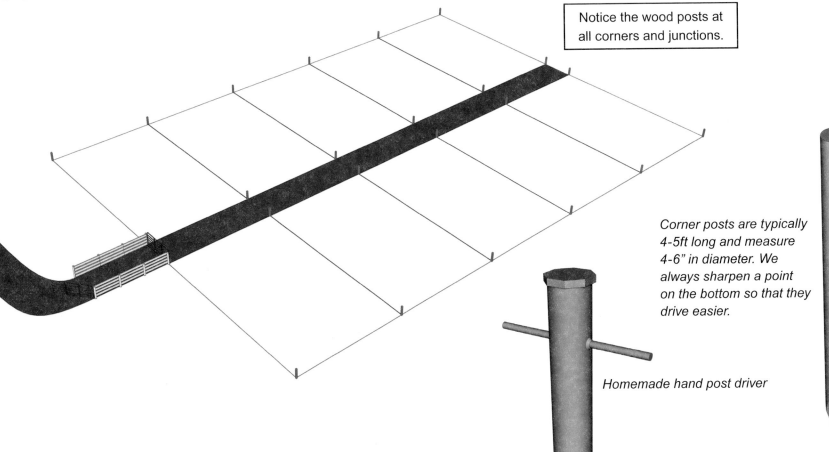

Notice the wood posts at all corners and junctions.

PIGS

Corner posts are typically 4-5ft long and measure 4-6" in diameter. We always sharpen a point on the bottom so that they drive easier.

Homemade hand post driver

Our standard fencing structure consists of wooden end posts and rebar line posts.

Step one is to locate and install all of the wooden posts. Wooden posts are harvested from locust, hedge apple, or cedar and are sharpened to a point on one end and driven into the ground with a post pounder. Diameters vary between 4-6" on a typical pasture post. End posts that support the load of a pipe gate or larger boundary fence (woven wire or high tensile) must be much stouter, but these

small diameter posts work terrific for our two-strand aluminum wire fence infrastructure. The smaller diameters and sharp points make driving the posts fairly easy. While we typically use a tractor mounted post pounder these days, we have a well-preserved artifact of years past.

It's a behemoth 50 lb. hand post driver made from 6" steel pipe. I set every post on this farm with that hand driver in my youth. Today we are blessed to only need it for fixing

the occasional broken post. I share that to say that with these small diameter wood posts on a wet day, one really strong person can set all of the ends and corners without any expensive machinery.

Having these sturdy ends and corners adds the right amount of integrity to the structure of our fence lines without adding excessive cost or landscape permanence to our management setup.

Orienting Gate Openings

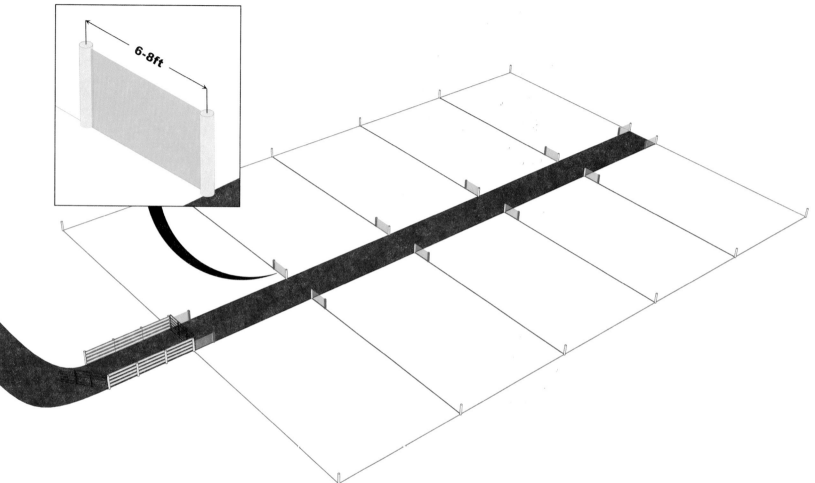

6-8ft

PIGS

We aren't finished with wooden posts just yet. We first need to decide where to place openings in the fence lines between paddocks that will be secured with moveable wooden gates. Why bother with gates? Why not just lift or roll up the electric fence?

Pigs are generally considered the smartest farm animal. As such, they're the slowest at developing trust in their caregivers. Whereas a cow learns quickly to trust me if I open an electric fence gate and call her, a pig is a different story. In just a few days cows quickly come to a call into a new paddock. They can see me open the electric fence gate and they trust me to open that gate for them.

Pigs are quite another matter. You can almost see their mental wheels turning. I open the electric fence gate and stand back, but they do not come. They look at me like "what kind of fast one are you trying to pull on me? I know that shocking wire was there; I don't see it but it must still be there. What you're holding in your hand must be a ploy, a trick. I don't trust you." And so it goes.

Meanwhile you get exasperated at them standing there looking at the opening, no wire, but refusing to go through. If you finally walk around and try to herd them, they whirl around and look at you like the enemy, buttocks resolutely toward the open gate, and finally break past you scattering in every direction except through the open gate.

Over time the pigs do indeed learn to trust. And nothing works like letting them run out of feed and then putting some feed on the ground in the new paddock. Or throwing some water through the gate on a hot day. Yes, you can do tricks and gradually win them over, but we've found it's much easier just to design a system that works easily from day one. That system is a physical gate.

This is so simple we wondered if we should include it in this book, but knowing our own epiphany when we came up with this, and hearing the ongoing laments of people trying to move pigs, we thought we'd include it. You can call this the sanity gate, or perhaps the don't-lose-your-religion gate.

Fortunately, pigs are not very tall, so these gates don't have to be high. However, they do need to be stout; which means if they were as high as cow gates, they'd be extremely heavy. But because they are short, you can still pick them up. Furthermore, they don't have to be exceptionally long, either. Six to eight feet is plenty long; these are not machinery access gates; they are simply portals of entry for the pigs to enter the next paddock. Remember, we have the ability to drive equipment over our fence because of the movable insulators on our line posts. For your own convenience, choose a gate opening width and standardize it throughout your entire operation. That way you only need one-sized gate that will work universally.

photo courtesy of Chris Slattery

Because these gates are physical, they don't have the stigma of an open electric fence wire. The pigs scratch on these gates; when removed, the opening is obvious. Unlike an electric wire, these gates have never shocked and hold no halo of hurt for the pigs. We simply open the gate and carry it to the far side of the next paddock. Don't worry, you can always beat the pigs there because as soon as they enter the new paddock, they go to work grazing and sniffing. Unlike cows,

which routinely walk first to the back of the new paddock and then graze back toward where they entered, pigs seem far less curious about their boundaries. Pigs only seem concerned about the here and now; what's in front of their face. They enter the new paddock and immediately slow down, giving you enough time to carry the gate to the back side and put it on the opening there. After all the pigs have entered the new paddock, you go back and carry the old back gate up to the opening you just opened. In other words, the front gate remains the front gate and the back gate remains the back gate. We'll illustrate the move and this gate procedure in the following pages.

This simple little physical gate was revolutionary for our handling pigs in rotated paddocks. We like wood not just because we have a band saw mill, but because wood is an insulator. That way if it happens to touch the electric fence, it won't short out. In their rubbing, pigs often scoot these gates an inch or two, so having an insulated gate builds forgiveness in the set up.

With pigs, nothing beats creating a situation in which the pigs want to go where you want them to go. That is true with all animals, of course, but pigs being intelligent and possessing an extremely low center of gravity, with a lot of strength, are the most critical for this truth. The physical gate is obvious to the pigs and they respond dramatically. A pig that responds dramatically – positively – to our actions is a wonderful partner.

PIGS

Line Posts

Line posts are spaced approximately 12ft apart.

PIGS

³/₈" rebar stake with two Dare insulators. Stakes range between 3-4ft in length.

We typically space these line posts approximately 12 feet apart, depending on how rough the terrain is. Do not skimp on these posts – the further apart they are, the harder it is to maintain optimal wire height.

For our line posts – the posts in between two ends – we prefer rebar with plastic insulators. Why rebar you might ask? Because it is cheap. We use ³/₈" rebar, and at the time of this writing, a 20ft stick costs less than $8. If you can get 6 posts out of it, you can do the math. Add two $0.25 cent insulators and you are in business for about $1.50 a post.

We are partial to the Dare brand round post insulators like the one pictured. The quality of the plastic is superior to others on the market and they have withstood the test of time here at Polyface.

Another bonus to the rebar is its knurled surface. The added friction allows the insulators to fasten more securely, and also aids in keeping the posts in the ground. We've been using them for 50 years and haven't had one rust out yet.

We like the quality of the Dare brand screw-on insulators.

Fence Wire

PIGS

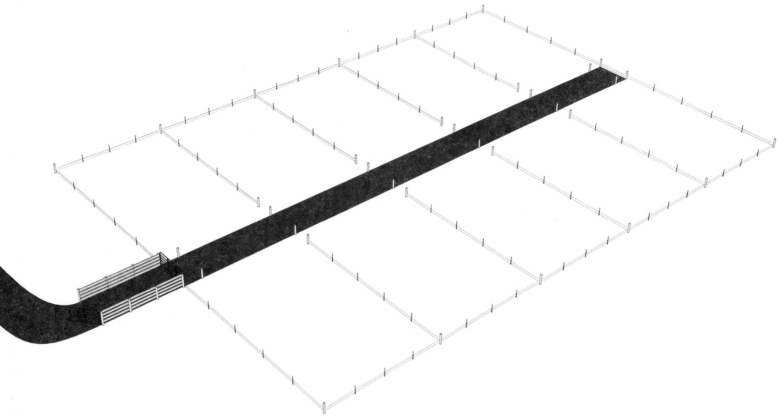

Our favorite permanent wire in all applications is 12-$\frac{1}{2}$ gauge aluminum. High quality polywire is great for portable cross fences, but for permanent installations, you want something strong enough to withstand wildlife pressure and something that will never degrade in sunlight. Aluminum is 4-5 times more conductive than galvanized wire, which means you don't have the resistance and voltage loss common in steel wire. Galvanized, no matter how well coated, will eventually dull down and flake off; not so with aluminum. It stays shiny, which means it's much easier to see for both pigs and wildlife.

Aluminum is also much lighter in weight, which means you can pull sags tight using fewer posts and without the need for corner braces. Those are huge positives. It's extremely malleable so you don't need special pliers or tools to work with it. It is more brittle than steel so it will break if you bend it back and forth only a couple of times. But that's really the only significant negative, which doesn't really count since we use it only for permanent installations.

No matter what wire you choose, none will be effective in containing pigs on pasture if they cannot see it. Pigs have especially poor eyesight. Fence lines must be kept clear so that pigs can easily see the wires. Our regular maintenance procedures include weed-whacking along fence lines and adjusting wire heights.

With the posts all in place, we then attach ceramic insulators to each post prior to routing our electric wires. The insulators isolate the electric spark and prevent it from grounding out on the posts. Reference appendix pages 538-549 for the finer details on how we install insulators and tie off fence wires.

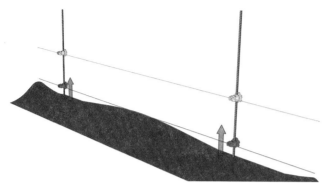

Adjusting wire height comes in handy when your pigs create berms along the fence lines. Pigs instinctually root and push material and it inevitably ends up along the perimeters. Unchecked, this material will short out your hotwire and eventually bury and topple it altogether. As part of our paddock prep, we walk the perimeter and adjust our insulators upward to achieve our desired wire heights. Over time, these berms stabilize and grow vegetation; they also catch any potential soil erosion in rain run-off situations.

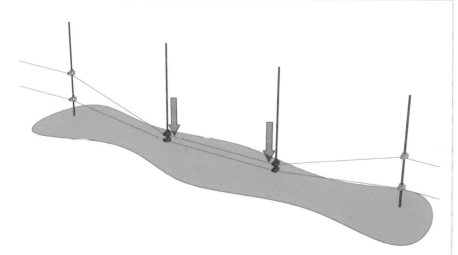

Another advantage to this line post arrangement is the flexibility to readily adjust the wire height. We do not like making a habit of it, but in a pinch, we have the option to drive our equipment right over our fences. We simply slide the insulators all the way to the ground and then reset them when finished.

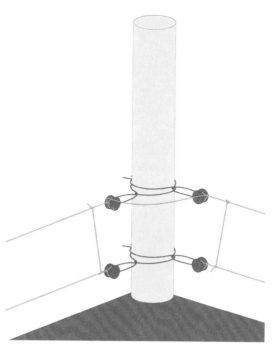

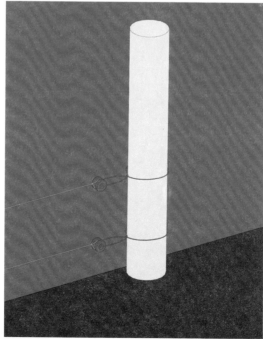

We fasten insulators to each post prior to routing the aluminum fence wire. These insulated connection points prevent the electric from grounding off on the posts. Pages 538-549 explains these connections in more detail.

PIGS

PIGS

Next let's talk about wire heights and configurations on pasture. We recommend that pasture paddocks have two wires to contain pigs under 150lbs. The lower wire should be ≈6" off of the ground and the second set at ≈12-15". We have separate acorn glen pastures up on the mountain that we use strictly to finish out larger pigs, and in these pastures we utilize only a single wire ≈12-15" high. We've found that these larger animals are trained well enough by this point to respect a single wire.

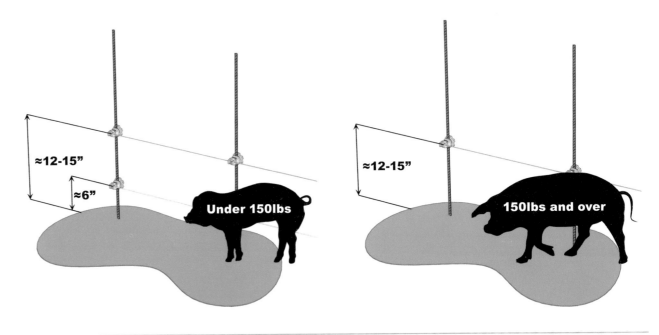

≈12-15" ≈6" **Under 150lbs**

≈12-15" **150lbs and over**

photo courtesy of Chris Slattery

Joel's Tip

A note on gate handles… We've tried every gate handle there is over the years. One-dollar handles, $10 handles and everything in between. If they have a spring, then they all have one thing in common: they all break! We've given up on searching for quality. We now use a basic yellow Dare handle for under $2 and plan on it breaking. If you have found the perfect, indestructible handle, then please write to us. We'd love to know about it! This is one notch above the cheapest gate handles, which are usually red; that's the only additional expense that seems worth it.

Electric Gates Along Roads

With 3 sides of the paddock completed, the only side left is the one adjacent to the road. For this front fence we do something a little differently. We still use rebar line posts and two strands of electric wire, but we install an electric gate handle to one end of each run of wire. This turns the entire front fence into a gate. Which end you place the gate handles on depends on the direction you will be entering/exiting. This becomes our regular means of accessing paddocks from the road with equipment. The gate can be as wide as we need it to be to service tractors, feed buggies, brush mowers, etc. If we need the entire distance, we simply pull the rebar posts out and put them back when we're done. Pretty simple.

It also serves as an additional means of moving the pigs between pastures. Say the pigs are halfway up the road and they are ready to be processed. Without this access point you would have to run your herd back through the wooden gate openings and through multiple pastures in order to get them to your corral. Been there done that – it's a lot of work. Yes, it can be difficult to get the pigs to cross that psychological barrier, but

we find that by this time they've developed enough trust in us that it isn't difficult like it is in the beginning. It's still much simpler than the alternative. Once the pigs are in your access road, you can yee-haw them into your corral or

other holding facilities.

We try to align all of our gates symmetrically so that they're opposite each other. It makes tractor work much simpler, even with such a narrow access road in place.

Joel's Tip

We try to align all of our gates symmetrically so that they're opposite each other. It makes tractor work much simpler, even with such a narrow access road in place.

PIGS

Wooden Gates

If you are already tallying all of the gate openings you have and lamenting the thought of spending weeks building gates for each opening, we want to ease your anxiety. If arranged properly, you will only need at most 3 gates per group of pigs, but can actually make do with just two gates. This sequence illustrates how we move our front gate first, followed by the back gate.

PIGS

1 Move the front gate ahead of the herd of pigs as they enter the new paddock. The pigs typically become so excited by the fresh forage that you have ample time to scurry ahead and set the new gate up before they get there.

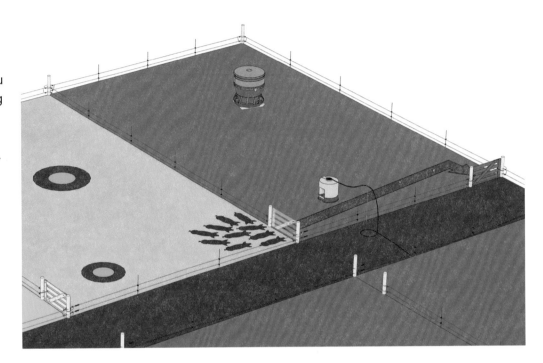

2 Once the pigs have all migrated to fresh pasture, you can advance the back gate and secure it on the new paddock opening.

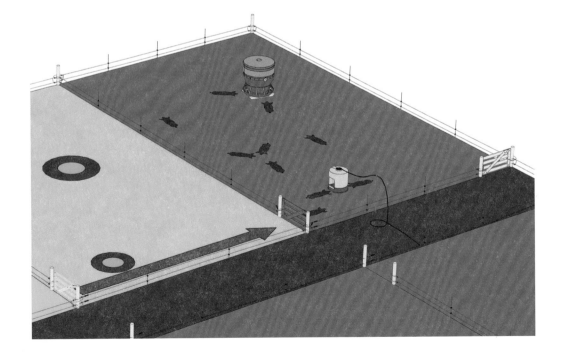

Gate construction is a simple, but crucial part of your pasture infrastructure. White oak is our preferred building material because of its strength and rot resistant nature. All planks are 1x6s or 1x4s, and we like to bolt the critical joints. The sandwich construction is the same concept as our cattle gates. Gates needn't be any more than 32" tall. The width will obviously depend on the size of your opening but 6-8ft is our normal range. Gates should overlap the posts slightly. Just remember, the wider the gate, the heavier it is. Refer to pages 252-259 for details on gate construction and apply those concepts here.

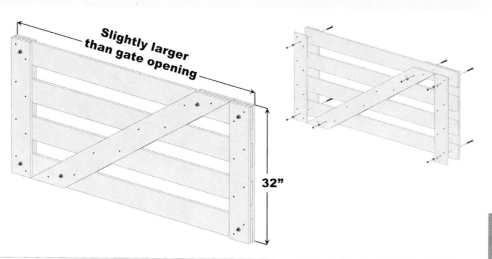

PIGS

Once you've constructed your gates, you will need a way to temporarily secure them. We've tried many different methods but prefer old fashioned chain and double-sided spring hooks. Wrap them around the gate and post somewhere in the lower half of the gate and get them as snug as possible. We recommend orienting the latch and any excess dangling chain on the side opposite of the pigs to prevent them from playing with it.

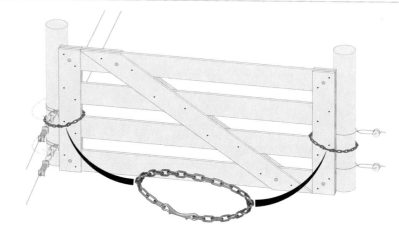

Lastly, we use homemade jumper cables to carry the spark across the gate to the fence on the other side. The reasons for doing this will make more sense as you continue reading. Our jumper cables consist of old extension cord material with an electric fence gate handle wired to each end. It's cheap, simple, and the extension cord material allows us to easily roll it up. The only way you can mess this up is to wire the handles to the wrong wire inside (there will be 2 or 3 to choose from). Make sure you wire each end to the same color wire and you will be good to go. You will only need one jumper per group of pigs.

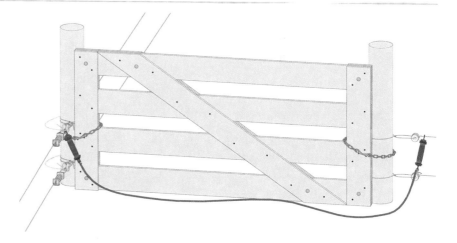

Routing Electric

PIGS

As we've stated, good training and a hot spark (more than 3,000 volts) are paramount to your success. With how low these fences are to the ground, it takes extra maintenance to ensure that fence lines are clear. Vegetation will quickly short out a fence. In order to ensure we have a good spark, we try to isolate our electricity to where we need it.

We design our fencing with jumpers so that you electrify only the paddocks in use. No sense energizing what's unnecessary. Depending on our perimeter fencing conditions, and proximity to other property/ residences, we may also keep the entire perimeter electrified all of the time. This is purely a judgment call.

If your pasture borders high dollar manicured estates, I cringe to think of what damage 50 hogs could do in one afternoon. If that is the case, I would secure my perimeters with quality woven wire and offset that with electric. Be safe, be respectful, be smart.

Here is a basic sequence of how we isolate spark with each paddock move. The jumper cables described on the previous page are often used to transmit spark across wooden gates where needed.

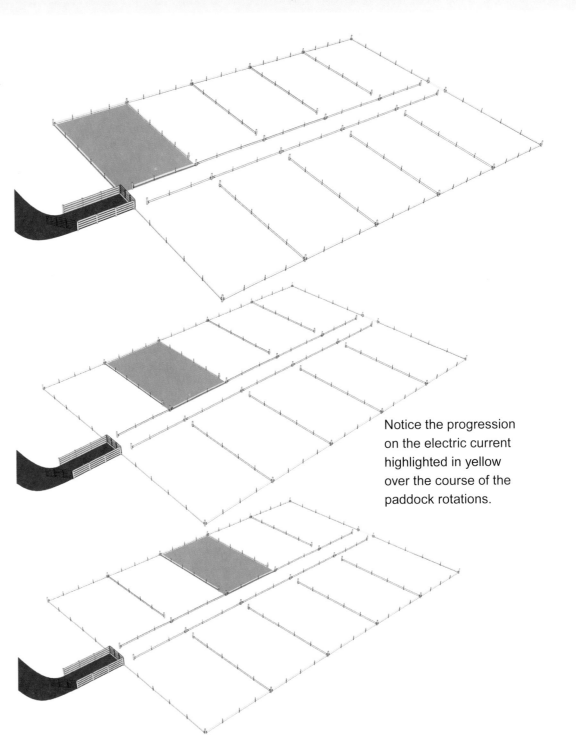

Notice the progression on the electric current highlighted in yellow over the course of the paddock rotations.

PIG HANDLING

photo courtesy of Chris Slattery

There are several different approaches to working pigs. The simplest method by far is always making the move seem like their idea. The best way to incentivize that is hunger. We try to make sure the pigs have an appetite when we are ready to move them. We also work with the innate nature of the animals. They are social, herd animals, so moving a group is always easier than moving one. Like cattle, we also sort the many from the few. In other words, it is easier to let 30 pigs back out of a corral, while holding a few back than to try and sort the two you want out of the 30.

Secondly, proper training to electric fence is the single most important aspect to a successful pasture operation. Follow our advice and make sure your hogs are well trained. Poor training will lead to an agonizing first and probably last experience with pastured pigs. With well-trained pigs, a person can easily complete simple moves without help by using some temporary bluff fences.

We would not, however, recommend trying to corral pigs under pressure with just electric fencing. It will not contain them. At best you will just have to catch a few. At worst, you'll have pigs that now realize they can get past the electric and you are done for. Simply said, pigs need stout PHYSICAL barriers for corralling and loading. Below we will walk you through several scenarios, sharing dos and don'ts and little anecdotes along the way.

Scenario 1
Trailer only = a bad idea

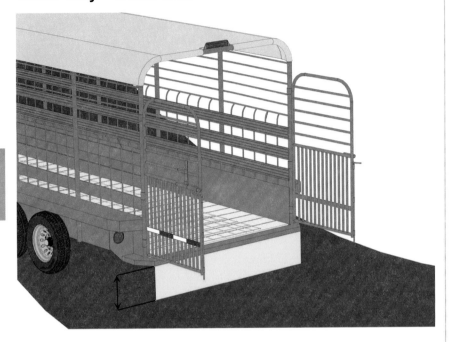

PIGS

Scenario 2
Trailer, sturdy ramp and pipe gate corral = a good start

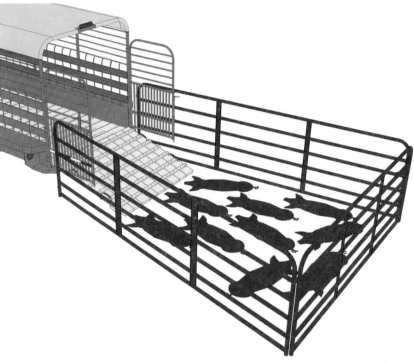

Anybody who has ever had pigs has pig loading stories. Years after the catastrophe, they are funny to tell and funny to hear. But at the time, they are not funny and some of the things you hear might not be nice.

Pigs have a low center of gravity. They don't like to get off the ground. They like to stay pretty glued to the earth. And they are smart. And distrusting.

Fortunately, most livestock hauling done locally today is in gooseneck trailers, which are pretty low to the ground. But unlike a cow, calf, goat, or sheep, pigs don't even like to jump up a foot. What is a simple little hop up for any other animal is a major impediment to a pig.

Their short little legs and rotund bodies aren't made for hopping up into things. Their eyes are so low they can't even see up there. What to you and I is clearly a low box, to them looks like the wall of a building. They don't even know it's open above.

Trust me on this one, don't even waste your time with a stock trailer if you don't have a ramp.

A good start to loading pigs would be a simple portable wooden ramp and a series of pipe gates chained together. The number and size of your gates ultimately depends on how many animals you have, but you can get the general idea here. Throwing some buckets of feed into the pen area and trailer can help coax hungry pigs to enter without a fuss. The less help you have, the more planning and forethought you will have to put into this. If you can manage it, an ideal scenario would be to set this area up inside their paddock days ahead of time and routinely feed the pigs inside. That way when you are ready to load, they will be conditioned.

If you decide to build a wooden ramp, the design is crucial to your success. First of all, it has to be stout. Any flexion in it and the pigs will not go up it. It also needs to be solid. (No gaps to see through.) Lastly, it has to have traction. We add rungs horizontally across it for their hooves to grip onto (similar to our poultry ramps). Details on wooden ramp construction can be found starting on page 339.

Scenario 3
Trailer, sturdy ramp, and semi-permanent corral = things are looking up

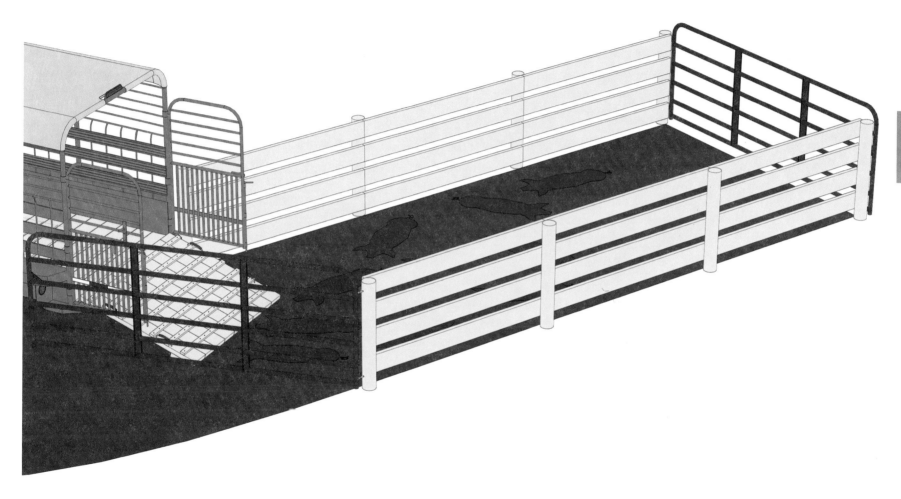

An even better infrastructure setup is to build a semi-permanent corral setup with a portable wooden ramp. On the farms where we have our larger pig pasture operations, we tend to build something along these lines. Since we try to design our pasture systems around a central access road, we utilize that road as an alleyway and build a corral at the entrance.

We simply drive right through the corral as part of our roadway when it is not in use. In some instances, we opt to install permanent pipe gates, while on other farms we bring them when we need them.

A basic permanent corral like this doesn't cost us very much to build and pays for itself quickly in not having to transport and setup enough gates to contain 30+ pigs every time we need to handle them.

Scenario 4
Trailer, earthen ramp, permanent corral and auxilary pipegates = bullet-proof pig handling

PIGS

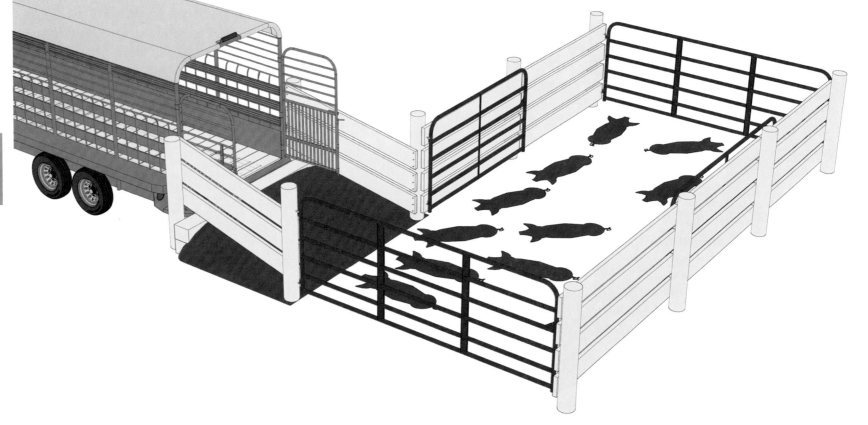

As mentioned previously, pigs don't like to walk on something hollow. They have an aversion to a ramp that sounds hollow underneath. Pigs want to stay on the ground, period. Enter the earthen ramp.

In this scenario, we take the semi-permanent corral concept one step further. Here we add an earthen ramp to mate with whatever transportation device we plan to use. Bigger the trailer, bigger the ramp. Thankfully, most stock trailers are low to the ground so it doesn't require much dirt or much material, but

it sure makes life easier when trying to load pigs.

I know it's easy to think you can just throw a couple of boards up, like glorified chicken ladders, and entice the pigs up there. Listen, don't do it. Don't even try. You'll be telling stories years from now. Take it from someone who has tried every shortcut known to farmers to get pigs loaded. 'Tain't worth it.

Build the freakin' ground-based ramp. You'll never regret it and your marriage might be saved in the process. Trust me on this. The

main thing to remember about loading pigs is this: you have one shot at it. If you mess up, you will not get a second chance. Short cuts never pay off. Due diligence in the setup is the only way to have an experience you'll want to repeat.

Lastly, for a most pleasurable loading experience, try adding some strategically placed auxiliary pipe gates to aid in guiding the pigs up the ramp and into the trailer. If you are a one- or two-man rodeo, a setup like this may very well be worth the investment.

Scenario 5
Yee-hawing (Pig Drive)

photo courtesy of Jessa Howdyshell

PIGS

This is by far the most rewarding method of pig transportation. It's become a ritualistic tradition that we always look forward too. We gather the apprentices and the grandkids; every one picks out a prodding stick and we head up the mountain.

We really only herd pigs that are approaching slaughter weight. The extra pounds make them slower and lazier – aka less ornery. We annually walk our pigs down from the mountain acorn glens in late fall.

It's a celebrated sigh of relief as we shut the last of the mountain pig pastures down for the winter. It is a long walk, and the more tired the pigs get, the more compliant they are.

By the time we've reached home, they are delighted to file into their warm winter quarters for a meal and some rest. It's important not to push a fat hog like that in hot weather though – or they could run into some cardiovascular issues.

Specialized Equipment

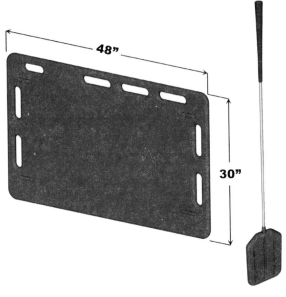

Pic 1

Pic 2

photos courtesy of Jessa Howdyshell

PIGS

Sort Boards

Another critical tool for handling pigs is the sort board. When under pressure pigs have a tendency to run through danger. In other words, their reactions tend to be opposite of our intentions. If you tap them on the butt, they back up and if you nudge their nose they will go forward. The best way to move a pig is to make them think it was their idea. That's where the sort boards come in. A pig cannot see any openings through the solid sort boards. To them it appears like an impenetrable barrier. When used in formation like picture 1 above, several handlers can successfully work a group of pigs into a pen or trailer.

Sure, you can try making your own sort boards, but nothing beats the strength and lightweight nature of a well manufactured sort board. This is one item we are not sorry for buying instead of making. Best $45 you'll spend on your pig operation. We buy ours from KANE Manufacturing and we like the 30" x 48" size for most applications, but also use the 30" x 36" for working pigs in narrow alleys such as in our livestock scale.

Sort Stick

Sort sticks are also a useful tool for handling pigs. Manufactured sticks often have rattles or some other noise-making medium on them. We have a few of these but more often than not we grab a tree branch in the moment when we need it. This is what we would consider a nicety, not a necessity.

Pig Moving Trailer

We do own a specialized trailer that hydraulically raises and lowers to transport pigs. (See picture 2.) This is by far the ultimate luxury that we have in our pig operation – and a justifiable one considering we run more than 800 pigs a year. We would not recommend spending that sort of capital upfront however. The beauty of it is that it lowers to the ground for loading, and then can be raised nearly 4 feet into the air, providing amazing ground clearance for tight spots, and offloading from large trailers or truck beds. Even with the trailer, we still need a corral loading area of some sort to pair with it.

Wooden Ramp

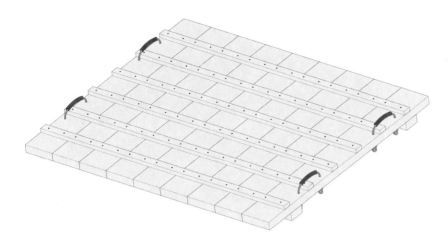

Hardware

DECK SCREWS

- ❑ **108x***   2"-½

- ❑ **54x***   3"

Length of screws may vary depending on thickness of lumber used.

- ❑ **≈32"**   Scrap Garden Hose (or plastic pipe)

- ❑ **≈8ft**   Braided Rope

Tools

- ❑ Circular Saw

- ❑ Power Drill

- ❑ Driver Bits (for screws)

- ❑ ½" HSS Drill Bit

- ❑ Marking Pencil

- ❑ Tape Measure

- ❑ Table Saw

PIGS

Wood Cutlist

We recommend using rough-sawn hardwood lumber for this project, preferably oak. Store bought softwoods will rot when exposed to the elements and may eventually break under the load, causing potential injury to livestock and handlers. Store bought wood is also smooth and slippery, especially when wet and/or muddy. Slippery surfaces are a huge deterrent for pigs, whereas rough-sawn lumber adds much needed grip for the pigs' hooves.

Wood scraps are highlighted in red.
Do not discard until project is completed.

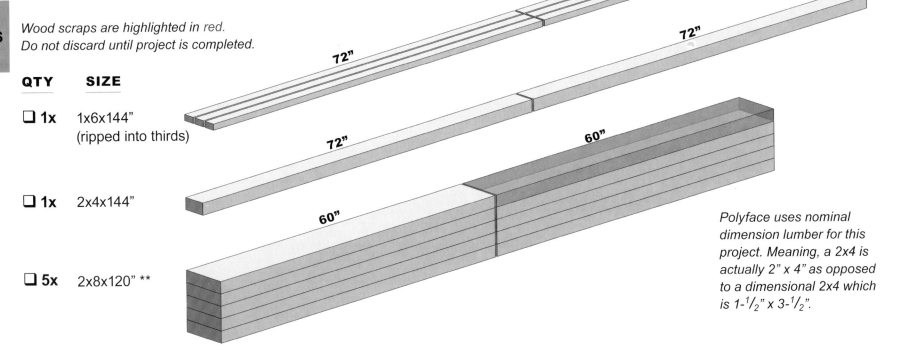

QTY	SIZE
☐ **1x**	1x6x144" (ripped into thirds)
☐ **1x**	2x4x144"
☐ **5x**	2x8x120" **

Polyface uses nominal dimension lumber for this project. Meaning, a 2x4 is actually 2" x 4" as opposed to a dimensional 2x4 which is 1-$\frac{1}{2}$" x 3-$\frac{1}{2}$".

**The ramp we illustrate is 60" long. Now, according to Temple Grandin, the maximum angle of your ramp should not exceed 25°, while it is ideal to shoot for 15°.

At 60" long, your trailer height can vary up to about 22" high. Any higher of a step and you will need to make a longer ramp.

On the contrary, if you only have a short step into your trailer you can always shorten this ramp as needed. That will make it lighter and easier to handle!

22" Maximum Height

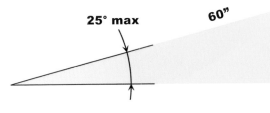

1

≈9x 2x8x60"

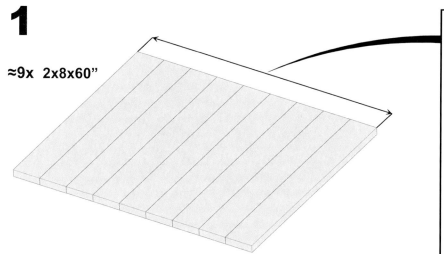

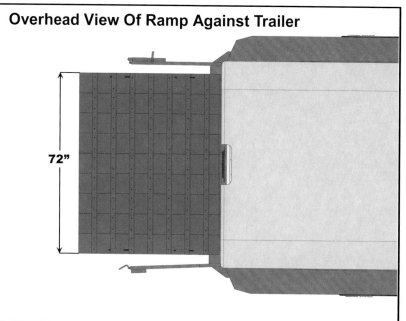

Overhead View Of Ramp Against Trailer

72"

The first step is to determine the width of your ramp. This will vary with the width of your trailer. Ours is approximately 6ft which works well with our trailers. Adjust your width accordingly.

PIGS

2

2x   2x4x72"
54x 3" Deck Screws

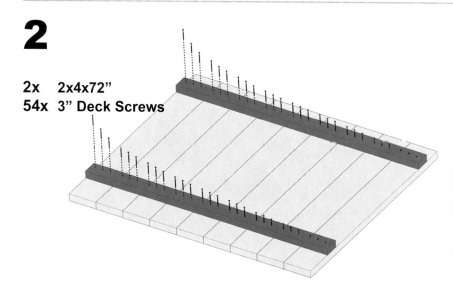

Fasten the 2x4 supports to the bottom of the ramp where shown. Use a liberal amount of appropriately sized screws. If using true 2" thick lumber, a 3" screw will work well. Check the other side for any screw tips that may have penetrated through and grind them down to prevent injury.

3

6x   1x2x72"
108x 2-½" Deck Screws

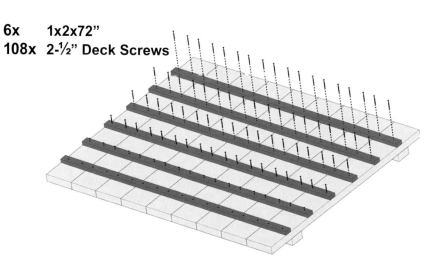

Flip the ramp right side up and space the horizontal rungs 8-10" apart. Fasten them with a liberal amount of appropriately sized screws.

4

Drill two ¹/₂" holes approximately 8-10" apart from each other. You will drill a total of 4 sets of holes in the locations shown.

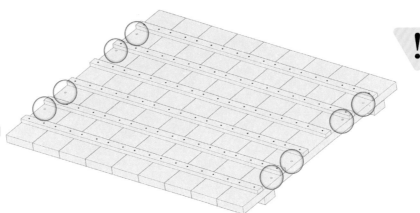

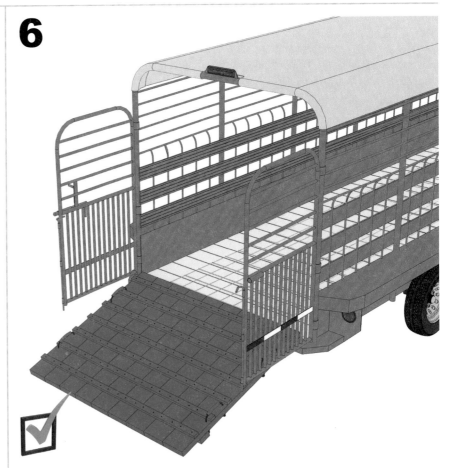

! This ramp is heavy and takes two people to safely maneuver it. For convenience we make some simple rope handles that are placed in ergonomic positions for the handlers. The number and location of the handles on your ramp may vary.

PIGS

5

4x   18-24" Lengths of Braided Rope
4x   6-8" Lengths Of Scrap Garden Hose Or Plastic Pipe

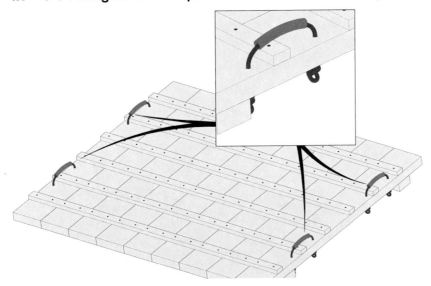

Tie a knot on one end of the rope and fish the other end from the bottom of the ramp and up through a hole. Next slide a piece of hose/pipe onto the rope and then fish it down through the second hole. Adjust the length to suit your liking and tie a knot in the loose end to secure the handle. Repeat the process for each of the four handles.

6

Your wooden ramp is complete (and ready to save your marriage).

Earthen Ramp

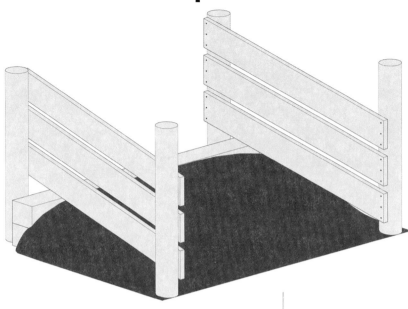

Tools

☐ Circular Saw

☐ Marking Pencil

☐ Hammer

☐ Tape Measure

☐ Marking Paint

☐ Post Hole Digger

☐ Digging Bar

Optional - Corral

☐ **2x**   12ft Pipe Gates

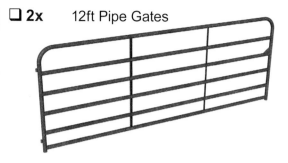

☐ **1x**   10ft Pipe Gate

PIGS

☐ **1x**   8ft Pipe Gate

☐ **8x**   Lag Screw Gate Hinges

Additional posts and fence rail material will need to be procured to build a corral, but quantities may vary depending on the size and shape. Refer to step 8 for details on size, and determine your lumber needs from there.

Materials

PIGS

QTY	SIZE
☐ ≈2x	8x10 Railroad Ties

Size and quantity will vary depending on how high your ramp needs to be. Length may also vary depending on trailer used. Round timbers are an adequate ok to substitute, but may create a gap between the trailer and ramp for legs to get caught in.

☐ ≈6x   2x10

Quantity based on width and number of boards used. Lumber does not have to be 10" wide, you can use whatever is available to you.

☐ 4x   8" Diameter Corner Posts

Posts should be approximately 8" in diameter and be set at least 30" into the ground for stability. Depending how tall your ramp needs to be, you may need longer posts.

☐ ≈3x   A few tractor bucket scoops of dirt

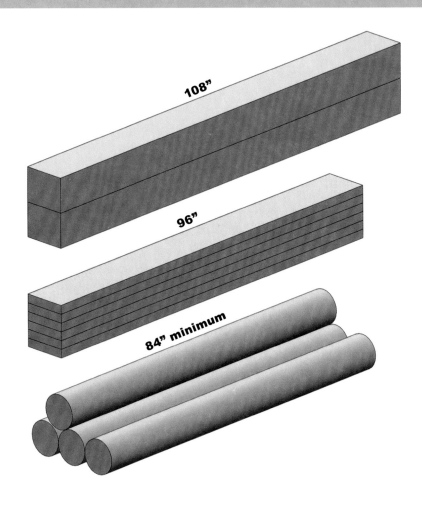

108"

96"

84" minimum

Hardware

☐ 36x   20D Galvanized Nails*

*Number and size of nails may vary depending on what lumber you use for the fence rails.

Joel's Tip

I recommend physically bringing the trailer you plan to use out to the pasture where you plan to build your ramp. Get a feel for how easily you can maneuver it into position and then set your posts precisely where they need to be. Also note that the rear end height of your trailer can vary quite a bit depending on the topography of the land it is sitting on. Having the trailer on-site will ensure that you set the elevation of your ramp correctly.

PIGS

2

2x   Railroad Ties

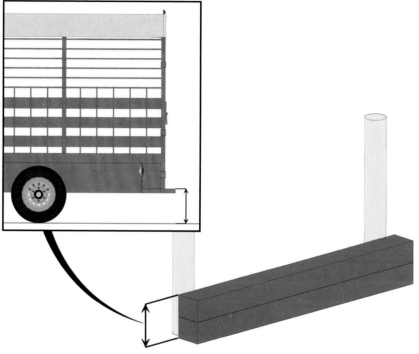

1

2x   Fence Posts

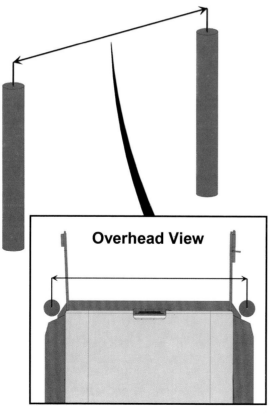

Overhead View

The first step in creating an earthen ramp is to determine the width of the posts. This will be based on the width of your stock trailer. Ideally you want the back end of the trailer to nest between your posts and be able to open the rear door(s) without obstruction.

The second step is to determine how high your ramp needs to be. This is also trailer specific.

With a height determined, you can now install the retaining wall. We've depicted square landscape timbers to use for the retaining wall, but round logs work too. They do not need to be anchored into the ground because there will be dirt on one side and the posts are holding them on the other.

3

2x   Fence Posts

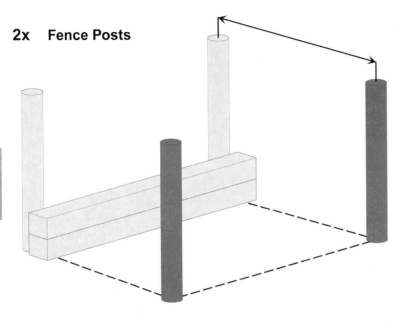

PIGS

1:4 Gradient Slope

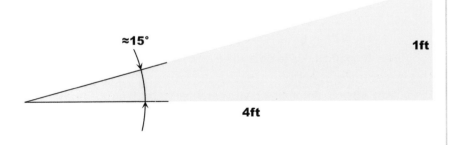

≈15°

1ft

4ft

This distance between the posts is based on how high your ramp is. According to Temple Grandin, the maximum angle should not exceed 25°, while it is ideal to shoot for 15°. A 15° angle approximately equates to a 1:4 gradient. In other words, for every 1ft of height, your ramp must extend 4ft away from the trailer.

4

1x   Pile of Dirt

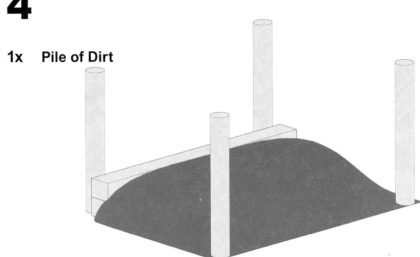

With your posts in place, back fill with dirt and shape the ramp. Be sure to pack it in well. Do not be afraid to leave the dirt a little high. After a few rains it will inevitably settle.

5

3x   2x10 Fence Rails
18x 20D Nails

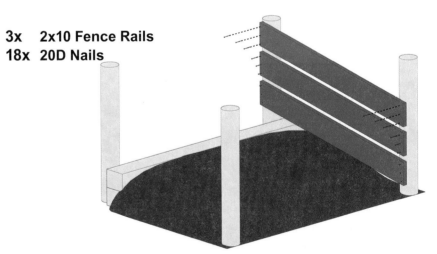

Fasten the fence rails to the ramp. The size and quantity may vary. Start your first rail close to the ground and end the top rail at least 32-36" up. These rails double as bracing to support the front posts.

6

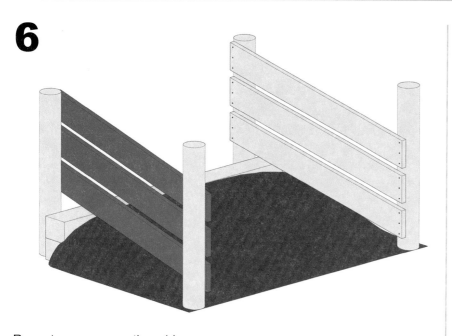

Repeat process on other side.

7

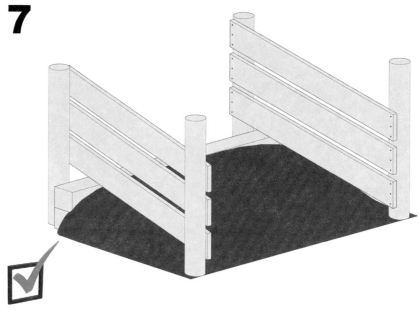

Your ramp is complete. Continue by building a corral around it.

8

Corral size and fence post spacings are not critical. Temple Grandin recommends approximately 6ft² per market-sized hog. This coral shown is approximately 25 x 13ft which is enough for approximately 50 pigs. If you plan to utilize gates to aid in sorting and loading, you will need to adjust the width of your corral so that a pipe gate, when mounted on hinges, can swing freely between the two sides. Refer to the loading sequence on the following pages to get a feel for how the gates are used and how they swing.

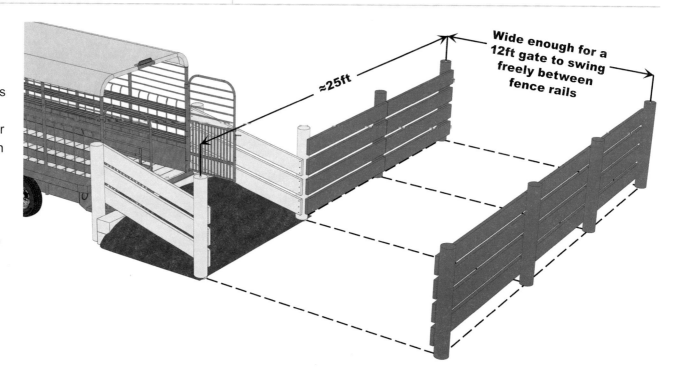

≈25ft

Wide enough for a 12ft gate to swing freely between fence rails

9

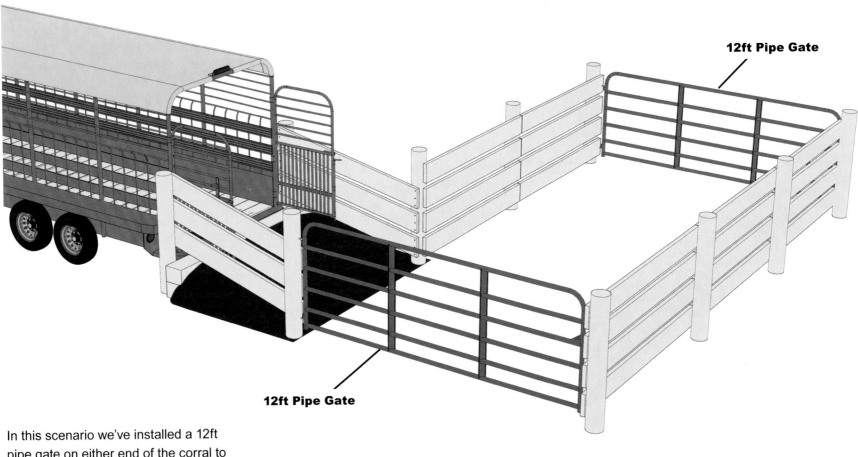

12ft Pipe Gate

12ft Pipe Gate

In this scenario we've installed a 12ft pipe gate on either end of the corral to secure the two ends.

10

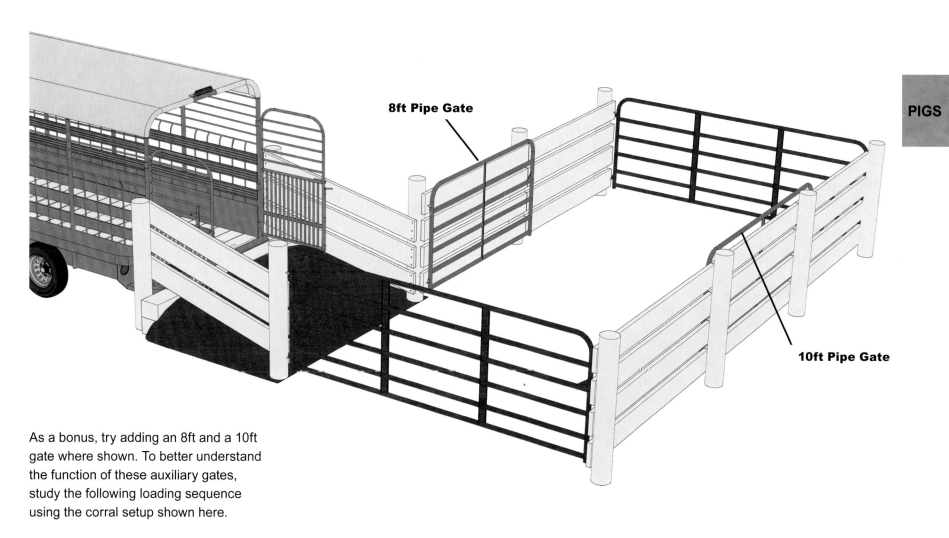

8ft Pipe Gate

10ft Pipe Gate

As a bonus, try adding an 8ft and a 10ft gate where shown. To better understand the function of these auxiliary gates, study the following loading sequence using the corral setup shown here.

Loading Sequence

1

Yee-haw pigs from pasture into the corral and lock them in with a gate.

PIGS

As mentioned before, this is an optional scenario, but definitely makes things easier when help is hard to find. I'm sure there are many other nifty corral ideas out there, and we'd love to hear what tricks have worked for you.

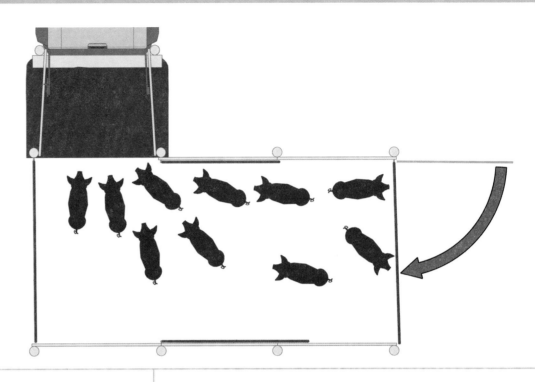

2

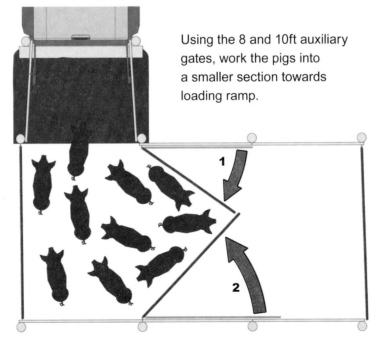

Using the 8 and 10ft auxiliary gates, work the pigs into a smaller section towards loading ramp.

3

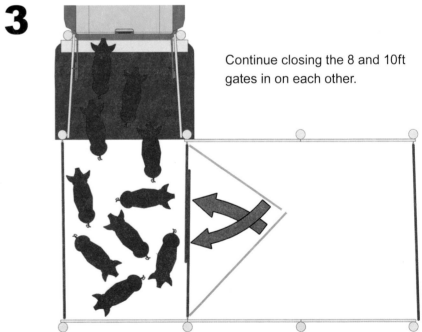

Continue closing the 8 and 10ft gates in on each other.

4

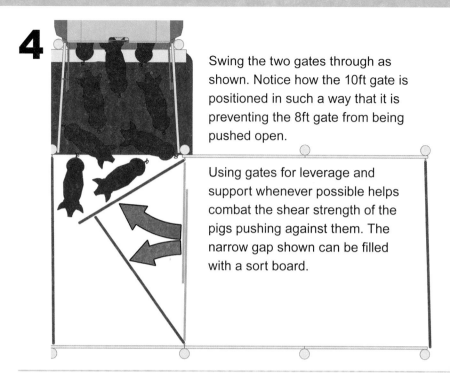

Swing the two gates through as shown. Notice how the 10ft gate is positioned in such a way that it is preventing the 8ft gate from being pushed open.

Using gates for leverage and support whenever possible helps combat the shear strength of the pigs pushing against them. The narrow gap shown can be filled with a sort board.

6

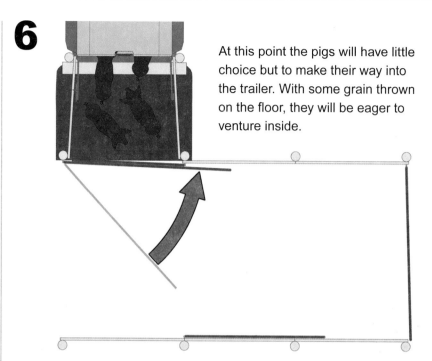

At this point the pigs will have little choice but to make their way into the trailer. With some grain thrown on the floor, they will be eager to venture inside.

PIGS

5

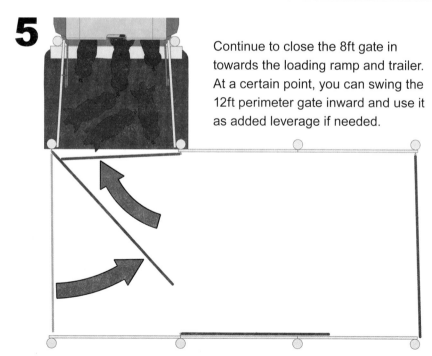

Continue to close the 8ft gate in towards the loading ramp and trailer. At a certain point, you can swing the 12ft perimeter gate inward and use it as added leverage if needed.

photo courtesy of Jessa Howdyshell

351

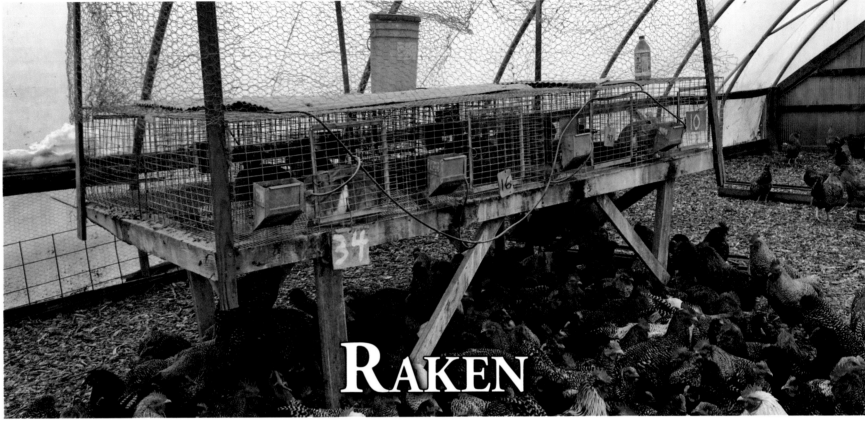

RAKEN

We elevated our rabbit pens by using a mobile hutch.   photo courtesy of Chris Slattery

When our son Daniel was 8 years old, he decided he wanted his own farm enterprise. He'd grown up looking at his uncle's rabbit hutches up in the rafters of the barn where we'd stored them when my older brother went off to college and shut down his teenage rabbit business. (In our family, we believe strongly in child entrepreneurship and encourage it from the womb.)

At about that time, a young couple in our church fellowship decided to move into the city and needed to find a new home for their three pet rabbits: two does and a buck. Daniel delightedly took the rabbits off their hands and began his enterprise that later turned into his 4-H project. One doe was a Dutch rabbit; another was a Californian and the buck was a New Zealand White. With those three breeds as a starting point, he developed a composite and retained his own breeding stock in a crude line breeding program.

For the first five years he suffered roughly 50 percent mortality as he worked through diarrhea caused by fresh grass. Most rabbits are so far removed from grass that they can hardly digest it anymore. Fortunately he didn't have to make a living with his fledgling enterprise so suffering these losses was more emotional than economic. He had a friendly banker.

Eventually, the genetics acclimated to forage and the mortality tapered off. By that time, he needed substantial housing to handle the 30 does in the breeding flock. That necessitated the birth of the "Raken" house. Raken is short for rabbit plus chicken. It's a perfect example of the permaculture stacking concept, whereby we use cubic footage inside a structure rather than just the floor square footage. The difference in output is remarkable.

The rabbits are housed safely above chickens that can roam freely in the space below. By either clipping wings or putting poultry netting barriers up, we keep the chickens off the top of the rabbit pens. (Chicken manure falling into the rabbit pens would not go through the floor and would create sanitation issues for the bunnies.)

Rabbits are fastidious by nature; they choose a toilet spot and go there all the time. Rabbit urine is extremely high in nitrogen (ammonia). That means when they repeatedly use the same toilet area it eventually gets wet, and the volatile ammonia vaporizes up into the rabbits causing respiratory problems.

Commercial rabbitries deal with this issue generally in two ways. One is to clean out every week. Some even put buckets under the toilet spots so the urine doesn't run everywhere. Talk about stinky. The other option is to raise earthworms underneath the rabbits.

We decided to go a different route: why not create compost aerated by chickens? We start with a foot or more of wood chips. Because the chickens scratch in this bedding, it stays aerated for good decomposition. The scratching also incorporates the urine into the bedding, and kind of kneads it in, which disperses the urine moisture out into the bedding.

The composting bedding grows lots of bugs which stimulates the chickens to scratch more. The decomposition utilizes the manure and urine, eliminating odors. You could go in there and eat

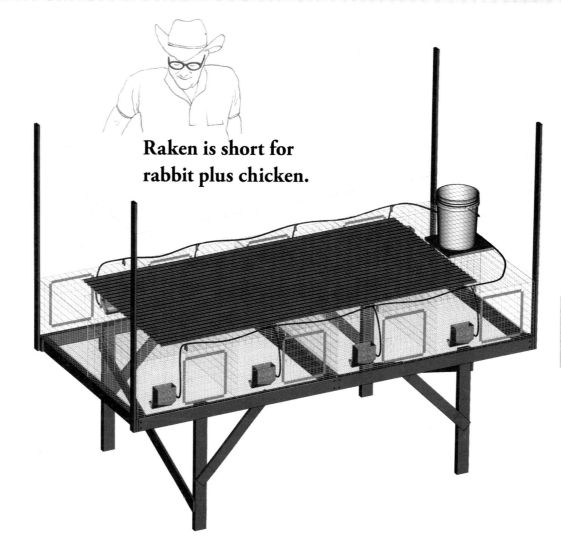

Raken is short for rabbit plus chicken.

a sandwich; it smells wonderful. The chickens also provide a gentle soothing background noise for the rabbits, which reduces their stress from sudden farm noises. Rabbits are highly excitable and of course they can hear really well.

If we take the chickens out for any reason, the Raken house begins to stink within a week. That's how important the chickens are. Because these laying hens do not get fresh pasture; we

throw in our food scraps and garden wastes, including weeds and lawn clippings. Chickens in housing like this need 3 sq. ft. per bird in order for their manure load to be low enough to incorporate into the bedding. At higher density, they tend to cap the bedding, which stops the scratching, which stops the aeration, which stops decomposition. To ensure proper aeration under the rabbits, a good rule of thumb is one

RABBITS

chicken per three rabbits. That bedding grows a lot of goodies that need discovery. Occasionally a rabbit pees on a chicken, but it's very seldom.

Initially, the Raken house was a simple shed on the end of our main farm equipment shed. But winters were brutal, freezing water and even some of the baby bunnies. Eventually we moved to two different structures; one for summer and one for winter.

Twice a year we have a rabbit moving day to shift living quarters. Summer quarters are a high-ceiling airy hoop house with white covering that stays cool. Lots of high ventilation keeps things comfortable. In the winter, all the rabbits move into the transparent hoop houses where we over winter the laying hens.

Although rabbits can breed and kindle year round, they definitely respond to daylight and express more fertility in the spring and early summer. The added light from the winter hoop house arrangements, increases both animal comfort and fertility.

It is key to protect rabbits from drafts, while still providing ample air circulation. With the hoop structures, ventilation occurs up around the top of the structure allowing plenty of airflow without a direct draft on the animals. Skirting boards fastened to the inside of the poles at the entrance hold the bedding in (see pages 438-439 for details). When we clean out, we pop off these boards to offer easy access for the front end loader.

By far and away, this bedding is the best compost we make on the farm. It never gets real hot and by definition it has two different types

photo courtesy of Jessa Howdyshell

of manure inputs. This material is ideal for side dressing leafy greens in the garden.

The Raken house is perfectly adapted to urban livestock as long as you don't have any roosters. It's quiet and highly productive, offering the potential for $10,000 per year out of a two-car garage. Who says we need factory farms to feed ourselves?

This permaculture stacking concept consists of two main structural components:
1. The wire rabbit pen
2. The mobile rabbit hutch

Wire Pens

Rabbits become easily stressed and are quite prone to diseases. In the wild, rabbits have a lot of room to run around and live fairly solitary lives. Some people raise rabbits in a colony, but our experience is that these are prone to disease because the rabbits are abnormally crowded.

Furthermore, they burrow into the ground or bedding to have their babies and by the time the bunnies come out, you can't match them up to the right does. From a management standpoint, to select the best does and bucks, you need to be able to track which bunnies go with which does.

Outdoor systems are problematic due to coccidiosis and escape. Rabbits dig so any containment needs to thwart all the possibilities of a furry Houdini. That's much harder than you can imagine, and requires burying something impenetrable at least 30 inches underground. Rabbits are highly prone to coccidiosis which is ubiquitous in the soil. While some radical animal welfarists may consider these wire pens unacceptable, they are the closest natural way we've found to mimic the singularity and hygiene in the wild. The pens can be cleaned for good sanitation. One other beauty of the Raken house is that the chickens provide a soothing background that minimizes stress.

Our wire rabbit pens are 60" x 30" x 16" high, divided in half so that each unit consists of two separate pens. We do not claim to have invented this rabbit cage design. The specifications of the cage and the materials used are fairly standard throughout the industry. We've adopted our own variants of commercially available designs over the years to end up with what we have today. This design works well for us and felt it worthwhile to share it with you.

We've also optimized material sizes to reduce waste and have provided a detailed shopping list to make things easier on you.

Notice the rabbit pens along the entire left side of this 20 x120ft hoophouse.

photo courtesy of Jessa Howdyshell

Mobile Rabbit Hutch

Over the years we've used numerous methods to suspend the rabbit pens. In the beginning, we hung them from the rafters with wire.

When we moved over to the hoop houses, though, the hoops did not offer enough strength to hang the pens. At that point, we shifted to a simple table type foundation. The final permutation is the completely mobile structure depicted in this book. As we've used different spaces for different things, we've wanted to put the rabbits in different places. Sometimes we even put a bunch of them in the barn. The entirely portable hutch gives us unlimited placement options.

This hutch structure is built with 2x4 lumber and is constructed like a table without a top. It is sized appropriately to support 4 pens.

We can pick the entire unit up either by hand or with extended forks on the front end loader, and move it outside or somewhere else. The whole set up is completely portable and scalable. We can add or subtract units as our herd numbers fluctuate. It makes clean out easy and also enables us to put pigs in the Raken house in the winter when the rabbits and chickens go over to the warmer hoop houses.

Tools

- ❑ 8" Tin Snips

- ❑ J-Clip Pliers

- ❑ Adjustable Wrench

- ❑ Flush Cut Wire Cutters

- ❑ Pliers

- ❑ Tape Measure

- ❑ Power Drill

- ❑ Drill Bit*

Drill bit will be used to create a hole in the bucket for water. It should be sized slightly smaller than the water tubing you are using. The objective is to squeeze the water tube though the hole, creating a seal that is water tight.

Materials

We recommend purchasing these supplies from the Klubertanz Equipment Company, Inc. and have listed their stock numbers in the materials list. Bass Equipment Company is another supplier we order from to purchase these supplies. (Their stock numbers will differ of course).

QUANTITY PER PEN

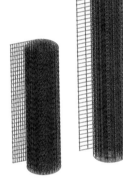

- ❑ **11ft**   1" x 2" x 48" 14 Gauge (galvanized <u>after</u> welding) (X664)

- ❑ **5ft**   1" x ½" x 30" 14 Gauge (galvanized <u>after</u> welding) (X663)

Quantities of wire are given on a per pen basis. Depending on how many pens you plan to construct, it may be cheaper to purchase an entire roll of wire.

We highly recommend paying extra for wire that has been galvanized <u>after</u> welding. Rabbit droppings are highly corrosive and lesser quality wire simply will not hold up.

- ❑ **2x**   Wire Spring Door Locks (Right Hinge) (ZKR1)

RABBITS

☐ **≈5ft*** $^{3}/_{16}$" Flex Water Tubing (A401)

☐ **2x** Standard Water Valves, $^{3}/_{16}$" Barb, Brass Cap (A102)

☐ **2x** Valve Clips (A108)

☐ **1x** 5 Gallon Bucket

☐ **≈1*** Barbed Tee, $^{3}/_{16}$" (A405)

☐ **≈2*** Standoffs, $^{3}/_{16}$" (A403)

☐ **≈170x** $^{3}/_{8}$" x $^{5}/_{8}$" J-Clips, Zinc Plated (R400)

☐ **8x** Galvanized Metal Door Frames (Z011)

☐ **≈3ft** 16-½ Gauge Tie Wire

☐ **1x** Nest Box (10x10x19) (N016)

The number of nest boxes you need will correspond to the number of does you plan to breed at a time.

☐ **2x** 5-½" Fine-X® Wire Screen Feeders (FX50)

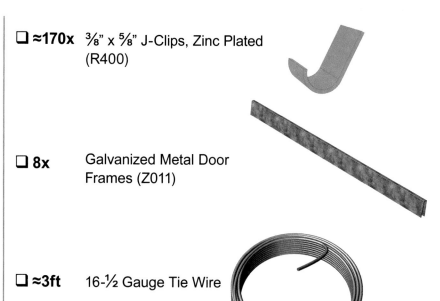

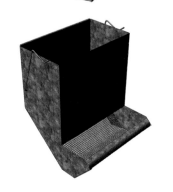

RABBITS

*Exact quantities of standoffs, barbed tees, and water tubing will depend on your setup. See pages 371-372 for details on water system layouts prior to purchasing.

Wire Cutlist

While there are certainly other ways to cut out the pen pieces, we have found this to be the most efficient use of material. By following this cutout arrangement on a 48" roll you will minimize waste and get 9 pens worth of material out of a 100ft roll.

RABBITS

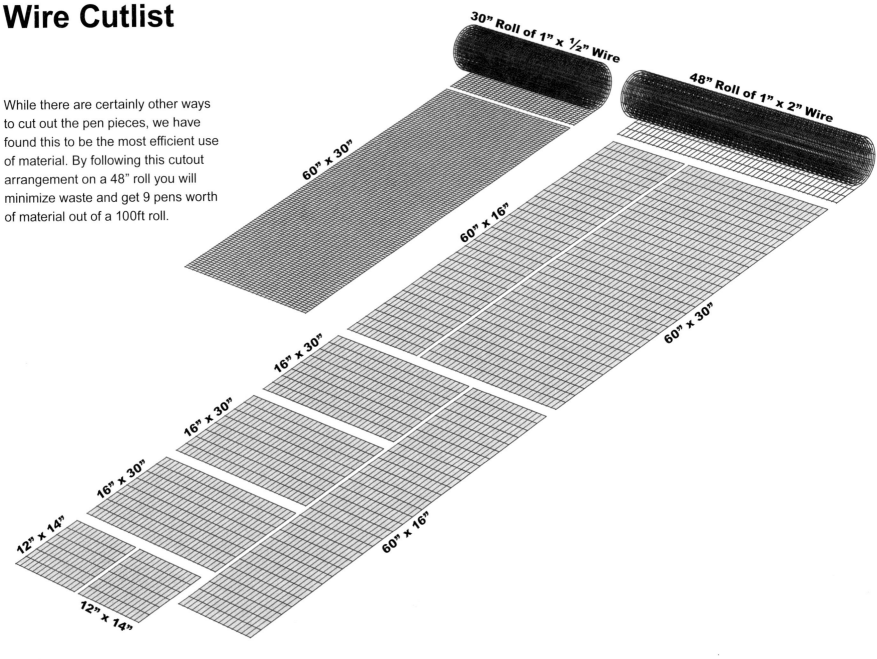

1

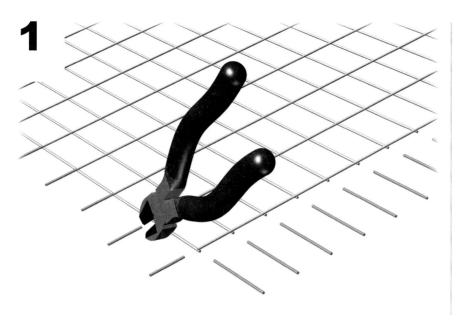

Eliminate sharp ends by trimming waste (shown in green) on all pieces.

2

1x   1x2" Wire (60" Wide x 30" Long)

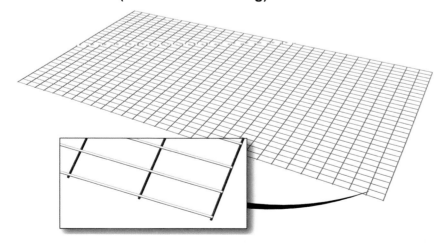

Orient with the highlighted portion of the wire on the bottom as shown.

3

2x   1x2" Wire (16" Wide x 30" Long)
4x   J-Clips

J-clip all four corners.

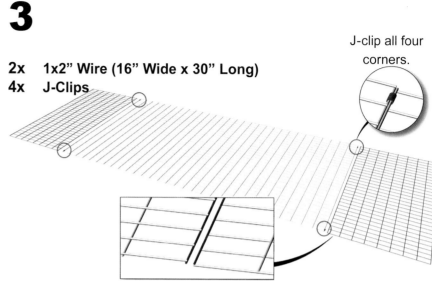

Pay attention to the orientation of highlighted wire for correct corner alignment with J-clips.

4

J-clip all four corners.

2x   1x2" Wire (60" Wide x 16" Long)
4x   J-Clips

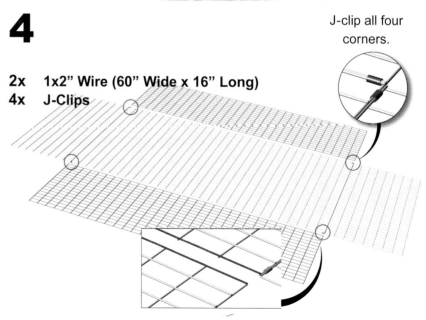

Pay close attention to the orientation of highlighted wire for correct corner alignment with J-clips.

5

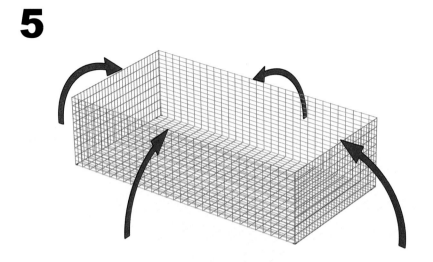

Fold all four sides up and together.

7

1x   1x2" Wire (16" Wide x 30" Long)
≈20x   J-Clips

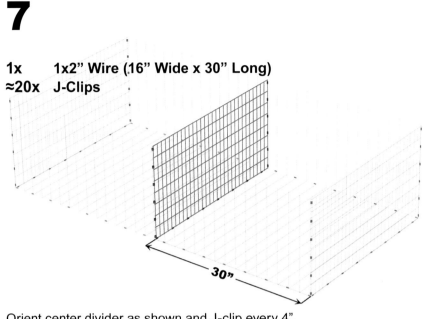

30"

Orient center divider as shown and J-clip every 4".

6

≈64x   J-Clips

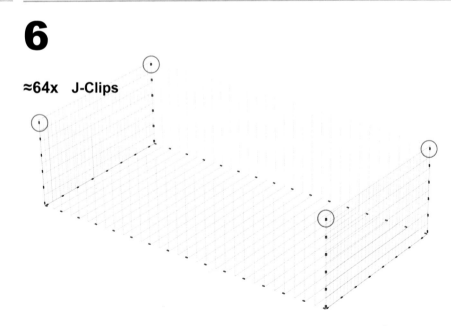

J-clip the top 4 corners first and then fill in with a J-clip every 4" as shown.

8

1x   1x½" Wire (60" Wide x 30" Long)

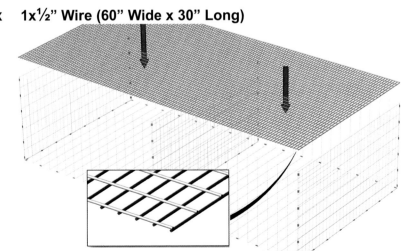

It is critical to orient the floor as shown with the ½" spaced wires facing into the pen because it prevents sores on the rabbit's paws.

9

≈48x   **J-Clips**

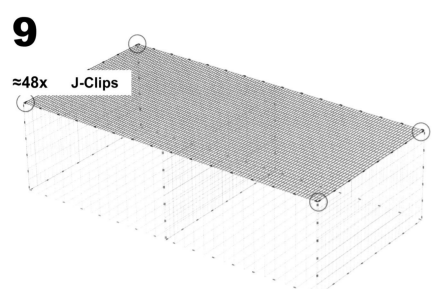

J-clip the 4 corners first and then fill in every 4" as shown. (You will not be able to fit the J-clip pliers into the bottom of the center divider. This will be addressed in step 12.)

11

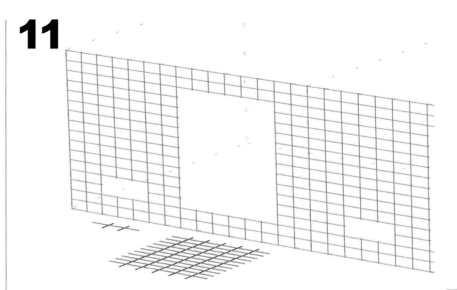

Repeat step 10 by cutting openings on the other side.

180°

10

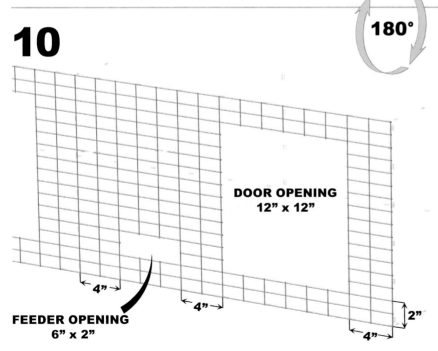

DOOR OPENING
12" x 12"

4"

4"

2"

4"

FEEDER OPENING
6" x 2"

Flip the pen over and cut openings in the front of the pen as shown.

180°

12

≈3ft   **16-½ Gauge Tie Wire**

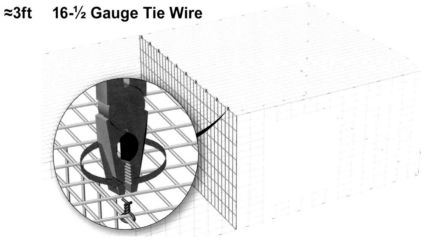

Flip the pen over and using 16-½ gauge wire, twist and tie the center divider to the bottom of the cage every 4" as shown.

13

4x   Metal Door Frames

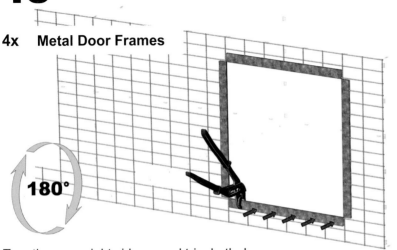

Turn the cage right side up and trim both door frames, crimping them with an adjustable wrench to hold them in place.

RABBITS

14

1x   1x2" Wire (12" Wide x 14" Long)
1x   Wire Spring Door Latch
≈6x J-Clips

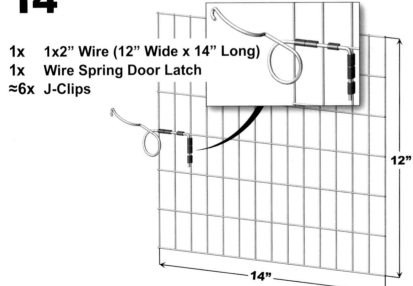

Attach door latch with J-clips.

15

≈5x   J-Clips

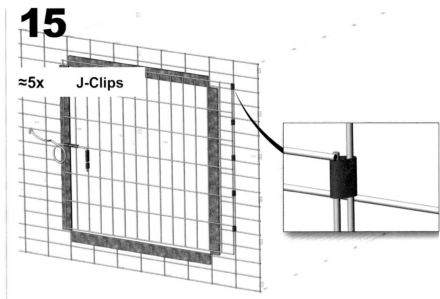

It is important to install the top J-clip as shown to prevent door from sliding downwards.

16

1x   5-½" Fine-X® Feeder

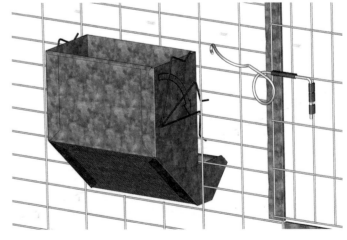

Install the feeder as shown.

17

Repeat steps 14-16 on the other side of the pen to attach the door latch, door and feeder.

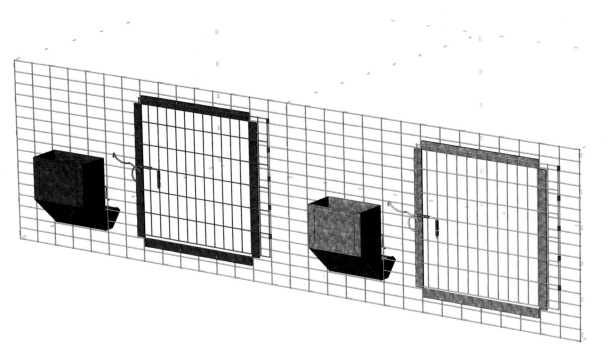

18

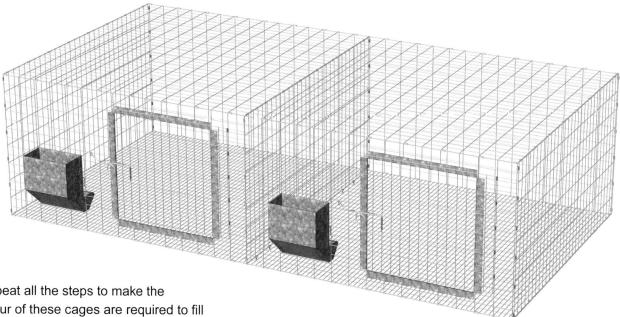

The rabbit wire cage is complete! Repeat all the steps to make the number of cages desired. A total of four of these cages are required to fill the mobile hutch featured next.

Mobile Hutch

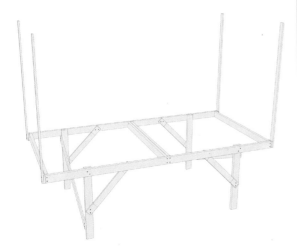

Tools

❏ Hand Saw

❏ Power Drill

❏ Flush Cut Wire Cutters

❏ Pliers

❏ 8" Tin Snips

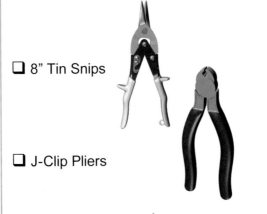

❏ J-Clip Pliers

❏ Speed Square

❏ Stapler

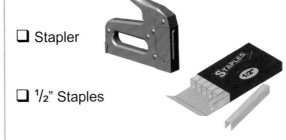

❏ ½" Staples

❏ Table Saw

❏ Driver Bits (for screws)

❏ Marking Pencil

❏ Tape Measure

Supplies

❏ **≈30ft** 36" Chicken Wire

This is an either or part. These will be used to secure the bottom edge of the chicken wire to the top of the rabbit pens in step 17.

❏ **≈20x*** ⅜" x ⅝" J-Clips (R400)

OR

❏ **≈5ft*** 16-½ Gauge Tie Wire

Hardware

☐ **32x**   1-⅝" Deck Screws

☐ **32x**   3" Deck Screws

Wood Cutlist

Wood scraps are highlighted in red. Do not discard until project is completed.

This project is built with dimensional lumber purchased from a building supply store. Meaning, a 2x4 is actually 1-$\frac{1}{2}$" x 3-$\frac{1}{2}$".

RABBITS

☐ **1x**   1x4x12 feet long (cut in half lengthwise)

64"   64"

☐ **3x**   1x4x12 feet long

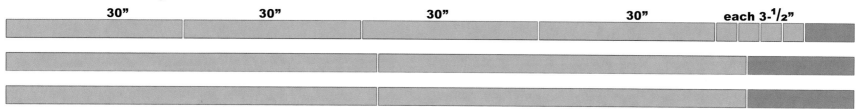
30"   30"   30"   30"   each 3-$\frac{1}{2}$"

☐ **4x**   2x4x10 feet long

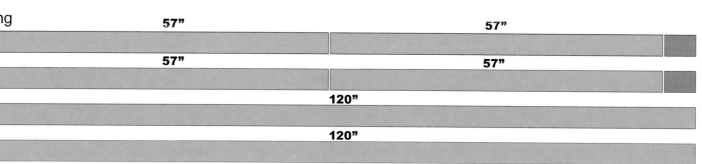
57"   57"
57"   57"
120"
120"

☐ **1x**   2x4x12 feet long

36"   36"   36"   36"

Build The Mobile Hutch Frame

1

1x   2x4x120"
2x   2x4x36"

Measure and position the 2x4 legs.

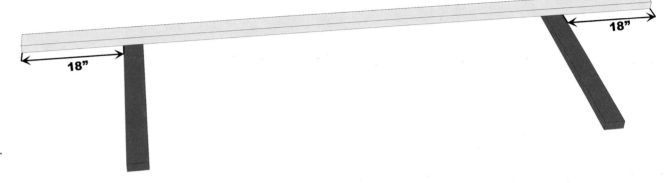

2

4x   3" Deck Screws

Square the legs and fasten them in place.

3

2x   1x4x30"
4x   1-⅝" Deck Screws

Install angle bracing as shown. If any material hangs over, be sure to trim it off.

4

4x   1-⅝" Deck Screws

180°

Flip over and secure braces to back.

5

Repeat steps 1-4 to create a
second side.

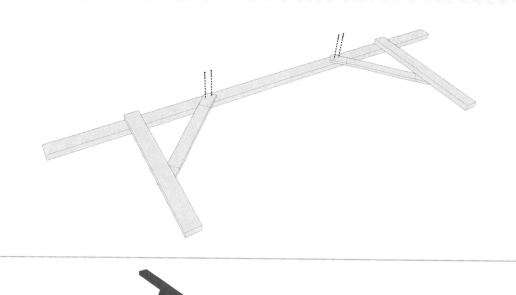

6

8x   3" Deck Screws
4x   2x4x57"

Install the 2x4 supports. Notice that the 2x4
supports are doubled up in the middle. This is
to ensure that the wire cages have an adequate
amount of surface area to rest on.

60"

7

4x 3" Deck Screws

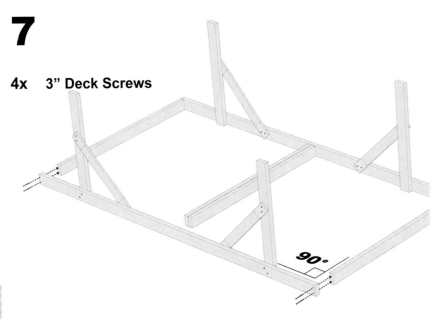

Screw the two ends of the second side in place.

8

4x 3" Deck Screws

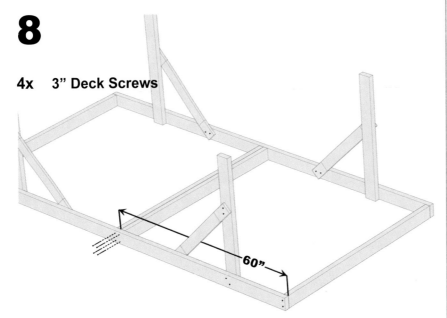

Measure and center the two middle members before fastening them in place.

9

1x 1x4x63"

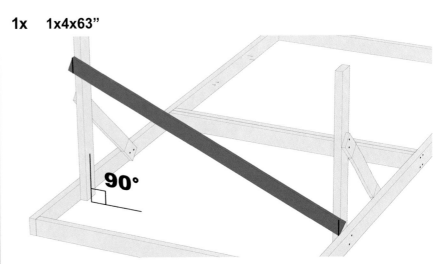

Position the piece and mark lines as shown (to miter and cut off excess).

10

Trim per your lines (removing ends shown in green) with a saw.

11

4x   1-⅝" Deck Screws

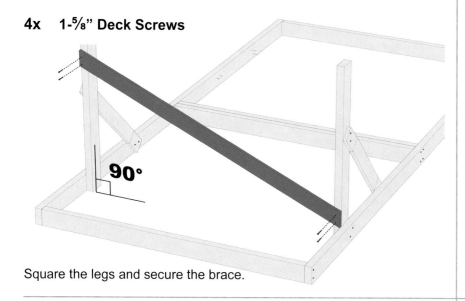

Square the legs and secure the brace.

12 Optional

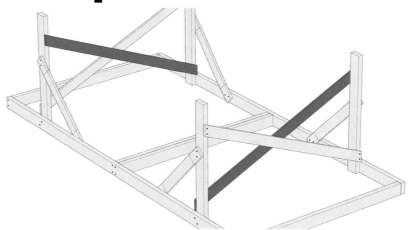

You can add additional diagonal bracing, but we find it adds more weight without gaining much extra stability. However, if you ever plan to run pigs underneath these hutches, the additional bracing is beneficial.

RABBITS

12

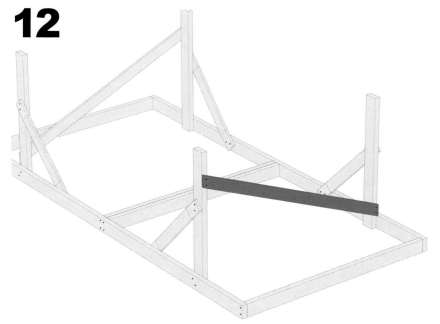

Repeat steps 9-11 for the brace on the other side.

13

2x   3" Deck Screws
1x   1x2x64"
1x   1x4x3-½ (does not need to be exact)

If this hutch will be used in a Raken setup we recommend preventative measures to keep chickens from perching on top. In addition to clipping a wing, we hang chicken wire above the rabbit cages to deter them from flying up.

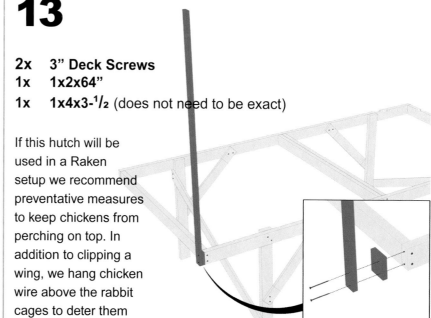

14

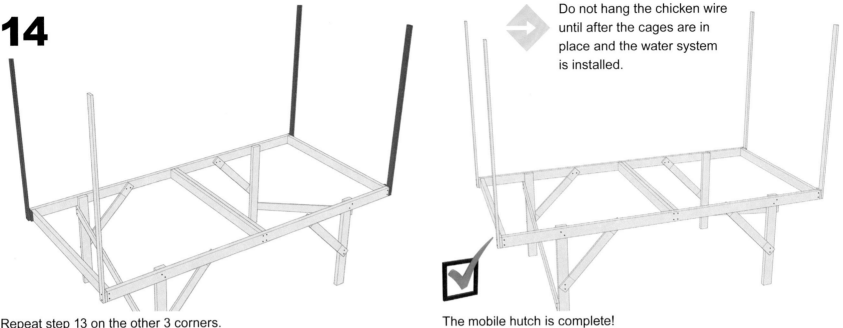

Do not hang the chicken wire until after the cages are in place and the water system is installed.

Repeat step 13 on the other 3 corners.

The mobile hutch is complete!

15

A mobile rabbit hutch is designed to support four wire rabbit pen structures for a total of 8 separate compartments. The hutches can be modified to hold any number of pens, but become more cumbersome and harder to handle the larger they are.

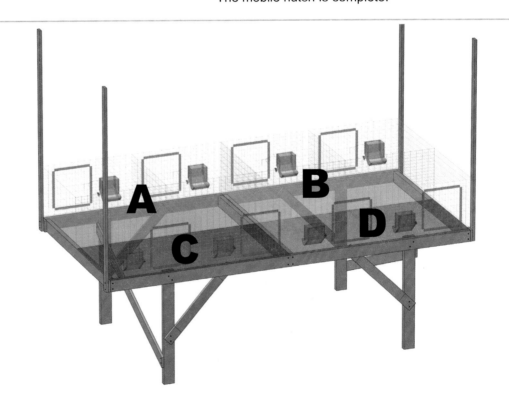

Water System – Basic Single Pen Setup

16

This setup uses:

1x	**5 Gallon Bucket**
5ft	**³/₁₆" Water Tubing**
1x	**Barbed Tee**
2x	**Standoffs**
2x	**Water Valves & Valve Clips**

While not shown here, we typically loosely place a lid over every bucket to prevent insects and debris from contaminating the water. Do not clamp the lid tight because it may create a vacuum that will prevent water from flowing through the system.

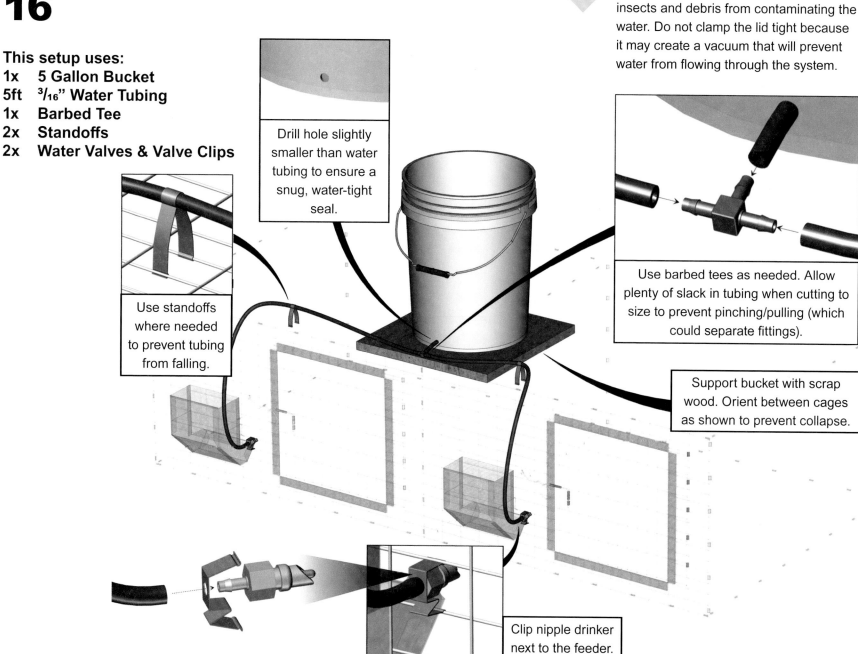

Drill hole slightly smaller than water tubing to ensure a snug, water-tight seal.

Use standoffs where needed to prevent tubing from falling.

Use barbed tees as needed. Allow plenty of slack in tubing when cutting to size to prevent pinching/pulling (which could separate fittings).

Support bucket with scrap wood. Orient between cages as shown to prevent collapse.

Clip nipple drinker next to the feeder.

RABBITS

Water System – Multiple Pen Setup

Using the same principles from the basic single pen water system, one can create a system to water an entire hutch.

This setup uses:
1x 5 Gallon Bucket
23ft $^3/_{16}$" Water Tubing
7x Barbed Tees
8x Standoffs
8x Water Valves & Valve Clips

RABBITS

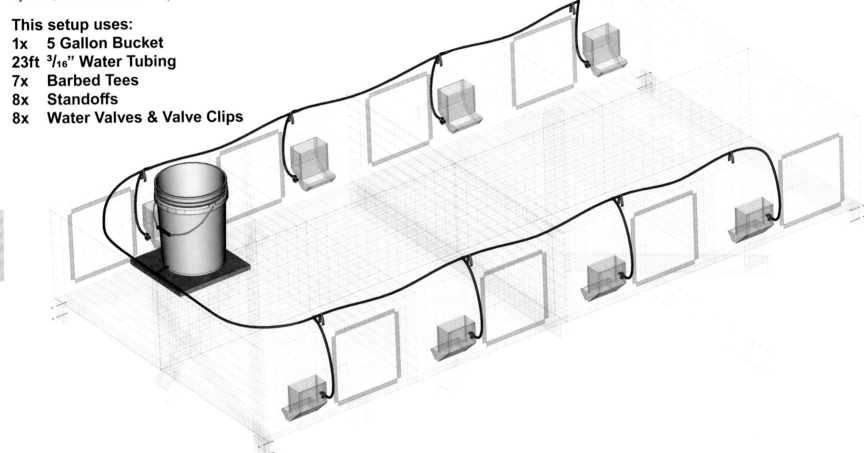

We use buckets as a water reservoir for two reasons: they are cheap and easy to monitor. When keeping animals, the first sign of problems is aberrant water consumption. If the bucket empties unusually fast, it probably means you have a leak somewhere. (If the watering system were hooked up to a continuous flow, you could have a huge swamp mess.) Water in the bedding is the fastest way

to bring in disease because it shuts down decomposition and offers a great pathogen growth medium. If the bucket empties unusually slowly, it could indicate sickness or a clogged system. The bucket offers a simple and cheap way to monitor all water usage.

In general, one 5 gallon bucket can service up to 30 pens. Obviously that will change based on how many does have litters,

time of year, size of rabbits, and type of feed.

As depicted above, we typically run one bucket per movable hutch (8 individual pens) so that everything remains self-contained and portable. Be sure to run the water tube right along the corner of the cage using the stand-offs, and offset it all the way down to the nipple to keep the rabbits from being able to scratch and gnaw it.

17

Since rabbits are nocturnal and like it cool, the sheet metal placed on top of the cages guarantee additional comfort. If a chicken happens to penetrate the protective netting boundary, it also keeps chicken poop from raining down on the rabbit cages. Chicken manure drops are much larger than rabbit manure pellets and therefore can clog up the small floor wire. (The floor wire must be small to keep baby bunnies from being able to fall through.)

Nest boxes can be metal or wood. We prefer metal ones because they are easier to clean and more durable. They can also take more weathering during storage. Our preferred nesting material is any kind of clean hay. It doesn't have to be especially high quality, but it should not be moldy. Put in plenty; the rabbits will pull out what they don't need.

Three foot chicken wire is used all the way around the top of the hutch. J-clip or tie wire bottom edge of the chicken wire to the top edge of cages and staple to upright risers.

If placing the hutch in a hoop house or similar shelter without shade, place roofing material in a manner to provide each pen with sufficient shade.

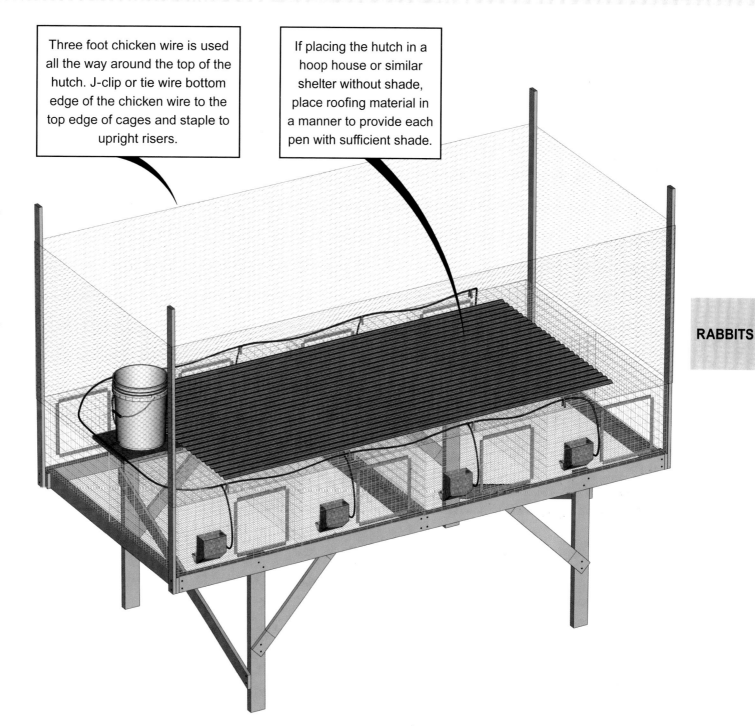

RABBITS

RABBITS

HAREPEN

photo courtesy of Chris Slattery

soil grew up over the wire so fast that the rabbits could dig out underneath the shelter--between its wall and the now buried wire.

The problem with putting wire like poultry netting on the bottom of the rabbit shelter was that the netting bent the grass over. Rabbits like to eat blades from the top; they don't like starting in the middle of the blade. Further complicating the situation, the rabbits would inadvertently bite the hard-to-see wires, which made them gun-shy about eating the grass. Biting on a wire didn't feel good.

We had several problems to solve:

1. Inability to dig out.
2. Complete mobility to new ground.
3. No bent over grass.
4. Break between soil and rabbit (for hygiene)

The solution was a floor made with ½" X 1" slats on 3" centers. This offered a nice wide 2" grazing access with no wire, so that the grass could flip back up in those areas in between. With this new design, the rabbits were able to graze roughly 70 percent of the grass in the shelter footprint. As an added bonus, the slats elevated the rabbits off the actual ground which kept them healthier. This, in conjunction with daily moves, helped

Perhaps no other design at Polyface has gone through more permutations than the pastured rabbit shelter. It was the first portable animal design my dad envisioned in the early 1960s. My older brother had rabbits and Dad wanted to figure out a way to move them along on pasture.

That initial trial morphed into our pastured poultry shelters. The problem with the big box

rabbit shelter was that we couldn't keep them from digging out. And if a rabbit gets out, it's not like a chicken. You can't just chase it down.

When my son Daniel took an interest in rabbits, we decided to try again. We placed poultry netting on the ground and let the grass grow up through it. Then we moved the shelter across it. That worked for a couple of years, but soon we had a buildup of coccidiosis and the

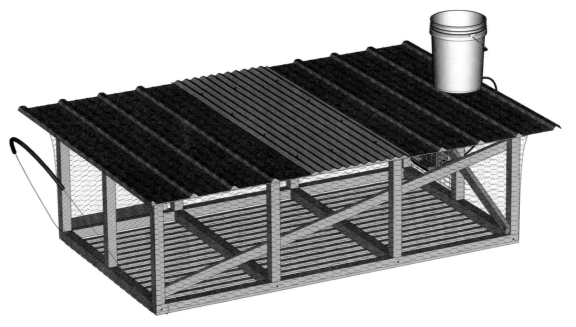

us reduce coccidiosis; perhaps the number one problem with rabbits – especially outdoor rabbits.

The only real negative is that you have to pick up the rabbits along with the weight of the shelter, but by making the shelters small and only putting one litter of bunnies in each one, it's all extremely light. We needed to keep each litter separate in order to record genetic performance and select new breeding stock anyway, so it didn't make sense to make large shelters that could handle multiple litters.

Of course the slats do rot over time, but not as fast as you might think. Since they're being moved every day, the constant sliding to each new spot prevents the slats from accumulating microbes that create rot. These small slats act like runners on a sleigh. They reduce friction with the ground making them surprisingly easy to move—even a small child can pull one of these.

Keeping them out of the grass in the off-season is also key to longevity. We stack the pens on top of each other to save room and store them on an old concrete pad behind our Raken building when not in use.

Because each of these is self-contained, they can be moved anywhere. This enables us to put the rabbits on brand new ground routinely,

which further reduces coccidiosis risk. We don't run them over the same ground more than twice a year. Ideally, you would rotate pasture locations from year to year in order to maintain sanitation. These shelters are ideal lawn mowers; rabbit droppings are small and fairly dry so they're more compatible up around the house area than, say, a cow.

To my knowledge, we've never had a serious predator attack on these rabbits, probably because rabbits are nocturnal so if something comes around, they run to the other side of the shelter. With a fleet of these, you can raise a lot of rabbits in a small area and save 30 percent of your feed bill. That's not pocket change.

After 30 years of genetic selection, Daniel's rabbits thrive on pasture. But even today, an occasional one will show signs of diarrhea. They

don't need lush pasture and definitely enjoy woody plants and weeds.

In general, outdoor field shelters should offer 3 sq. feet per rabbit. The slats cover 30 percent of the ground so only 70 percent of the forage is actually available. The rabbits will eat roughly 75 percent of their diet in forage, which is a huge feed savings and of course makes a rabbit with better texture and taste than commercial hot house rabbits.

We move the harepen every day. Unlike chickens, the rabbits are nocturnal so the best time to move them is in the evening, going into the active night. We wean at 6 weeks and process fryers at 12 weeks. That's slower growth than the commercial industry standard, but what's time to a rabbit? In our climate, they can be outside from the first of April until about Thanksgiving.

Tools

☐ Hand Saw

☐ Power Drill

☐ Driver Bits (for screws)

☐ Flush Cut Wire Cutters

☐ 8" Tin Snips

☐ Marking Pencil

☐ Hammer

☐ Tape Measure

☐ Speed Square

☐ Stapler

☐ ½" Staples

RABBITS

Supplies

We recommend purchasing rabbit supplies from the Klubertanz Equipment Company, Inc. and have listed their stock numbers in the materials list. Bass Equipment Company is another supplier we order from to purchase these supplies. (Their stock numbers will differ of course).

☐ **≈4ft**   ³/₁₆" Flex Water Tubing (A401)

☐ **1x**   Standard Water Valve, ³/₁₆" Barb, Brass Cap (A102)

☐ **1x**   Valve Clip (A108)

☐ **≈25ft** 24" Chicken Wire

☐ **7x**   Fencing Staples

☐ **1x**   1" x 2" 14 Gauge Wire (24" x 14")

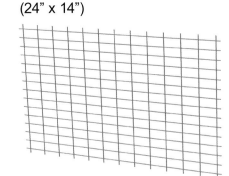

☐ **≈12ft** 12-½ Gauge Medium Tensile Wire

☐ **≈3ft** Scrap Garden Hose (or plastic pipe)

❑ **1x**   5 Gallon Bucket

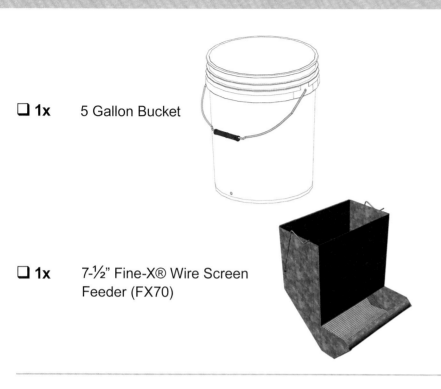

❑ **1x**   7-½" Fine-X® Wire Screen
Feeder (FX70)

DECK SCREWS

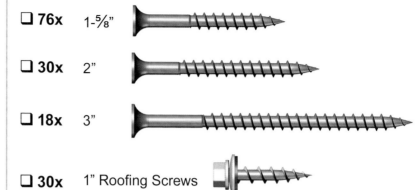

❑ **76x**   1-⅝"

❑ **30x**   2"

❑ **18x**   3"

❑ **30x**   1" Roofing Screws

Use 1" Roofing screws wherever possible. There will be areas where longer screws are needed to properly fasten, but it is good practice to minimize the number of screws that penetrate through the wood and create hazards for livestock and handlers.

Metal Cutlist

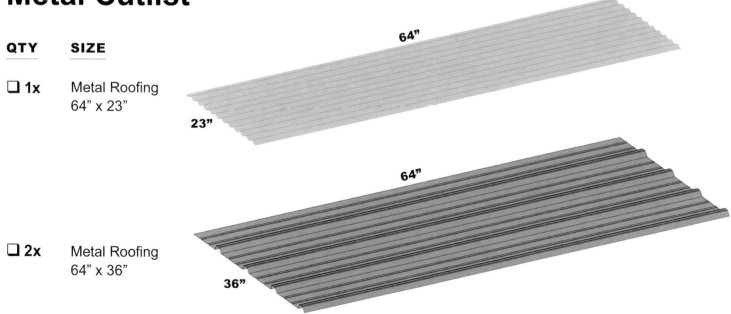

QTY	SIZE
❑ **1x**	Metal Roofing 64" x 23"
❑ **2x**	Metal Roofing 64" x 36"

64"

23"

64"

36"

Wood Cutlist

Wood scraps are highlighted in red.
Do not discard until project is completed.

☐ **7x** 2x2x8 Feet Long

24"	24"	
24"	24"	24"	24"
24"	24"	24"	24"
84"	
84"	
44"	44"	
44"	44"	

☐ **3x** 1x2x8 Feet Long

51"	
51"	
85-½"	

RABBITS

☐ **6x** 1x3x8 Feet Long

Polyface uses nominal dimension lumber for this project. Meaning, a 2x4 is actually 2" x 4" as opposed to a dimensional 2x4 which is 1-$\frac{1}{2}$" x 3-$\frac{1}{2}$".

48"	47"
48"	47"
50"	21"
50"	21"
84"	
84"	

☐ **16x** ½x1x8 Feet Long

84"

We recommend making the $\frac{1}{2}$ x 1" floor slats from white oak for durability and rot resistance.

Build The Side Walls

1

1x   1x3x84"

Measure and mark the three lines listed.

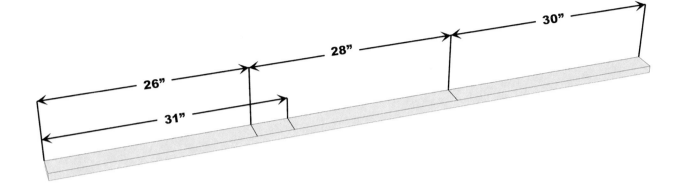

2

1x   2x2x84"

Measure and mark the two lines listed.

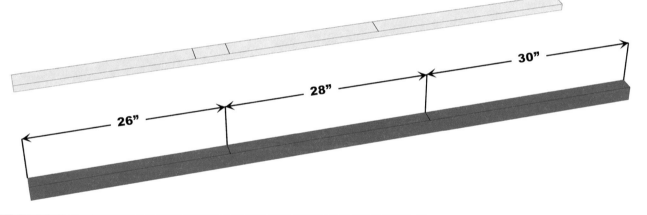

3

4x   3" Deck Screws
4x   2x2x24"

Position the two middle uprights to the right of the marks. Fasten them in place with screws.

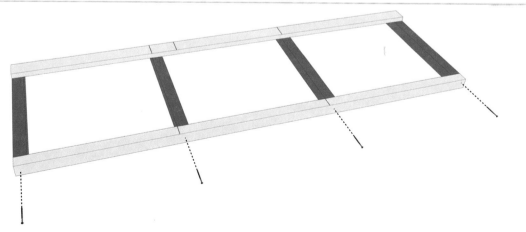

4

4x   2" Deck Screws

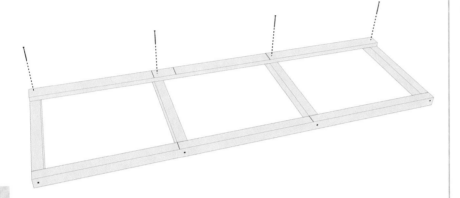

Fasten the top piece to the uprights.

5

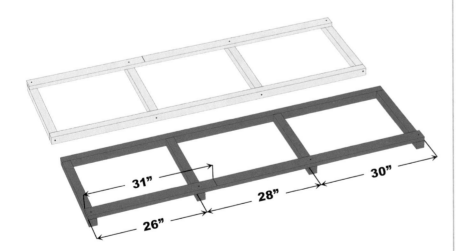

Repeat steps 1-4 to create a mirrored side as shown.

Build The End Walls

6

1x   1x3x50"

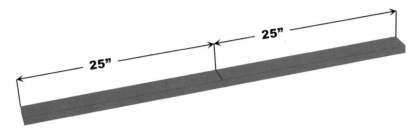

Measure and mark the center.

7

1x   2x2x44"

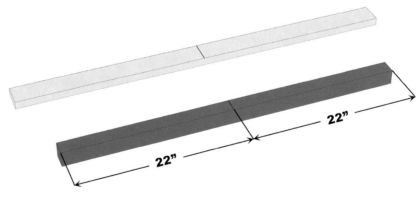

Measure and mark the center.

8

1x 2x2x24"
1x 3" Deck Screw

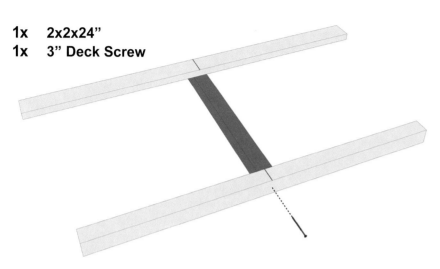

Center the upright on the marks and fasten in place.

10

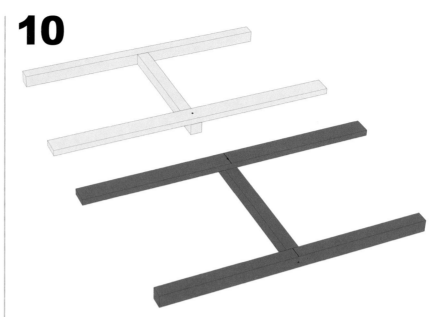

Repeat steps 6-9 to create a mirrored side as shown.

9

1x 2" Deck Screw

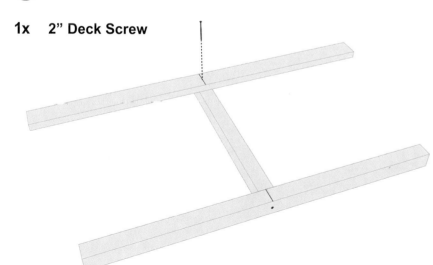

Center and fasten the top to the upright.

11

4x 2" Deck Screws

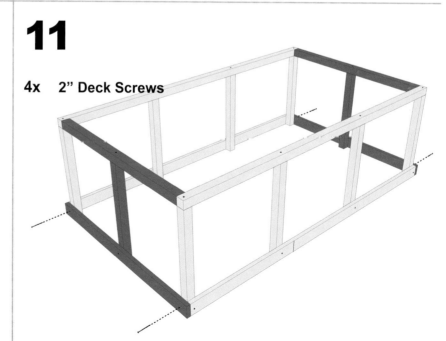

Assemble the four sections together as shown.

RABBITS

12

4x   3" Deck Screws

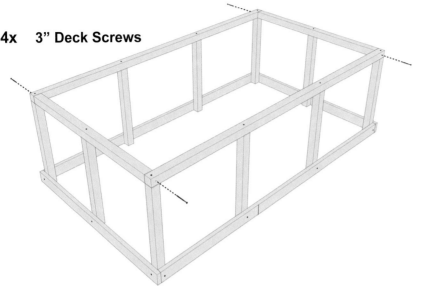

14

4x   2" Deck Screws
1x   1x2x85-$^{1}/_{2}$"

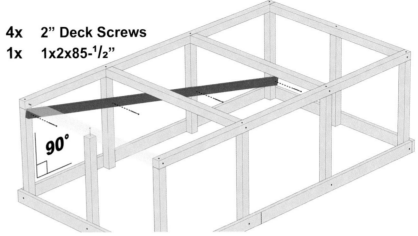

Square the long side and install diagonal brace. We typically only brace one of the 2 long sides to decrease weight. There isn't a lot of stress on these structures to merit additional bracing.

13

4x   3" Deck Screws
2x   2x2x44"

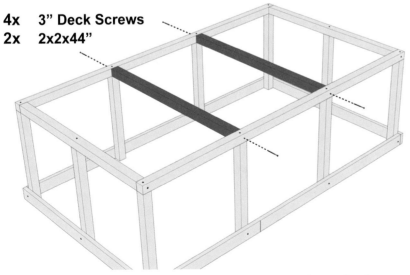

Align the 2x2 supports with the vertical uprights and fasten in place.

15

3x   2" Deck Screws
1x   1x2x51"

Square the short side and install the diagonal brace.

16

3x   2" Deck Screws
1x   1x2x51"

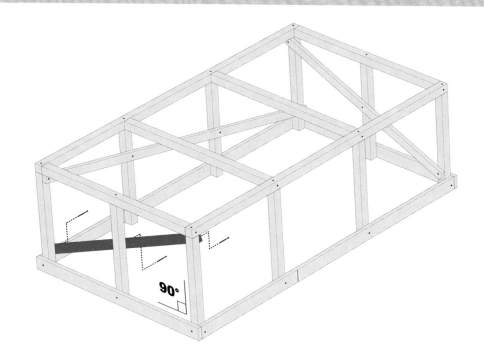

Install a diagonal brace on other short side, if desired.

17

64x   1-⅝" Deck Screws
16x   ½x1x84"

Cut 2" spacer blocks in order to create uniform spacing between the floor slats. Fasten the first slat in place.

Predrill holes to prevent splitting, and countersink screws to ensure that screws are flush on the bottom. (Any screw protruding out can get hung up while sliding the pen on the ground and make it harder to pull.)

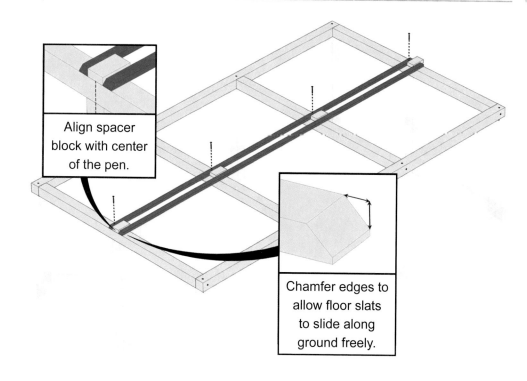

Align spacer block with center of the pen.

Chamfer edges to allow floor slats to slide along ground freely.

18

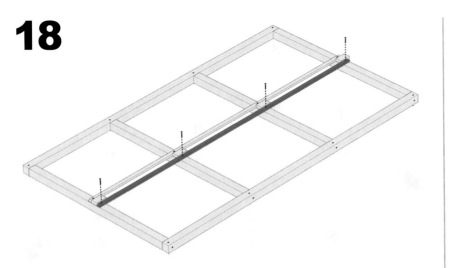

Fasten the other slat in place.

20

Continue moving spacer blocks and fastening the slats, working your way outwards.

19

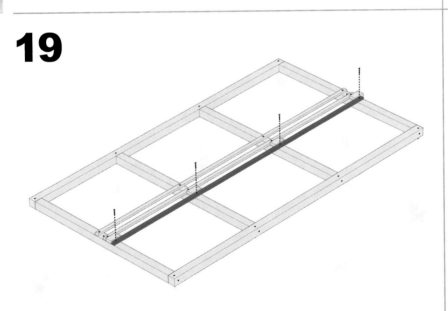

Move the spacer blocks and fasten the third slat.

21

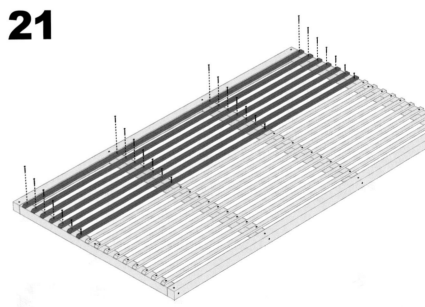

Fasten slats on the other side, working your way outwards.

22

2x   **2" Deck Screws**
1x   **1x3x48"**

180°

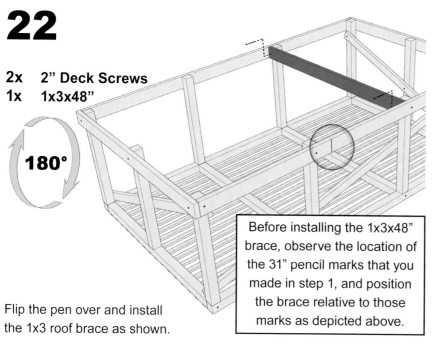

Flip the pen over and install the 1x3 roof brace as shown.

> Before installing the 1x3x48" brace, observe the location of the 31" pencil marks that you made in step 1, and position the brace relative to those marks as depicted above.

24

≈25ft of 24" Chicken Wire

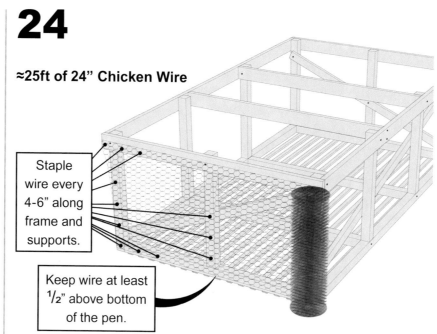

> Staple wire every 4-6" along frame and supports.

> Keep wire at least ¹/₂" above bottom of the pen.

RABBITS

23

4x   **2" Deck Screws**
1x   **1x3x48"**

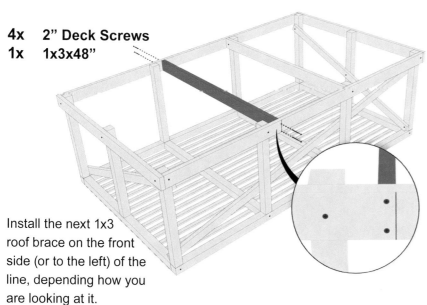

Install the next 1x3 roof brace on the front side (or to the left) of the line, depending how you are looking at it.

25

Fastening chicken wire is easier with two people: one person unrolls while applying tension and the second person staples.

Continue stapling all the way around the pen. Overlap wire at start/stop point by 1-2" and cut from roll using tin snips.

26

≈10-12ft   **12-½ Gauge Wire**
≈3ft   **Scrap Garden Hose
or Black Pipe**

Loop wire around frame and through the chicken wire at the front of the pen. Make a proper knot using the tying techniques illustrated here.

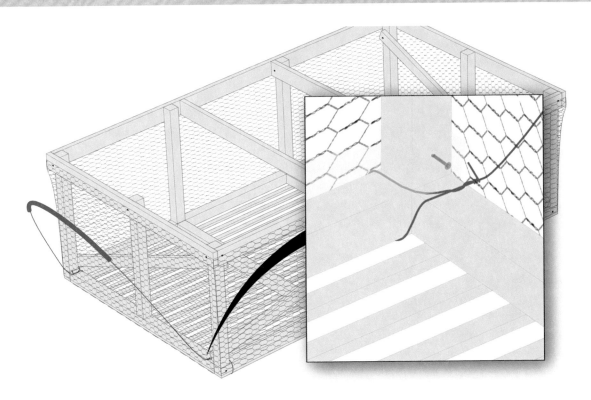

RABBITS

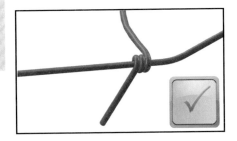

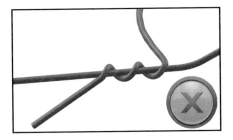

The length of the handle is ultimately determined by the stature of the individual moving the pen. Naturally a taller individual will need a longer handle. However, a good starting point would be 4-6" above the top of the pen when pulled straight upwards. By leaving plenty of extra wire on the tail, you will be able to lengthen it as needed. See the photo to the right for reference.

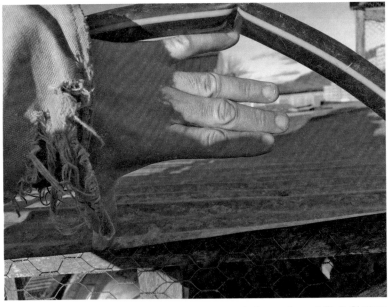

photo courtesy of Chris Slattery

27

10x   1" Roofing Screws
1x   36x64" Roof Panel

Orient the first roofing panel as shown. It should be flush with the 1x3 support installed in step 22 and overhang evenly everywhere else.

 Use 1" roofing screws wherever possible. There will be areas where longer screws are needed to properly fasten, but it is good practice to minimize screws that penetrate through the wood and create hazards.

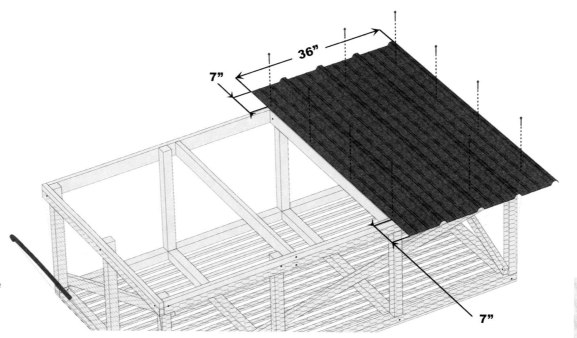

28

10x   1" Roofing Screws
1x   36x64" Roof Panel

Repeat the installation process for the second roofing panel. Panel should align flush with the 1x3 installed in step 23 and overhang evenly on the other sides.

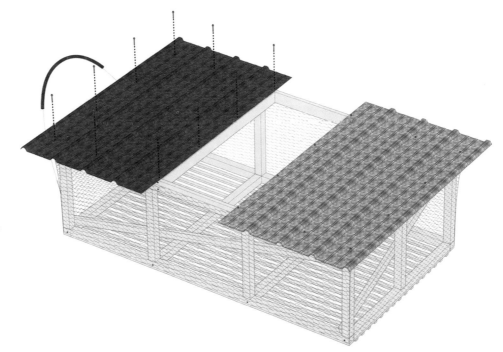

Assemble The Lid

29

2x 1x3x47"

31

4x 1-⁵⁄₈" Deck Screws

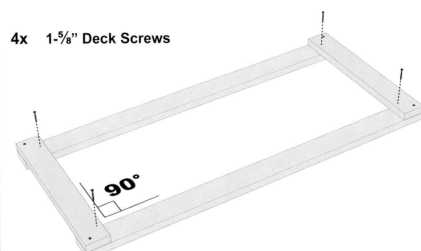

Square up the frame and add a second screw to each corner.

30

4x 1-⁵⁄₈" Deck Screws
2x 1x3x21"

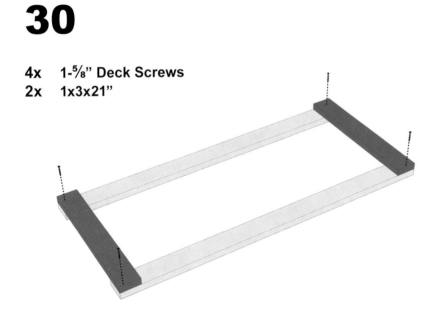

Place a single screw in each of the four corners.

32

4x 1-⁵⁄₈" Deck Screws
2x 1x3 Scraps

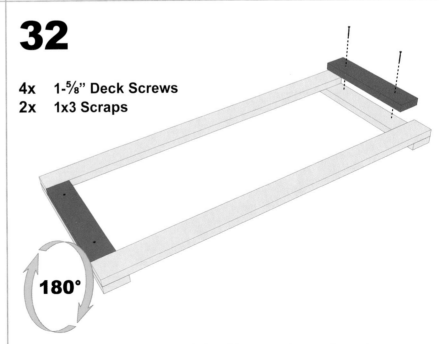

Turn over and fill the short ends in with scrap pieces of wood.

33

10x   1" Roofing Screws
1x   23x64" Roof Panel

Center the metal
roofing panel on the
wooden frame.

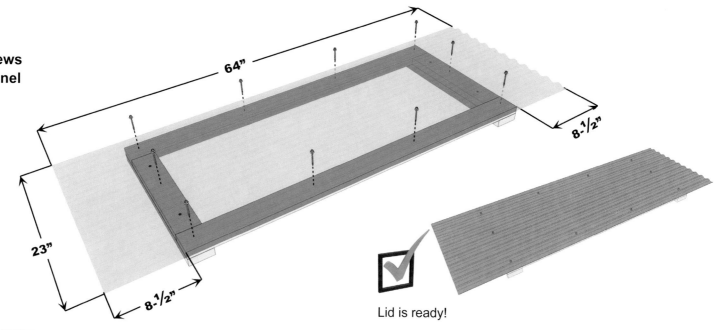

64"

8-1/2"

23"

8-1/2"

Lid is ready!

34

Install the lid in the middle of the pen
opening as shown. We do not use any
hinges or latches to hold the lid on,
and have never had any issues with it
coming off.

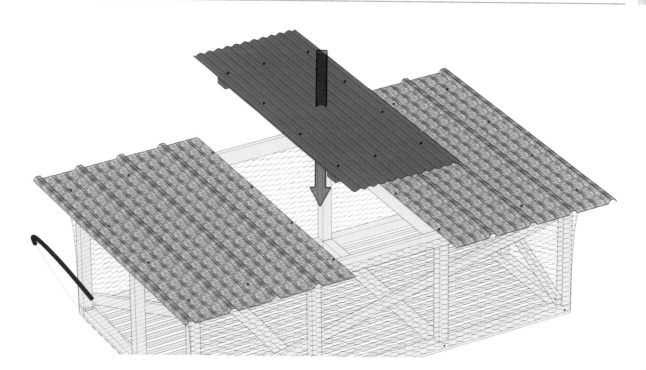

35

1x 24" x 14" Wire

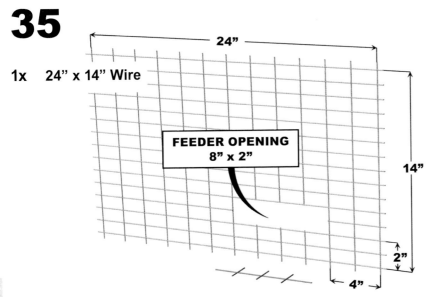

24"

14"

FEEDER OPENING
8" x 2"

2"

4"

RABBITS

Using a scrap piece of 1"x2" wire, cut an opening for a 7-$\frac{1}{2}$" Fine-X® feeder.

37

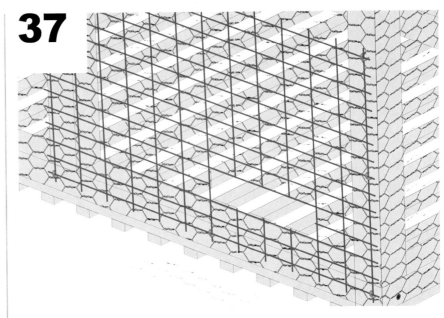

Cut chicken wire for the feeder opening.

36

7x Fencing Staples

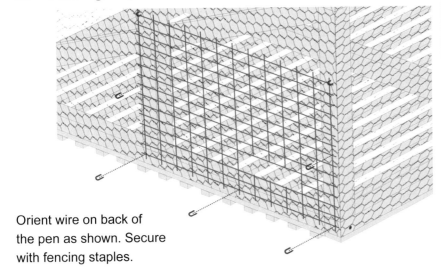

Orient wire on back of the pen as shown. Secure with fencing staples.

38

1x 7-$\frac{1}{2}$" Fine-X® Wire Screen Feeder

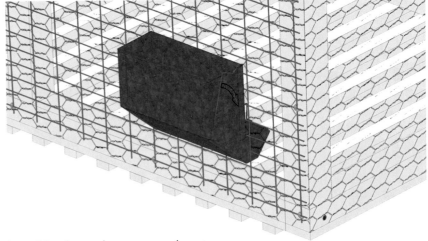

Insert feeder and secure as shown.

Install The Water System

39

1x 5 Gallon Bucket
≈4ft ³/₁₆" Water Tubing
1x Water Valve
1x Valve Clip

We use buckets for the same reasons as we do in the indoor rabbit pens and broiler pens. Monitoring water consumption is an excellent way to gage bunny health. We typically fill the water buckets half full and we mount them on the back side of the pen so that we do not have to lift that additional weight by the handle.

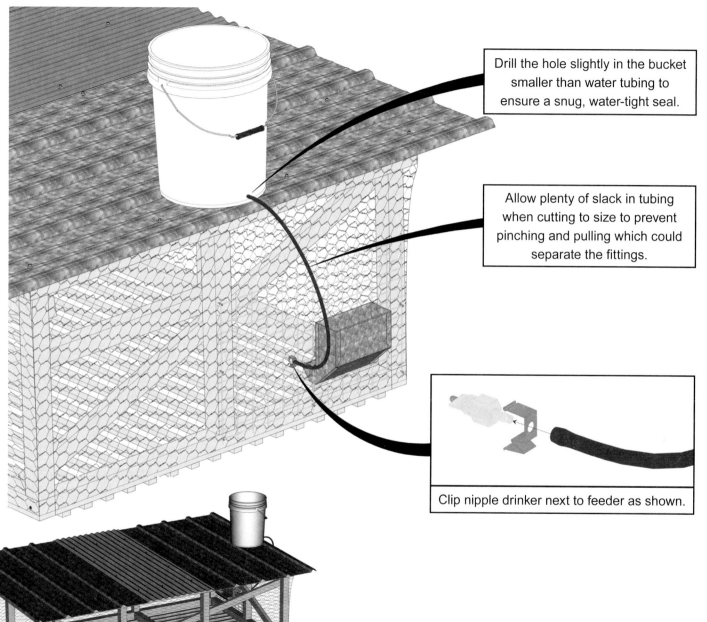

Drill the hole slightly in the bucket smaller than water tubing to ensure a snug, water-tight seal.

Allow plenty of slack in tubing when cutting to size to prevent pinching and pulling which could separate the fittings.

RABBITS

Clip nipple drinker next to feeder as shown.

The harepen is complete!

GOBBLEDYGO

TURKEYS

photo courtesy of Jean Shutt

When we started raising turkeys many years ago, we tried them in our regular broiler shelters, but it was problematic for a number of reasons.

1. The turkeys couldn't fully extend their heads when they were full grown.
2. When they got big, they broke up the light wooden braces.
3. They wanted more grass.
4. They liked to walk a lot more than chickens.

The next permutation was an extremely large hoop structure (see picture 1) surrounded by three sets of 164ft (50 meter) electrified poultry netting from Premier. This portable structure is 30ft wide and 50ft long, but it suffers all the regular problems of portable hoop structures. The ground-close bracing is awkward for both people and turkeys, and it would routinely hit turkeys and maim them. Not a good thing. Furthermore, the size made it cumbersome in the field. Our fields are rough, with lots of ditches and unevenness.

After a couple of years trying to work with the hoop design, we realized just how tough these turkeys are once they get past 6 weeks old. We'd see them perfectly happy standing out in rain and snow when they could easily go inside the protective cover. As a result of these observations, we designed and built this more modest and simple structure that we've now used for many years. Turkeys are much hardier than chickens. Although animal welfare certification probably would not accept this structure, it has been extremely adequate and the turkeys have thrived with it. The only critical shelter the turkeys seem to need is shade, and this provides

This old mobile hoop structure is no longer in use.

photo courtesy of Wendy Gray

Joel's Tip

We use 80% shade cloth because that seems to most closely approximate the variegated shade of a tree. If you look at shade under a tree, it's generally not opaque, but mottled, or kind of blotched because every particle of sunlight is not blocked by a leaf. That actually is more enjoyable for the animals. Less than 80% is not shady enough. The shade cloth is permeable but turkeys don't mind getting wet; they kind of like it on a hot day. The main benefit of the permeable cloth is to let the wind through. Even in extremely high winds (80 mph) the cloth catches no wind and simply flutters a bit. Grommets strategically placed down the edges and corners offer strong fastening points. You can also purchase grommet kits so you can install them yourself. Any major horticultural supplier has shade cloth. Most will custom cut it for you, but it's always cheaper to find a dimension they sell standard. The more you buy, the cheaper it is. We always fold them up and store them in a shed over the winter to reduce ultraviolet breakdown and weathering.

TURKEYS

better shade than a water-proof structure because the shade is mottled, resembling a tree. The other production protocol we wanted was plenty of access to pasture. Turkeys consume far more green material than chickens.

Like the Millennium Feathernet, the Gobbledygo is only half of the equation. The shelter works in conjunction with the electrified netting. We use Premier brand nets. Turkeys are the one animal Premier won't stand behind on these nets, but we've found that the secret is the right window of exposure.

The inconvenient truth is that the turkeys never understand the electric spark. Every other animal develops an aversion to it, but turkeys don't seem to. They're perfectly happy to stick their heads through the netting to get a grasshopper on the other side. You can see them jerking from the shock, but they don't seem to mind. It's unbelievable.

But turkeys absolutely do love far more ground and eat three times as much grass as a chicken. Giving them lots more room cuts down on feed costs and makes a far better eating experience. If you put the turkeys in the netting when they're too small, they walk through the holes and get out. So our current protocol is to raise the turkeys in a brooder for 3 weeks. Then we put them in the regular broiler shelters for the next 3 weeks.

At roughly 6 weeks, we put them in the electrified netting. At that point, they're too big to go through the holes but not big enough to bowl it over. They learn to stay in not because of the spark, but because they don't think they can walk over it. As they get bigger, of course, they could easily walk over the flimsy netting, but by that time they're well acculturated to the limits of their world.

The netting, then, is the mobile control. Now what about shelter? That's where the Gobbledygo comes in. It's really nothing more than a V-truss on an axle. Unlike chickens, turkeys don't mind getting wet. Their fragile time is in the first couple of weeks. Once they get up to 6 weeks they become almost bullet proof. We've

all 3 photos courtesy of Jason Pope

TURKEYS

had them take 10 inches of snow and hurricanes; they're almost indestructible unless they get extremely cold.

They don't need a solid roof; a nursery shade cloth works just fine. Like the cattle Shademobile, the purlins fit into notches on the main stringers, making the whole contraption easy to disassemble and tow up the public road. Ditto for the perches.

The Gobbledygo works fine for up to 400 turkeys, which is our favorite batch size. Using three nets and moving every other day, this offers a quarter acre every other day. At more than 400 in a flock, the birds escalate bullying. The occasional bullied bird in the 400-bird group can be pulled out and separated for protection. Most of them perk up and do fine, although they will always be a bit smaller than the rest of their mates. Most of the turkeys happily sleep on the ground. Some prefer perches; we offer enough perches to accommodate about half the flock. Depending on size, about 12 inches per bird is plenty of perch area.

The one significant design feature that's easy to miss is the bow-truss under the long tongue. Because the 3 inch pipe tongue is long and has the weight of the superstructure on it, the tongue is real bouncy. By putting a bow truss underneath it and actually bending it slightly upward, that long pipe stays rigid.

Although this design was developed for turkeys, this shade structure can be used for all animals: cows, sheep, pigs, goats. It's extremely versatile and lightweight. A couple of people can move it by hand. Wind is never an issue because the shade cloth is permeable. Obviously with other animals you don't need the perches.

When we're using it in the field, we usually hitch it to the rear end of a bulk feed cart, forming a two-vehicle train. Our favorite axle for the money is a mobile home axle. Many farm auctions have a couple axles for sale.

Now for a couple of production items. Turkey poults are far more fragile than chicks. They need a higher temperature, higher protein,

and more ambient temperature consistency. We've tried brooding them in the same facility as chicks, and even brooding them with chicks, but have gradually abandoned that and gone to dedicated brooders that we clean thoroughly before the first batch of the season.

Finally, most poultry books highly discourage running chickens and turkeys in proximity. The reason is the dreaded black head and most poultry experts will have you quacking in your shoes to ever let a turkey and chicken be near each other. We've been raising both for decades and have never had black head. We run them in the same pastures and adjacent to each other. We've never had an issue. I think the studies are all done in confinement situations, which are quite different than well managed mobile pasture situations.

Few things are as beautiful as a flock of 400 full-grown pastured turkeys on green grass in late summer lounging around the Gobbledygo. Get out your camera, folks.

Tools

- ❑ Power Drill

- ❑ 6 and/or 8ft Ladder

- ❑ Welder

- ❑ Circular Saw

This project requires a tractor or hydraulic loader. We tailored the assembly process to minimize the tractor time needed. If you do not own a tractor, borrowing one is a plausible option because it won't take long to set the prebuilt wooden structure on top of the wagon chassis.

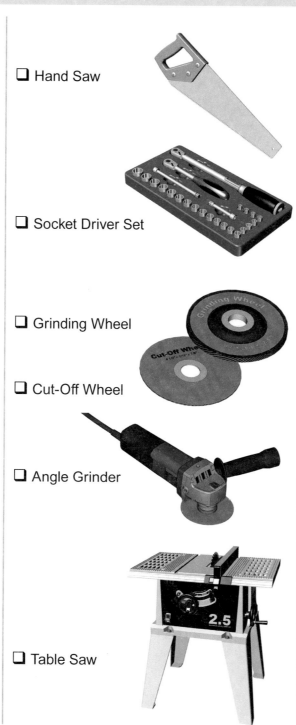

- ❑ Hand Saw

- ❑ Socket Driver Set

- ❑ Grinding Wheel

- ❑ Cut-Off Wheel

- ❑ Angle Grinder

- ❑ Table Saw

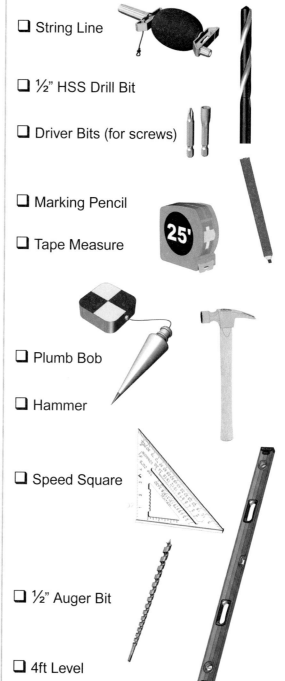

- ❑ String Line

- ❑ ½" HSS Drill Bit

- ❑ Driver Bits (for screws)

- ❑ Marking Pencil

- ❑ Tape Measure

- ❑ Plumb Bob

- ❑ Hammer

- ❑ Speed Square

- ❑ ½" Auger Bit

- ❑ 4ft Level

TURKEYS

Hardware

DECK SCREWS

☐ **50x**   2"

☐ **20x**   3"

NAILS

☐ **20x**   20D Galvanized Nails

☐ **2x**   8D Framing Nails

BOLTS

☐ **16x**   ½-13 x 3"
Carriage Bolts

☐ **12x**   ½-13 x 5" Carriage Bolts

☐ **4x**   ½-13 x 7" Carriage Bolts

TURKEYS

☐ **32x**   ½"-13 Hex Nuts

☐ **32x**   ½" Washers

Supplies

☐ **1x**   Shade Cloth 16'x32',
80% Shade

☐ **1x**   Trailer Jack (see note A)

Notes:

A.)   The trailer jack we prefer is one with a tube mount configuration. The small steel tube can be easily welded to a chassis and has proven to be a very durable design. In 2020, this jack could be found at Tractor Supply stores for less than $60. (Part # 11630699)

Wood Cutlist

Wood scraps are highlighted in red.
Do not discard until project is completed.

QTY	SIZE
❑ **2x**	2x6x8ft
❑ **1x**	1x6x8ft
❑ **6x**	1x6x12ft
❑ **6x**	2x8x12ft
❑ **4x**	2x10x16ft
❑ **9x**	2x4x16ft
❑ **14x**	2x4x12ft
❑ **4x**	2x4x10ft
❑ **1x**	2x4x8ft

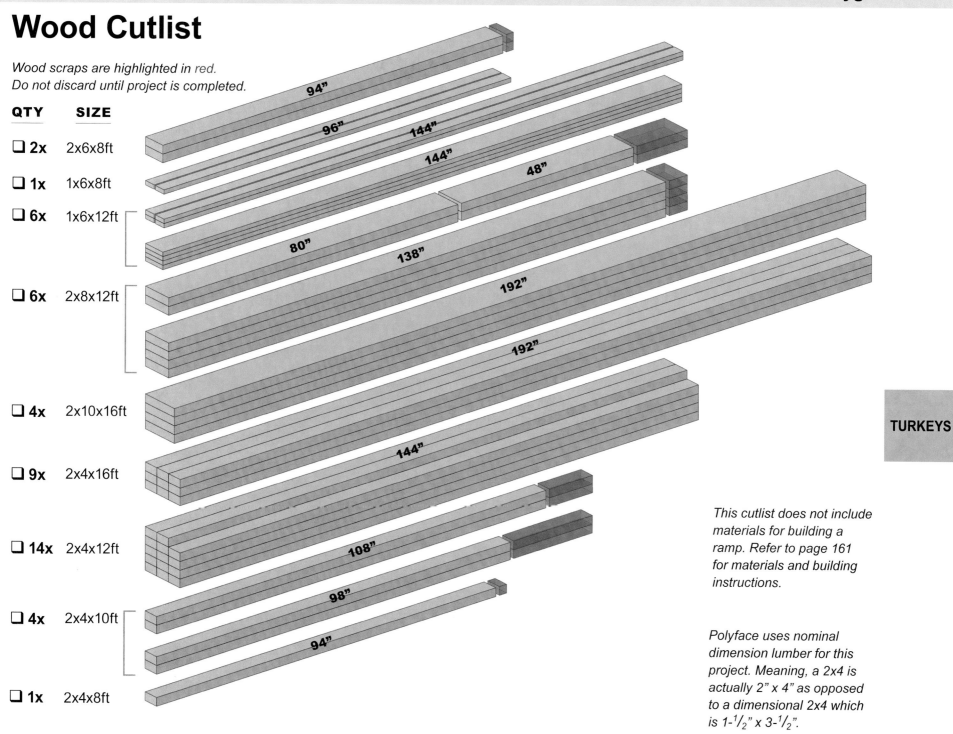

TURKEYS

This cutlist does not include materials for building a ramp. Refer to page 161 for materials and building instructions.

Polyface uses nominal dimension lumber for this project. Meaning, a 2x4 is actually 2" x 4" as opposed to a dimensional 2x4 which is 1-1/2" x 3-1/2".

Metal Cutlist

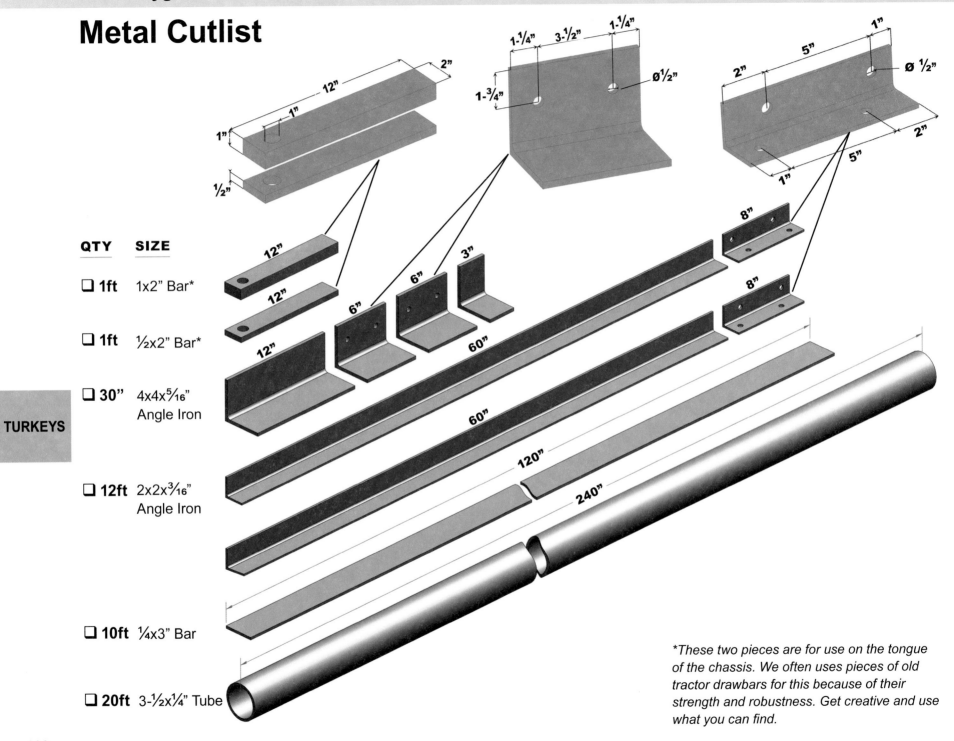

TURKEYS

QTY	SIZE
☐ 1ft	1x2" Bar*
☐ 1ft	½x2" Bar*
☐ 30"	4x4x⁵⁄₁₆" Angle Iron
☐ 12ft	2x2x³⁄₁₆" Angle Iron
☐ 10ft	¼x3" Bar
☐ 20ft	3-½x¼" Tube

These two pieces are for use on the tongue of the chassis. We often uses pieces of old tractor drawbars for this because of their strength and robustness. Get creative and use what you can find.

1

The frame structure is sized to fit on a 106" wheelbase mobile home axle. If the axle you have is narrower, then you will need to adjust your frame width dimensions accordingly.

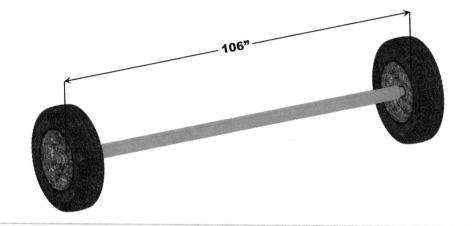

2

1x   3-¹/₂ x ¹/₄" Tube, 20ft Long

Cut a notch in the pipe to fit against the axle. We did not provide a template pattern because the radius of the cut will vary based on diameter of axle used.

TURKEYS

3

Center the pipe on the axle and square it.

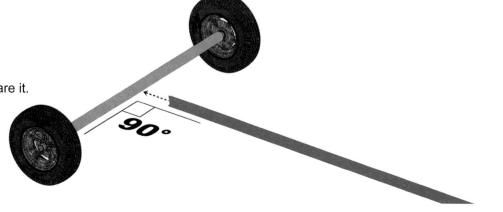

4

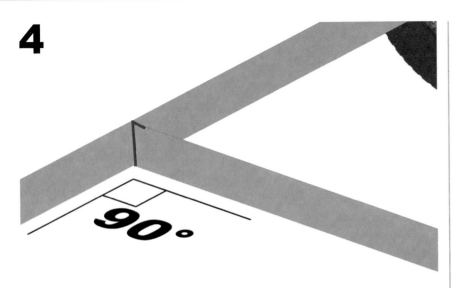

Weld the joint all around.

6

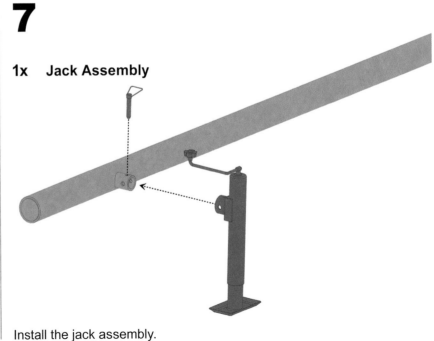

Weld the joint all around (shown in red).

5

1x   Weld-on Mount For Trailer Jack

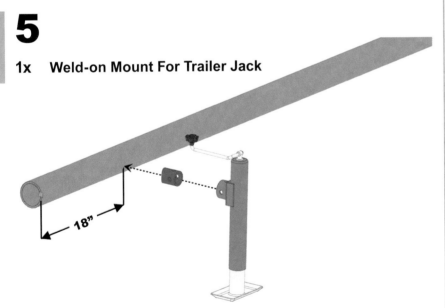

Measure approximately 18" back from the front of the pipe and weld the jack mount bracket in place.

7

1x   Jack Assembly

Install the jack assembly.

8

1x   1x2x12" Bar

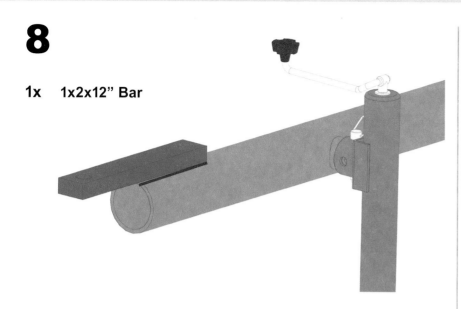

Orient top of hitch so that the hole protrudes about 6" from the end of the pipe. Weld joints on both sides.

9

1x   ½x2x12 Bar

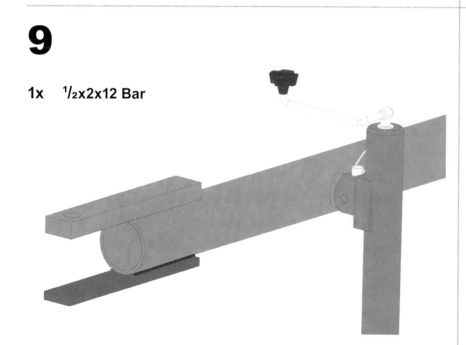

Align the bottom hole with the top hole and weld the joints on both sides.

10

2x   4x4x⁵⁄₁₆" Angle Iron, 6" Long

3-½" min

3-½" min

90"

Maintain at least 3-½" minimum between the inside of the tire and the bracket. The center dimension is 90", but may vary based on the track width of the axle used. Remember, if your center dimension is NOT 90" then you will need to adjust your frame width dimensions accordingly.

11

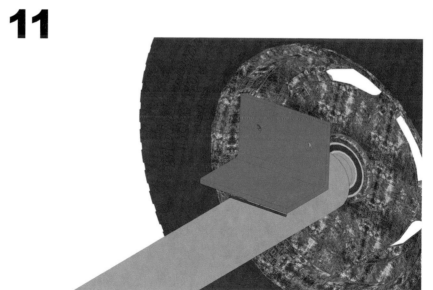

TURKEYS

Center and level the bracket on the axle as best as you can. Weld joints on all sides. You may need to remove tires if they are in the way of welding.

12

1x   2x2x³/₁₆" Angle Iron, 60" Long

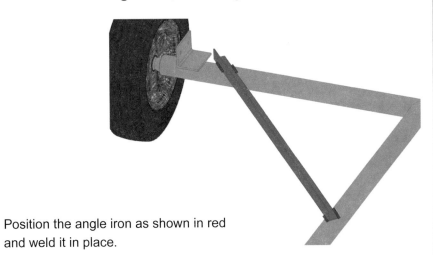

Position the angle iron as shown in red and weld it in place.

TURKEYS

13

1x   2x2x³/₁₆" Angle Iron, 60" Long

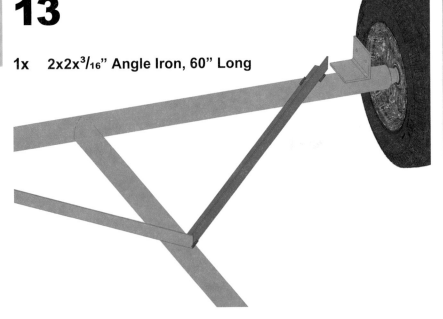

Repeat process on the other side.

14

1x   4x4x⁵/₁₆" Angle Iron, 12" Long

a)

Measure and mark the middle of the angle bracket.

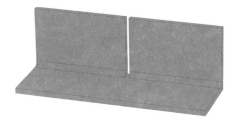

b)

Cut only through the top portion as shown.

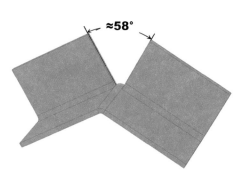

c)

Bend at cut to approximately 58°. You may need to bring this to a metal shop if you do not have the means to bend this part.

d)

Drill four Ø½" holes approximately where shown.

15

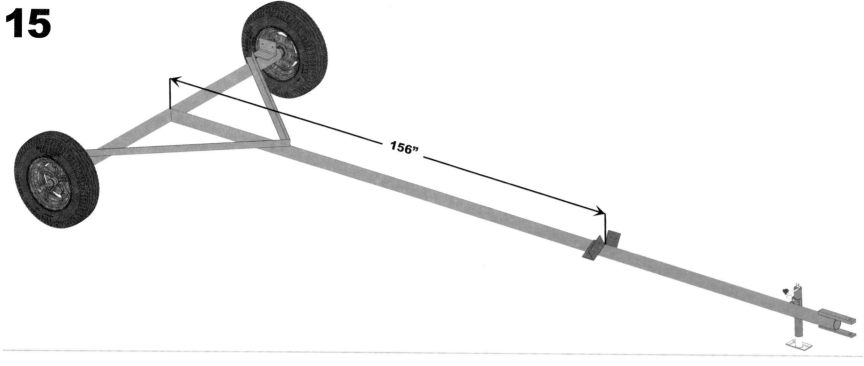

156"

16

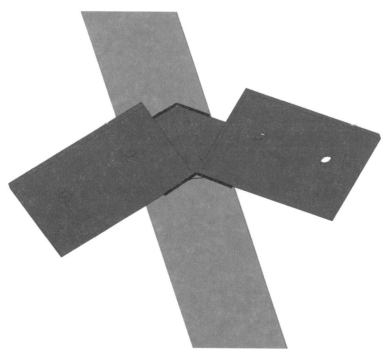

Level the bracket as best as you can on the pipe and weld the joints on all sides.

17

1x   ¼x3x120" Bar
1x   4x4x⁵/₁₆" Angle Iron, 3" Long

Joel's Tip

In this step we are fabricating a make-shift bowstring truss (or bow-struss) for short. Picture an archers bow. The string applies tension on the bow. It's sort of the same principal here. The ¹/4" x 3 flat stock steel welded to the pipe with some applied tension, results in a stiffening of the entire structure. I cannot speak to the physics involved, but I know from experience that it works!

The key here is to weld the flat stock onto the pipe first, with a bow in it that is slightly smaller than the wedge (angle iron) you will be placing in between. After the ends are welded, you pry the gap open wide enough to slip the wedge in place. This "preloads" the structure and removes excess flex from the long steel pipe.

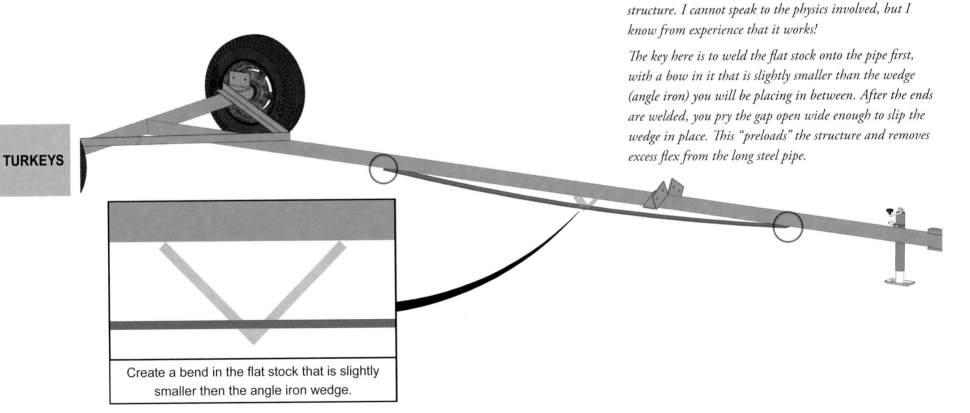

TURKEYS

Create a bend in the flat stock that is slightly smaller then the angle iron wedge.

With the help of another person and/or some clamps, position the flat stock approximately where shown and weld both ends.

18

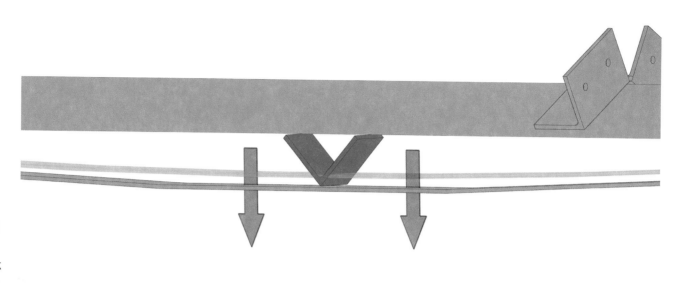

Using a crowbar, digging bar, or some other form of leverage, pry the gap open wide enough to slide the wedge in place. If the wedge is loose, or if you cannot insert the wedge, you may need to cut a weld and adjust your bend accordingly. With the wedge firmly in place, tack weld it to the pipe to prevent it from moving.

19

TURKEYS

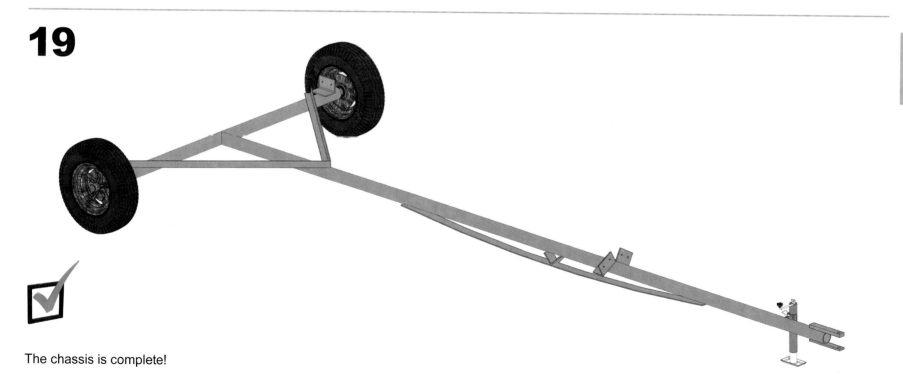

The chassis is complete!

20

1x   2x10x192"

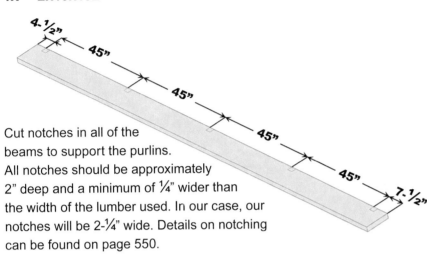

4-½"
45"
45"
45"
45"
7-½"

Cut notches in all of the
beams to support the purlins.
All notches should be approximately
2" deep and a minimum of ¼" wider than
the width of the lumber used. In our case, our
notches will be 2-¼" wide. Details on notching
can be found on page 550.

22

1x   2x10x192"

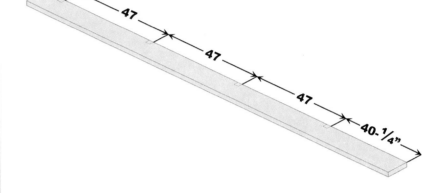

10-¾"
47
47
47
40-¼"

Keep track of the boards as you
notch them. It <u>does</u> matter what
goes where in order to ensure
that everything lines up in the
end! It may help to label your
boards after each step number.

21

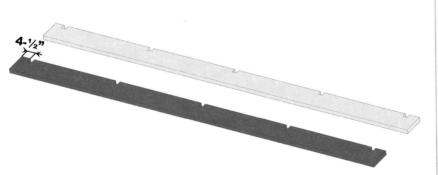

4-½"

Repeat step 20 to create a duplicate.

23

10-¾"

Repeat step 22 to create a duplicate.

24

1x   2x8x138"

65°

Miter one end of the 2x8.

25

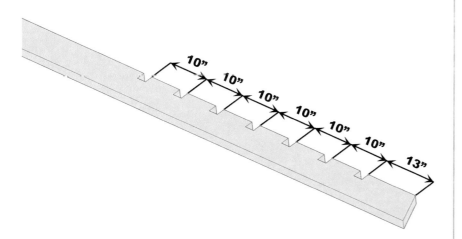

10"
10"
10"
10"
10"
10"
10"
13"

Space 7 notches 10" apart, starting 13" from the mitered end of the board.

26

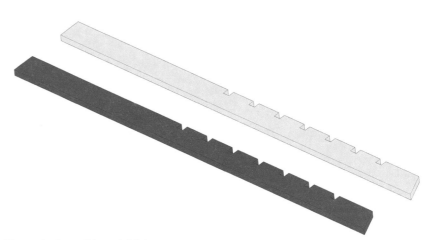

Repeat steps 24 and 25 to create notches on the next beam.

27

TURKEYS

1x   2x8x80"

Place scrap wood blocks on the ground to elevate the boards for the next step.

28

2x10x192 from step 22

2x10x192 from step 20

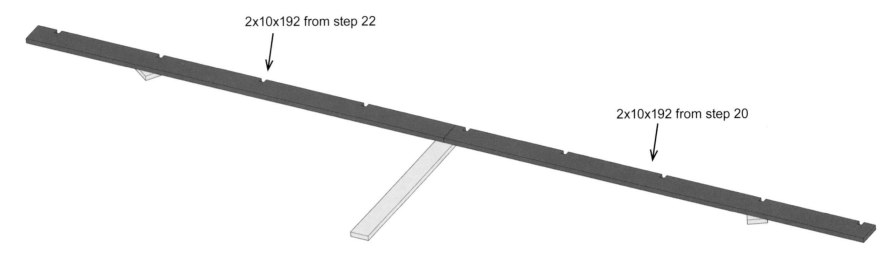

29

2x   20D Galvanized Nails

Place one nail in the corner of each of the two boards to tack them in place and prevent members from moving during orientation.

90°

90°

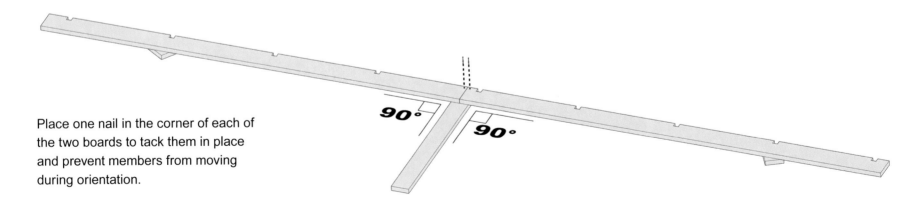

30

2x   8D Framing Nails
1x   String Line

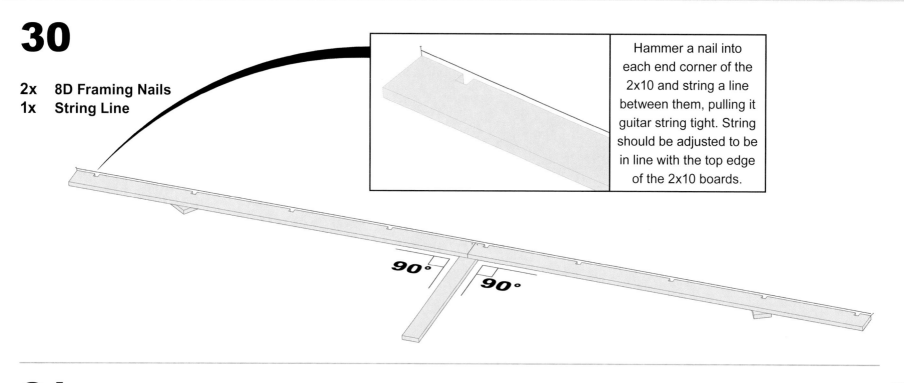

Hammer a nail into each end corner of the 2x10 and string a line between them, pulling it guitar string tight. String should be adjusted to be in line with the top edge of the 2x10 boards.

90°   90°

31

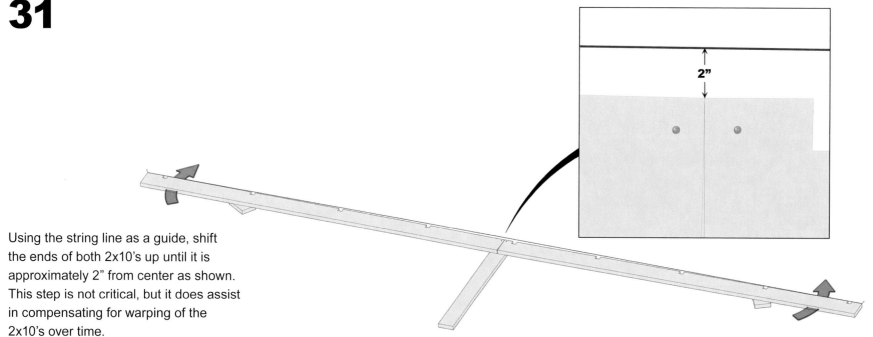

2"

Using the string line as a guide, shift the ends of both 2x10's up until it is approximately 2" from center as shown. This step is not critical, but it does assist in compensating for warping of the 2x10's over time.

32

201-½"

201-½"

Measure corner to corner as shown to ensure that everything is symmetrical. Our measurements are ≈201-½", but that is not important. What matters is that whatever your measurement is, it is equal on both sides. Adjust the 2x8x80 upright as needed.

33

2x   2x8x138 From Step 25

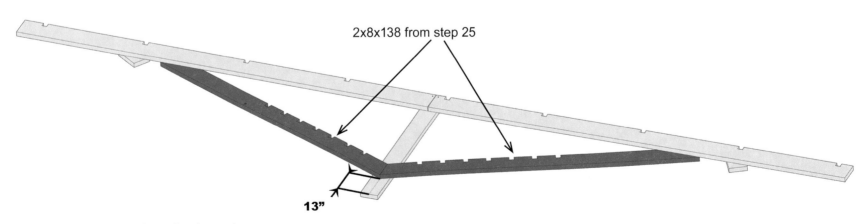

2x8x138 from step 25

13"

Place the braces approximately where shown.

34

4x 20D Galvanized Nails

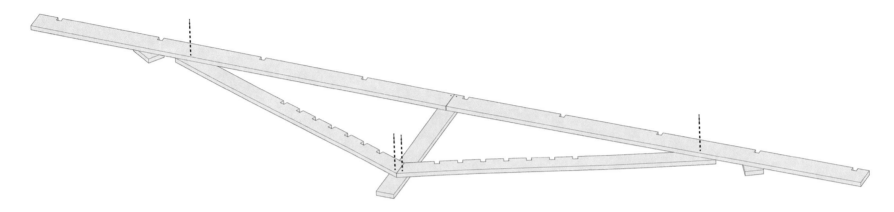

Tack the braces in place with nails.

35

TURKEYS

4x 20D Galvanized Nails
1x 2x8x48

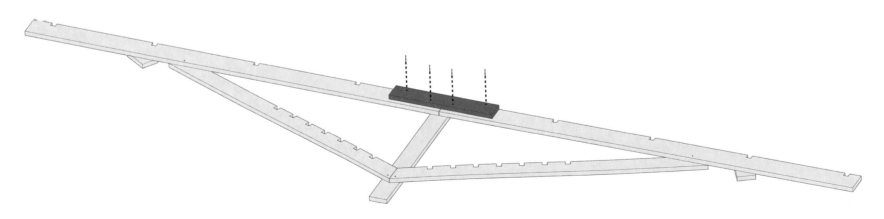

Tack the reinforcing scab in place.

36

6x ½-13 x 5" Carriage Bolts
6x ½" Washers
6x ½-13 Hex Nuts

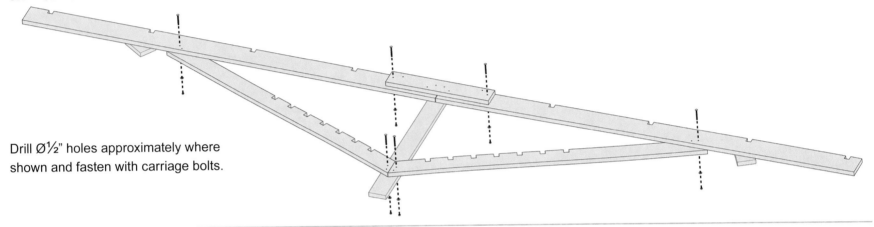

Drill Ø½" holes approximately where shown and fasten with carriage bolts.

37

TURKEYS

2x ½-13 x 7" Carriage Bolts
2x ½" Washers
2x ½-13 Hex Nuts

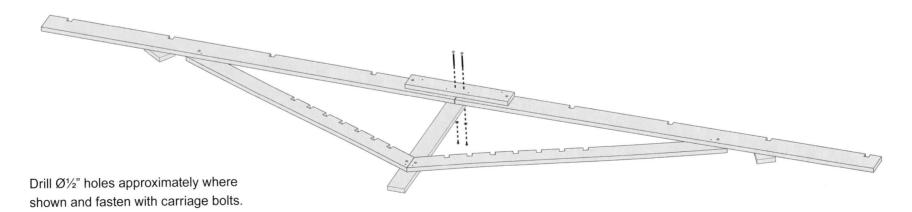

Drill Ø½" holes approximately where shown and fasten with carriage bolts.

38

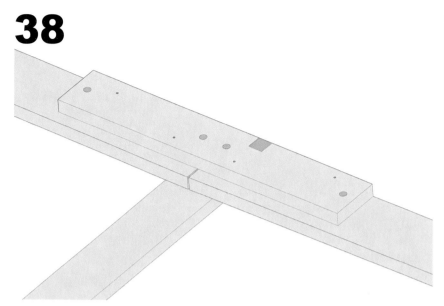

Cut a notch into the scab piece that corresponds with the notched 2x10 below it.

39

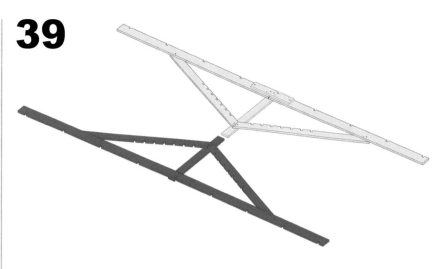

Repeat steps 24-38 to create the opposite hand version. The orientation of the notches are crucial to ensure that the purlins will line up during final assembly.

40

! These sides are heavy and you will need some help turning them upright and sliding into position. We recommend assembling these as close together as possible like the orientation shown in step 39.

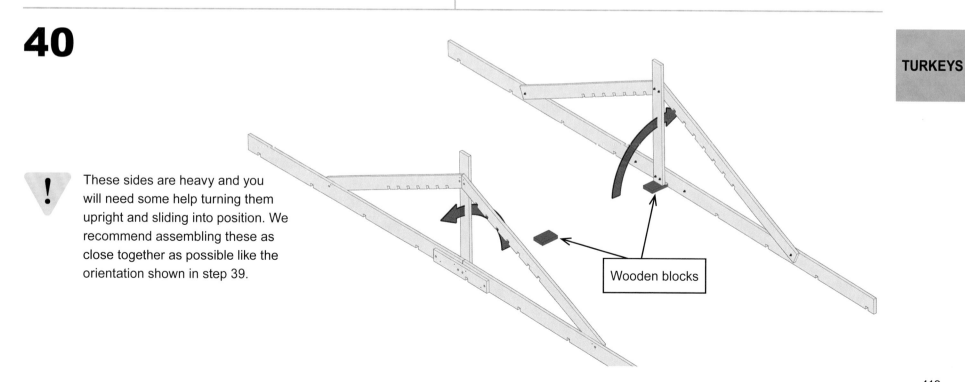

Wooden blocks

41

Move the sections to an appropriate distance from each other.

43

1x   2x6x94
4x   3" Decks Screws

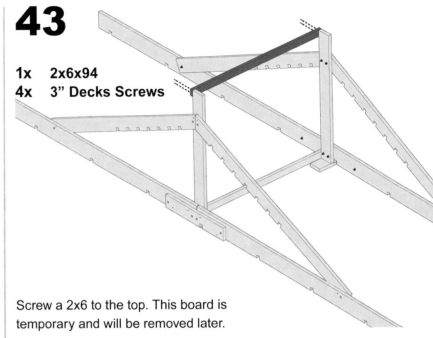

Screw a 2x6 to the top. This board is temporary and will be removed later.

42

1x   2x4x94
4x   3" Decks Screws

Screw the first 2x4 cross-member in place.

44

1x   2x4x108
4x   3" Decks Screws

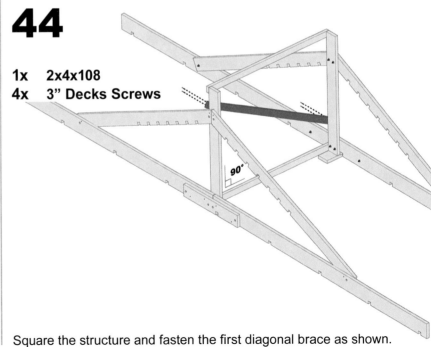

Square the structure and fasten the first diagonal brace as shown.

45

1x   2x4x108
4x   3" Decks Screws

Repeat process to secure the second
diagonal brace.

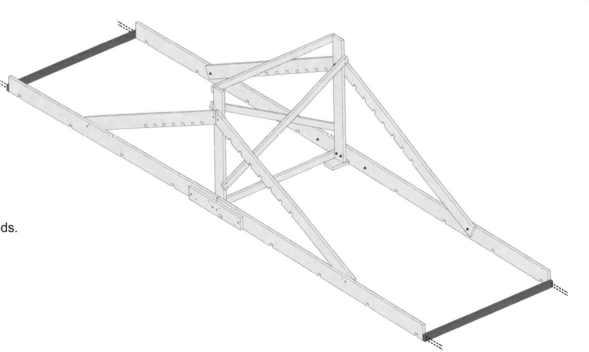

46

2x   2x4x98
8x   3" Decks Screws

Fasten the cross braces to both ends.

47

2x 1x6x144
12x 2" Decks Screws

Square the structure and install additional bracing as shown.

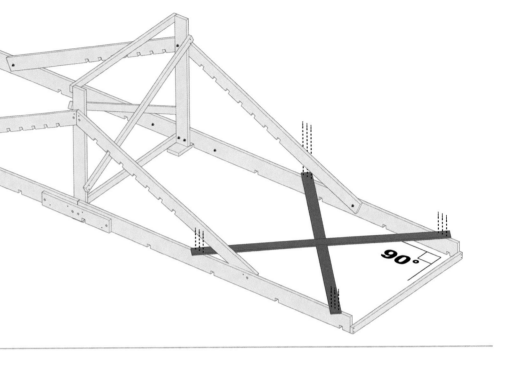

48

2x 1x6x144
12x 2" Decks Screws

Square and install the additional bracing
on the other side.

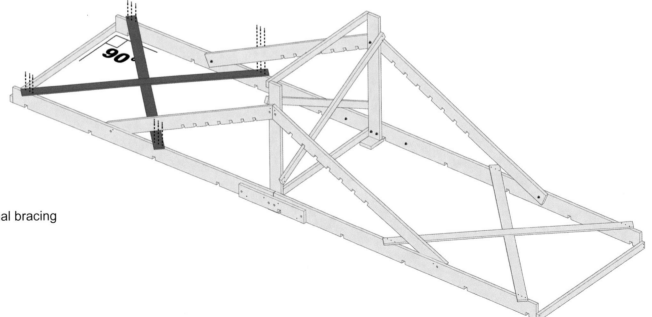

49

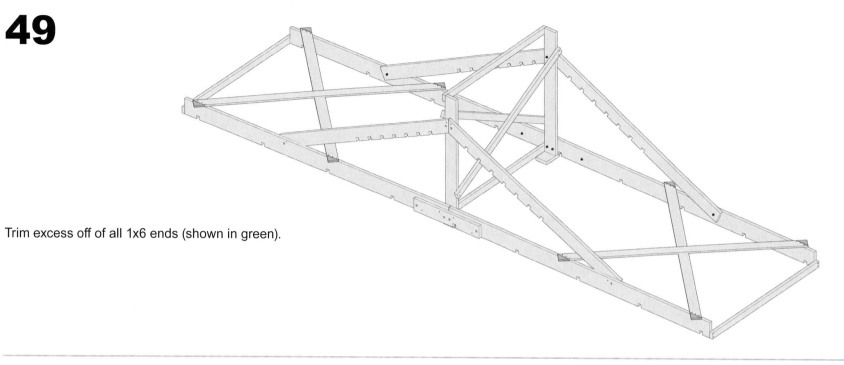

Trim excess off of all 1x6 ends (shown in green).

50

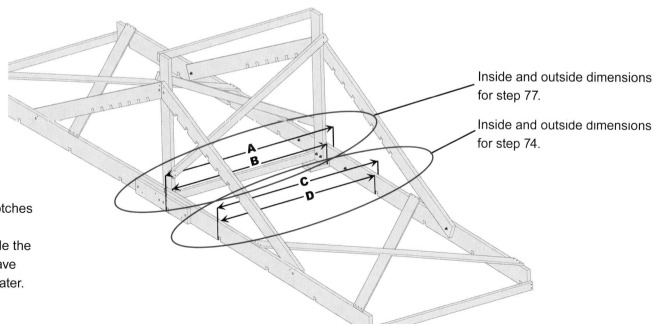

The purpose of this step is purely to make your life easier later in the assembly process when cutting notches in the purlins.
Record these dimensions now while the structure is on the ground. It will save you from having to climb a ladder later.

Inside and outside dimensions for step 77.

Inside and outside dimensions for step 74.

51

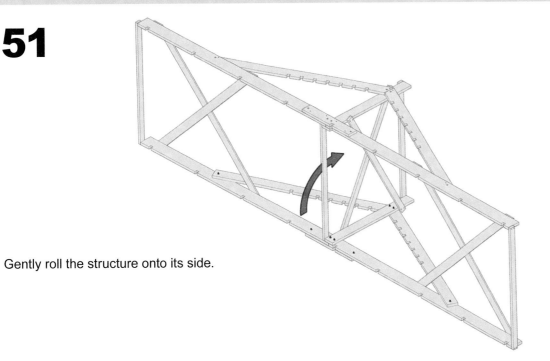

Gently roll the structure onto its side.

Joel's Tip

You will need a tractor and/or an army of Amish men for the following steps. When made out of oak, this structure weighs close to 2,000lbs. Carefully tip the structure onto its side and then over again as shown below.

52

TURKEYS

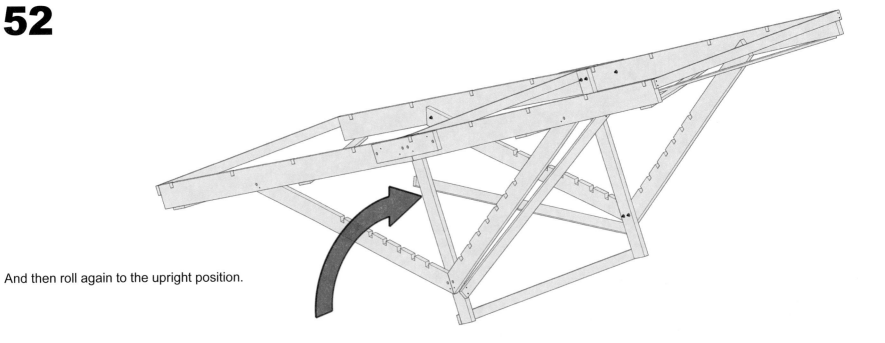

And then roll again to the upright position.

Attach To The Chassis

53

Roll the chassis right up against the front of the structure.

54

Using a tractor, lift one side of the structure up and maneuver the chassis underneath. Orient as shown in step 55. Screw and/or clamp a block to prevent the structure from sliding off. Also chock the wheel on that side as well.

55

3x   **Scrap Wood Blocks**
3x   **3" Decks Screws**

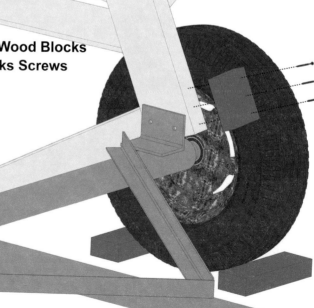

TURKEYS

56

Repeat the process on the other side by lifting it with a tractor and then aligning the chassis underneath the structure. Screw/clamp another block and chock the wheels to prevent anything from rolling away.

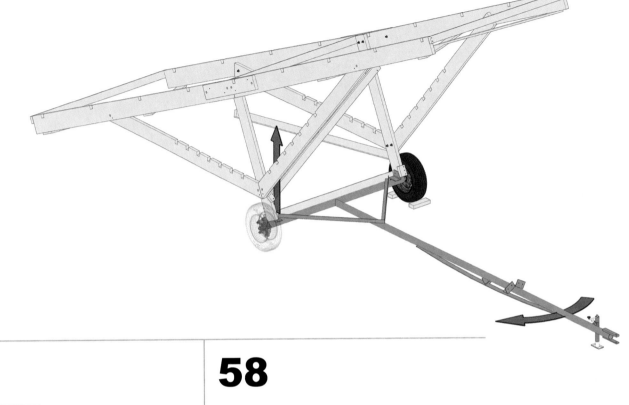

57

3x   **Scrap Wood Blocks**
3x   **3" Decks Screws**

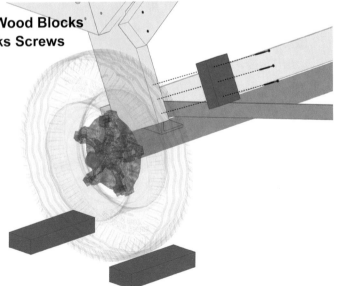

58

With the temporary blocks in place and wheels chocked, carefully tilt the structure upright.

59

4x   2x4 Brace Posts
4x   3" Decks Screws

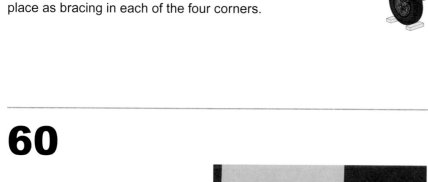

Measure the height of the structure in the front as well as the back to determine what "level" is. Use some 2x4s and temporarily screw them in place as bracing in each of the four corners.

60

4x   ½-13 x 3" Carriage Bolts
4x   ½" Washers
4x   ½-13 Hex Nuts

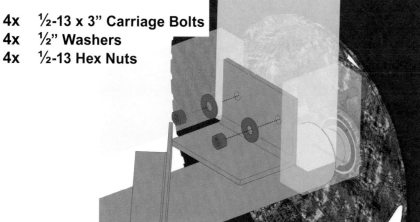

Center the 2x8x80 oak board on the angle iron bracket, and using the existing holes in the bracket as a guide, drill Ø½" holes through the wood. Secure with the carriage bolts. Repeat on other side as well.

61

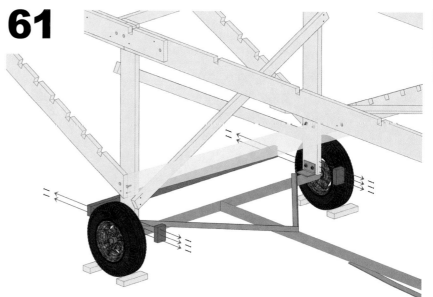

TURKEYS

After both sides are bolted and secure, remove the temporary wood blocks (shown in red) and the 2x6x94.

62

Mark the center of the front 2x4x98 as shown. Then, using a level or plumb bob, align that mark with the center of the chassis. It doesn't have to be perfect, but the closer it is the easier the following steps will be. Brace the structure as needed to hold it in that position.

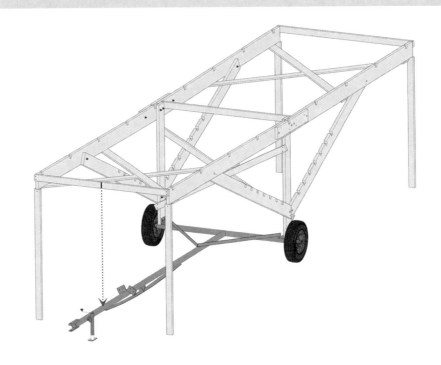

63

1x   2x6x94

Using a helper, hold a 2x6x94 roughly in the orientation that it will be mounted and scribe a mark as shown in red.

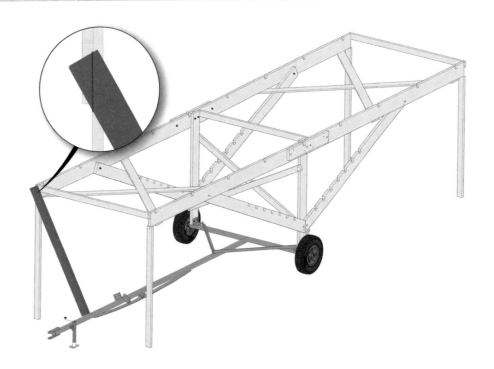

64

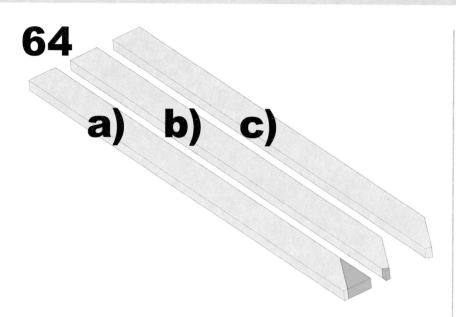

a)   b)   c)

Miter the board using the scribed line from the previous step. If there is a narrow point remaining, you may trim that back slightly as shown.

65

1x   3" Decks Screw

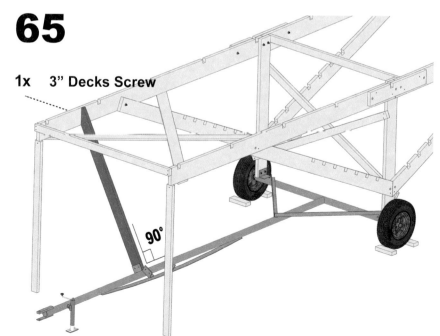

90°

Square the 2x6 with the chassis and tack the top in place with a screw.

66

2x   ½-13 x 3" Carriage Bolts
2x   ½" Washers
2x   ½-13 Hex Nuts

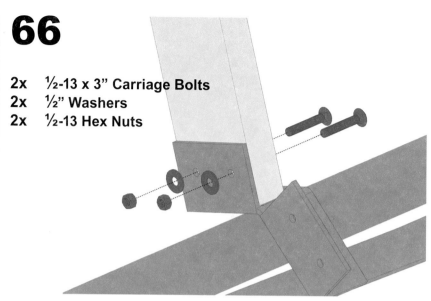

Using the existing holes in the bracket as a guide, drill Ø½" holes through the wood. Secure with the carriage bolts.

67

1x   2x2x³/₁₆" Angle Iron, 8" Long
4x   ½-13 x 3" Carriage Bolts
4x   ½" Washers
4x   ½-13 Hex Nuts

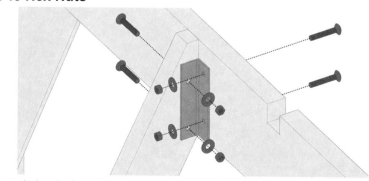

Using the existing holes in the bracket as a guide, drill Ø½" holes through the wood. Secure with the carriage bolts.

TURKEYS

68

Repeat steps 63-67 for other side.

69

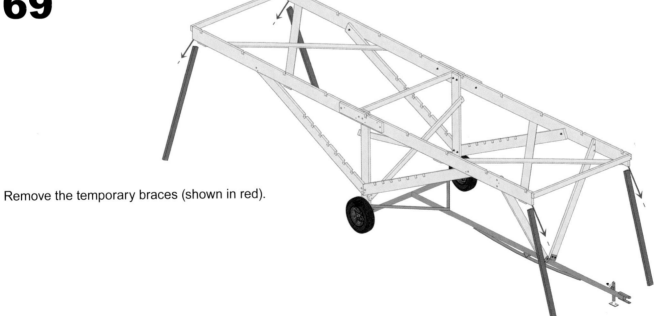

Remove the temporary braces (shown in red).

Build The Roosting Bars

70

Measure for the notches on the roost bars. Because of the way the angle braces get closer together the higher you go, record dimensions at both the widest and narrowest locations and size notches appropriately.

The widest dimension on the outside, F, will be at the bottom as shown. The narrowest dimension on the inside, E, will be up at the top.

71

1x   2x4x144

Mark center on the 2x4x144, and using the dimensions from the previous step and mark out the notches. Notches should be centered on the board. Be sure to oversize the notches at least ¼" to account for any warping or swelling that may occur.

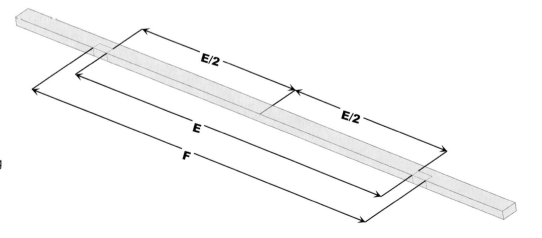

72

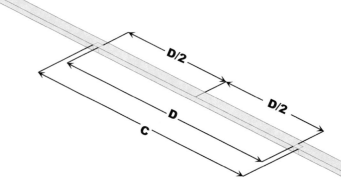

Using your first completed roost bar as a template. Mark and cut 13 more.

74

Create The Roof Purlins

1x   2x4x192

Mark the center on the 2x4x192. Using the dimensions from step 50, mark out the notches. Remember to oversize the notches slightly to account for swelling and warping.

TURKEYS

73

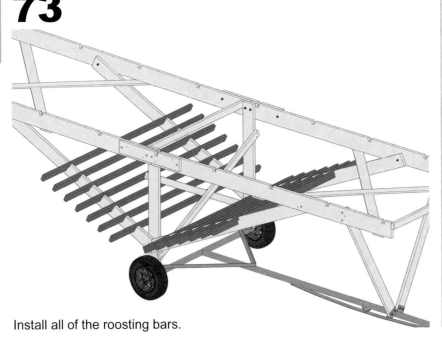

Install all of the roosting bars.

75

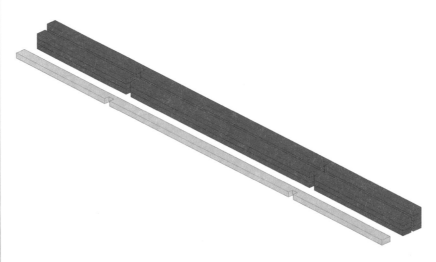

Using your first completed purlin as a template, mark and cut 7 more.

76

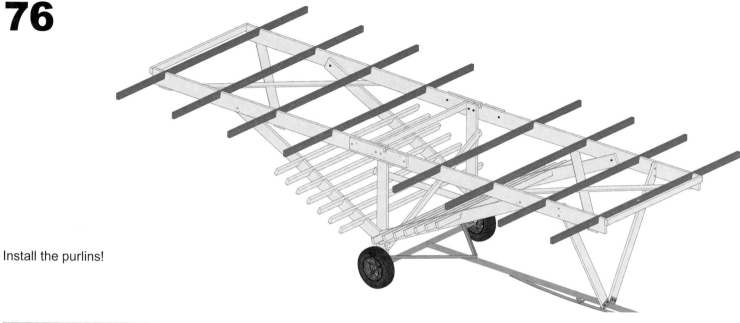

Install the purlins!

77

1x   2x4x192

TURKEYS

The final purlin will be slightly different because there are two pieces of wood that the notch must accommodate. Using the dimensions from step 50, mark out notches for the last purlin. Oversize notches slightly just like in previous steps.

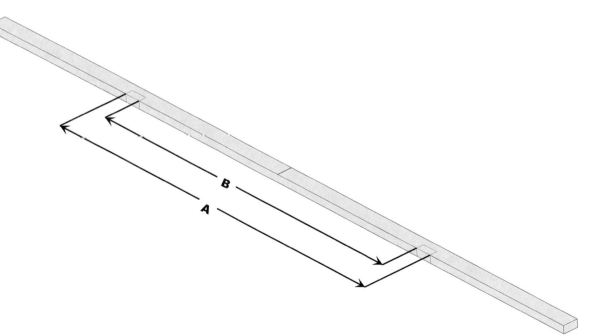

78

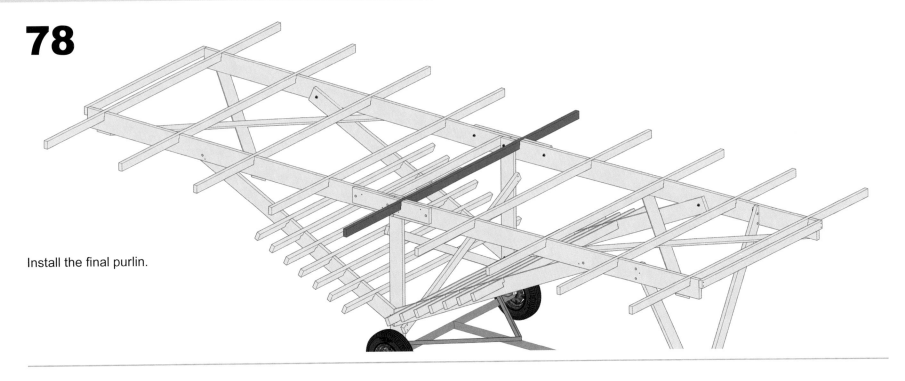

Install the final purlin.

Install The Shade Cloth

TURKEYS

79

1x   16x32ft Shade Cloth

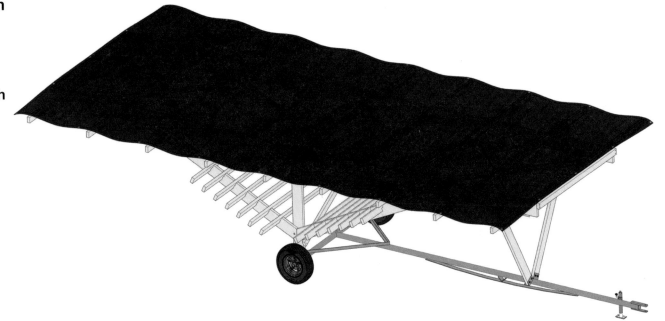

80

2x 1x3x96
6x 2" Decks Screws

With the shade cloth positioned appropriately, start by fastening one end with the 1x3 strip and screws. Pull the opposite end gently to straighten out any wrinkles and secure in the same manner.

There are 101 ways to secure shade cloth, but we've found this to be the best. The premise of this concept is to secure the cloth using "clamping force" rather than knots and string tied through the grommets. How often have you seen tarps tear and fail right where you've tied them? That's because all of the forces are directed through the single knot location. On the contrary, by clamping the shade between two pieces of wood with a screw, you distribute the forces throughout the entire area in contact with the wood.

Once it has been initially installed, it is incredibly simple for a single individual to unfold and re-attach it. If you are methodical, you can even find and reuse the same holes in the shade cloth from year to year.

To un-attach, simply back a screw out of the hole, slide the shade cloth out from between the wood pieces, and then retighten the screw, leaving the wood strips in place. Repeat all the way around.

To re-attach, unfold the cloth as demonstrated on pages 536-537 and then, finding the previous years hole, back each screw out, slide the shade in between the wood pieces and retighten the screw to clamp the shade cloth in place. Start at each end and then work your way down the sides last.

With diligence and care, you can get many years out of this piece of shade cloth.

TURKEYS

photo courtesy of Jessa Howdyshell

81

4x   1x3x144
16x 2" Decks Screws

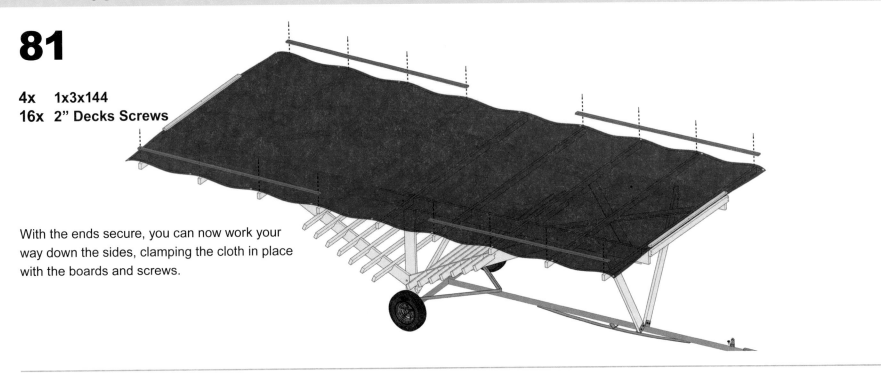

With the ends secure, you can now work your way down the sides, clamping the cloth in place with the boards and screws.

Build A Ramp

82

The ramp design we use for the Gobbledygo is the same ramp used for the Millennium Feathernet. See page 161 for construction details.

83

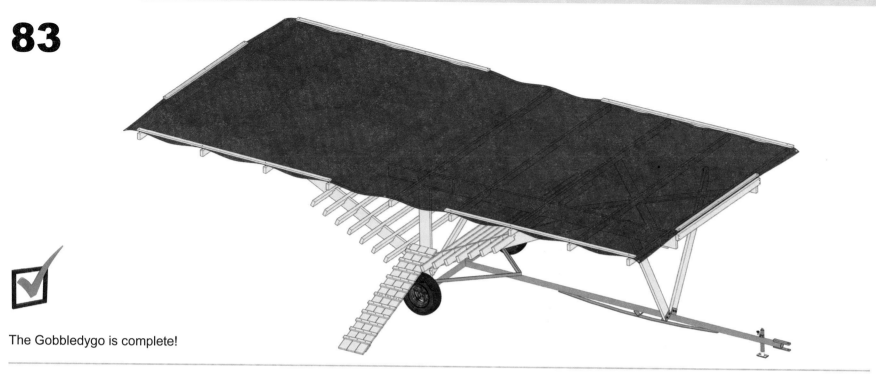

The Gobbledygo is complete!

Add A Bulk Feed Buggy

84

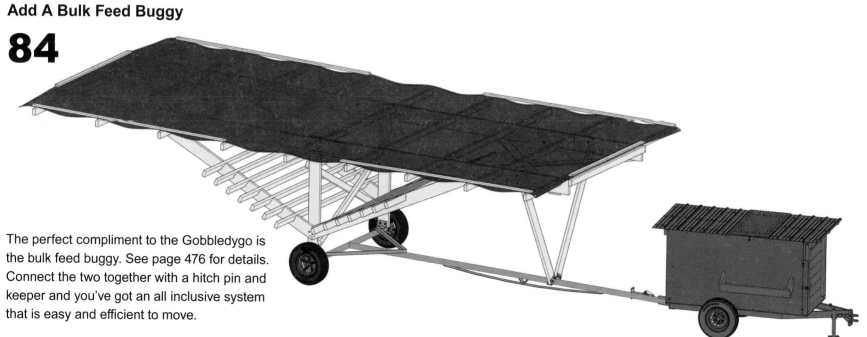

The perfect compliment to the Gobbledygo is the bulk feed buggy. See page 476 for details. Connect the two together with a hitch pin and keeper and you've got an all inclusive system that is easy and efficient to move.

photo courtesy of Jessa Howdyshell

When we began running the Eggmobiles, we initially tried to winterize them. Because they are purposely made to encourage lots of air flow, they are horrendously uncomfortable for the birds in the winter. We put plywood sheets on the floor and then threw in a bunch of wood chips in order to cozy it up. Even though we did our best, we could not make it comfortable. Water froze; the birds completely stopped laying, huddled around like a bunch of unhappy prisoners. Clean out was

no picnic either once weather warmed up. All that collected and caked poop on the plywood panels and trying to get them out through the doors.

After a couple of winters and some cogitating, we decided to build a hoop house. For the record, I love hoop houses. Plastic should be used for three things: hoop houses, water pipe, and shade cloths. Hoop houses offer passive solar heat but they're also fairly cheap to build. If all you want is covered space, few designs beat the

simple hoop house. Dollar for dollar, you can get a lot of coverage pretty cheap.

In the winter, few habitats are as comfortable as a hoop house. Even in sub-zero temperatures, these structures always warm up to 70 degrees F. We had a blizzard one year that filled the narrow space between the first two hoop houses all the way to the top–12 feet. It piled up all around them and turned them into igloos. But inside, the birds were cozy and comfortable. In fact, the piled up snow snugged down the structures to make them immune to the winds that inevitably follow change of weather after a blizzard.

Many insurance companies won't cover hoop houses due to their perceived fragility. The trade-off is that in most areas you don't need a building permit and they don't add to your property taxes. Hoop houses have a lot going for them. We built the first two nearly 30 years ago and over the years added three more bigger ones. They're beautiful and versatile.

What this is not:

This is not a step by step tutorial of how to construct a hoop house. The internet is saturated with how-to's on this topic already.

What this is:

This section is a compilation of tips/tricks and notable wisdom we have gained throughout our years of experience with sheltering animals in hoop house structures.

Choosing a building site

When choosing a building site, two of the most important aspects to consider are wind direction and watershed. Ideally, the house should be oriented long side parallel to your region's dominant trade winds. This will ensure adequate ventilation in all seasons. For us that means orienting the structure in the east-west direction.

Second, this hoop house needs to be dry on the inside. If you build it in a low spot or on the side of a hill, you will be dealing with perpetual water issues. Pick a high spot to build, and if on the side of a hill, build swales to divert surface water from flooding your dry bedding. Wet bedding leads to sick animals. You also want any water collecting off of the roof to shed away from the structure. We violated these rules with the first two we built and eventually dug French drains alongside to drain off water. This is nothing more than a ditch about 18 inches deep filled with about 3 inches of number 5 gravel, then perforated (leaky) drainage pipe, then filled up to soil level with the same gravel. Water runs fast through the gravel into the pipe, protecting the interior from dampness.

Sizing your hoop house

Determining the size of a hoop house is based on several factors:

1.　Lot size

Obviously, the topography of your land and the square footage you have to work with will play a key role in how large you can make your hoop house. Excavation work can easily exceed the cost of the hoop house itself. It may be cheaper to build two or more smaller structures as opposed to one large hoop depending on your landscape. Do your homework and price out all of the options. Generally, economies of scale

photo courtesy of Kate Simon

WINTER HOOP HOUSES

cheapen structures if they're bigger, unless they get so big they need extra trusses and expensive supports.

2.　Access

While we aim to minimize the amount of machinery needed in our winter housing, animals still need food and bedding. Be sure that you have ample access to at least one (but ideally both) ends. If you anticipate feed deliveries directly into

photo courtesy of Jessa Howdyshell

your feed storage bins and/or pig feeders, be sure that those large trucks have decent all-weather access. Grading and solid footing for vehicles will pay big benefits when spring thaw turns the ground into mush.

Make sure your access doors allow entrance with whatever equipment you have, both in height and width. The higher and wider the doors and structure the easier it is to maneuver front end loaders and other equipment.

3. Animal capacity

Chickens need 3 square feet per bird and ideally flock size should not exceed 1,000. Stress includes mass, density, and time factors. At less than 3 square feet per bird, they can't scratch their bedding fast enough to keep up with the manure load and it caps over.

Pigs should have 15-20 square feet per animal. Rabbit hutches are a standard 30" x 30" per doe or buck.

4. Water

Another fundamental requirement for winter housing is access to reliable fresh water, even in the coldest of conditions. Be sure to install freeze-proof water. In our poultry-only hoop houses, we like to install our water hydrants inside of the hoop house itself. On our pig houses we typically install the hydrants along the exterior, outside of the pigs' reach. We also make it standard protocol to unhook and drain water hoses if the temperature will drop below freezing. True frost free waterers are extremely expensive whether they use deep earth temperatures or electricity to keep from freezing. Those installations are not portable. We prefer a simple frost-free hydrant and cheap, portable waterers so we can move them around the bedding and avoid wet spots. Animals dribble where they drink.

5. Cost/maintainability

We've used economy-model hoop kits, realizing that if a blizzard comes, we may need to go out with a stick covered with some rags to poke the plastic and help the snow to slide off. In 25 years I think we've had to do this twice. The cost of a blizzard-proof structure might pay in Minnesota or Buffalo, New York, but in more temperate areas some strategic labor is the cheaper way to go.

Hoop House Coverings

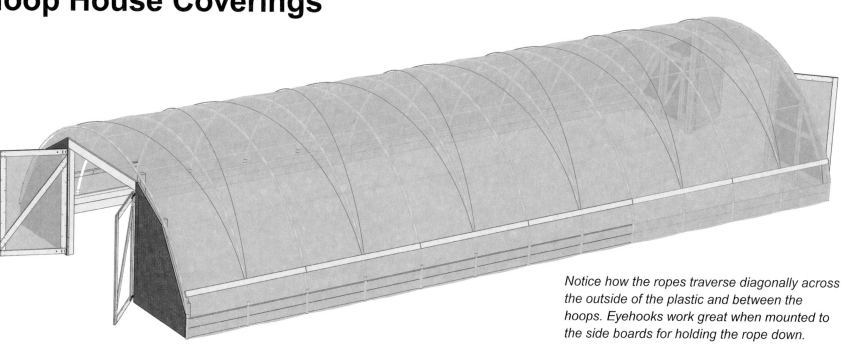

Notice how the ropes traverse diagonally across the outside of the plastic and between the hoops. Eyehooks work great when mounted to the side boards for holding the rope down.

The most important factor to ensure covering longevity is eliminating "flap". Covering slack enables it to flap in the wind. This stresses the joints and stretches the covering even more; thus exacerbating the problem until it ultimately fails. Once it starts to rip at one point, it catches wind and turns into a kite. Been there and done that. Therefore, flapping plastic = bad!

Believe it or not, temperature changes can often cause things to stretch and shrink as well. Fortunately, we've found that the key to keeping the covering taut in all conditions lies in a rope that crisscrosses diagonally over the top of the hoop house. Just a small nylon rope arranged like the picture above that is tied off with a little bit of tension pulling down on the

plastic will do the trick. Nylon is critical to keep the rope from weathering.

Make sure you install the covering on a calm and preferably warm day. Plastic expands in heat, so the warmer the day of installation is, the tighter the fabric will stay. In winter's cold, it'll get as tight as a drum, but that's what you want because plastic also gets more brittle when it's cold. Tight plastic won't wiggle and crinkle as easily.

When we pick our covering moment, we unroll the plastic along the long edge and use C-clamps with two wood blocks to squeeze the plastic every 20ft or so. We tie ropes to the C-clamps, toss the ropes over the skeleton, and get some friends to come by for a hoop house raising. We all begin pulling on our

ropes from the one side while two or three helpers with rag-covered poles shepherd the plastic up and over the skeleton. In literally 5 minutes the structure goes from bones to skin; it's one of the most amazing and wonderful things to ever do. The first thing is to affix the ends. After that, we affix the edges. A 1x6" purlin at 4ft above the ground is the pull point. We precut 1x3" boards and lay them along the ground at the ready. After screwing these pinch or sandwich boards to affix one side, we all go to the other side and stand on the plastic to pull it as tight as possible while another couple of people install the other sandwich board.

A simple vertical zig-zag rope with screw eyes on the purlin and down at ground level keep the skirt from flapping. This way we can

WINTER HOOP HOUSES

both photos courtesy of Chris Slattery

raise and lower that skirt edge for increased ventilation in the summer. While all of this may sound somewhat crude (it is) we've had covers last 10 years with virtually no maintenance. While that is not a century, for hoop houses that's a nice record.

To enjoy full season use, shade cloths can make a hoop house as comfortable as a shed. With the sides rolled up 3 feet, the cross breeze makes it downright inviting. We not only grow plants in our hoop houses during the summer; we even have events and parties in them.

Sides

Since our system revolves around deep bedding, we need a way to contain it and our animals. We also need a way to prevent livestock from coming in contact with the delicate plastic covering. As you could probably guess, this plastic, as durable as it may be, is no match for curious fowl, let alone mischievous swine. Chickens can peck a hole in the plastic in no time.

To protect the plastic, start with a perimeter of 16-24" high baseboards all the way around the structure; we prefer to use rot resistant oak for these features because of their strength and longevity. These baseboards are what keep the bedding inside. We then, depending on the livestock we are attempting to contain, secure chicken wire, hog panel, or both above the baseboards. The wire sides also double as lattices for climbing vegetables in the warmer months. We never bother to clean the vegetable vines off them in the fall;

the animals do that just fine.

For the record, we do not encase the side pillars in concrete like most specifications demand. Knock on wood, we've never had a wind yank these anchor pipes out of the ground. Perhaps one reason is because we lock them together with our bedding base boards. All that linkage certainly strengthens the whole. To pound the pipe columns in the ground, we now use a regular fence post pounder on the tractor. In the early days, we welded a 1 inch thick piece of flat bar on a pipe that would sleeve over the pillar pipes; that gave us a big nail head to pound with a sledge hammer. If gale force winds come, we close the windward end to keep the house from becoming a box kite.

Hoop House Endwalls

Joel's Tip

Today's hoop house kits typically come with plastic that is 4-6mils thick and is rated for 1-4 years of service. All of the options can be overwhelming, but our favorite covering is 11 mil laminated material from Northern Greenhouses. It lasts up to 15 years so it's worth the extra 30% initial cost versus regular hoop house plastic that only lasts 3 years or so. The laminations keep it from tearing easily because it's kind of like a plastic six-pack tie.

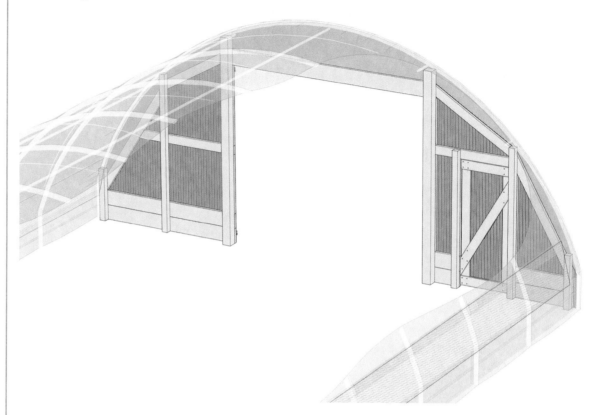

Hoop house manufacturers provide all sorts of options for endwall kits. We, however, choose to save a buck and build our own. We simply dig holes and set posts to form our wall uprights. We then fasten them to the hoops using provided brackets and hardware (consult hoop provider for these accessories). Then we roughly frame in our openings and fasten the excess overlapping covering material to it. Lastly we sheet the ends with metal.

When sizing your openings, make them wide and tall enough to access with whatever equipment you use/plan to use. We recommend making it as large as possible because you can always frame it in to be smaller later. Ours are tall enough to fit wagon loads of hay. A 30ft wide structure is even tall enough to offload dump trailers of bedding material inside.

We also highly recommend inserting a person door on either end, especially when using this structure for poultry. It allows easy access for feeding and egg collection, and keeps birds and heat inside. A latch for the outside of the person door is recommended in addition to a handy self-closing device we will elaborate on later in this chapter.

WINTER HOOP HOUSES

Hoop House Doors

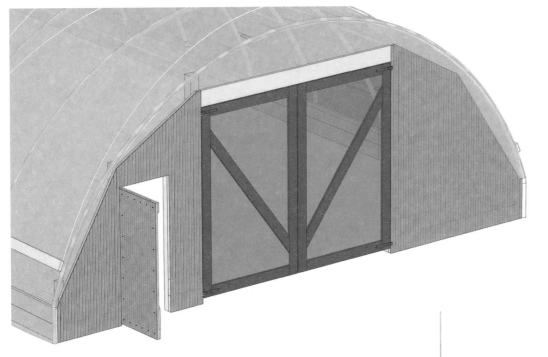

Doors are an absolute must for ventilation and we've experimented with different styles. We have sliders, hinged barn-style doors, and even overhead garage style doors. It all amounts to what we had available at construction time. Remember our mantra — use what you've got and get creative. Chickens and pigs certainly don't care, nor do we.

We will say that hinged doors are cheaper to make than sliding doors, but sliders are more durable and are less prone to breaking off in the wind. When we construct hinged doors, we like to sandwich our extra covering material between a 1x6 door frame because it makes for a lightweight, yet functional door. Gate hinge straps and lag hinges work well on large doors like this. These doors can afford to be lighter weight because they do not handle any animal pressure. Depending on what animals you are sheltering, you'll either have heavy metal gates or chicken wire screens on the inside of these doors to contain livestock.

WINTER HOOP HOUSES

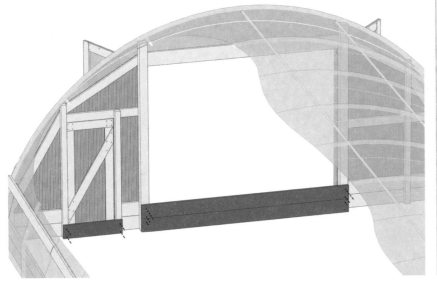

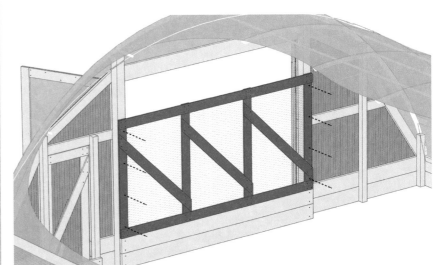

For a house full of poultry only, be sure to fill it with at least 1 foot of bedding material and then secure baseboards across the doorways as shown. (See diagrams on the bottom to the left.)

We then erect a simple screen structure across the main opening made from a 1x6 frame with chicken wire stapled across it. Depending on the breed of bird, 6ft tall chicken wire may be enough. With flightier fowl, it may need to be taller. Another option is to clip a wing on every bird as you bring them inside, which will also help tremendously if you plan to have rabbit hutches inside as well.

photo courtesy of Jessa Howdyshell

Inner Partitions

Divider walls are an excellent way to partition sections of the structure for different animal groups. Obviously for larger stock like pigs you would need to substantially beef up this divider, but for poultry this is really all you need. Place a baseboard on the bottom to keep the chicken wire off the bedding. Chicken wire 6-8 ft off the floor is typically high enough. If you are dealing with flightier fowl (heritage chickens, guineas, etc.) you will need to extend mesh all the way up to the roof plastic.

Similar to the end walls, secure several upright posts to the hoops above and bury into the ground. Chicken wire can be cut to fit and tied with wire to the hoops on the perimeter. Also plan to install an access door for personnel to travel between sections easily.

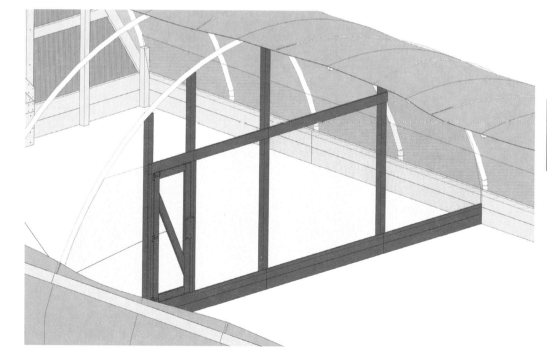

WINTER HOOP HOUSES

439

Self-closing Access Door

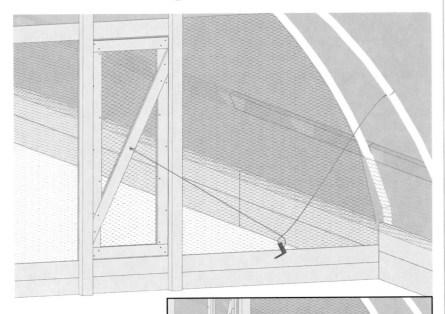

Feed Storage

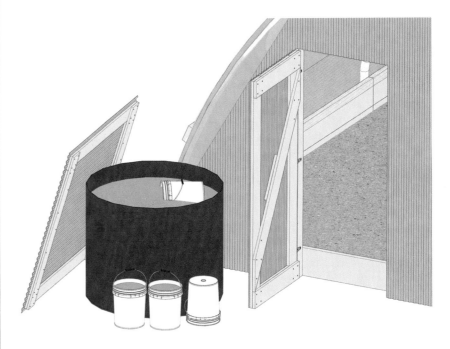

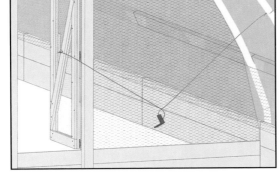

This is a cheap, slick way to create a self-closing door, which is worth its weight in gold after the first time you forget to close a door behind you and you spend the next hour catching chickens outside. It is especially awesome when your hands are full and all you need to do is carefully push through the door while it closes itself.

This setup consists of a length of 12-½ gauge medium tensile wire and a ≈5lb scrap of metal. Simply tie one end to a central location on the door using an eye-hook or other means. Tie the other end up on a hoop allowing it to sag about 1ft or so from the floor when the door is closed. Then, attach the weigh to the lowest point of the loop. As you open the door the weighted wire raises up, and then gravity does the work at pulling it back closed. It's really that simple!

We utilize old 1000 gallon steel tanks cut in half for grain storage. We build simple lids from 1x6 wood frames sheeted with metal. Heavy enough to not blow away yet light enough for one person to slide off easily. We place them next to the doors on each end and fill them full directly from the feed truck. Then all we have to do is bucket it in when needed. It's brilliantly simple and cheap. No tractors to start on the coldest of mornings. Oh, and we also use these in the summer to store feed out in the fields for the broilers.

A Typical Hoop House Layout

In the scenario on the next page we've divided the hoop house into two sections: one-third and two-thirds. We try our best to keep individual flocks or hatches of birds separate, especially in a contained environment. We will split flocks as needed but we never combine. Flock separation eliminates a lot of fighting.

Notice how we arranged the mezzanine platforms with the feeders on top. There are dozens of possibilities for your operation based on your own needs. If we have small pigs running around you will not need poultry drinkers because the chickens can drink out of the pig water tanks. You may find that chickens are flying up and over divider walls if the mezzanines are placed too close. The solutions are simple if this becomes a problem: either move the mezzanines or add height to the divider.

The mezzanines allow stacking multi-species in the house to utilize the cubic area rather than just the floor linear area. The rabbit hutches stacked above the chickens offer the most conducive ingredients for composting and deep bedding bug growing. The extra manure load and urine stimulate biological activity in the deep bedding, which in turn draws the chickens to do more aerating, which stimulates more biological activity.

If we raise pigs under the mezzanine tables, we place all the fragile poultry feeders and waterers on the mezzanine to protect them from the pigs. The pigs cannot be bigger than about 80 pounds or they will eat an occasional chicken. The chickens never learn about their danger. Pigs and chickens seem to be the most

compatible critters in this stacking scenario, perhaps because they are both omnivores.

We do throw in spoiled meat or carcasses left over from hunting to give the chickens fresh animal protein when they don't have access to pasture insects and worms. Finally, sheep also work well on the floor under the mezzanine. Chickens like to pick at the hay fed to the sheep; the sheep spread their manure and urine around the best. The pigs pick a toilet spot that should be fed a steady topping of carbon. This is a good place to feed some junky hay to the pigs; they eat half and poop on half. Otherwise things can get smelly. Except for this aspect of the pigs, generally we don't need to re-bed during the 100 days of hoop house living as long as we've put enough carbon in there to start. If something caps over, we might take a forked potato hoe or rake to stir things up and uncap it.

The beauty of this system is that we can rearrange things throughout the winter season. Feeders and waterers can be relocated to spread the manure more evenly and nest boxes can be arranged to better suit the laying patterns of the hens.

Also notice how the ramp and step from our Feathernet can be reused inside. Bulk feeders typically migrate from Eggmobiles and Feathernets as do the drinker tubs. It's all about maximizing the use of limited infrastructure.

We have two hoop houses 20ftx120ft and three that are 30ftx120ft. That length seems to be about as long as you can go with natural ventilation. If you go longer, the air flow slows down too much before it gets to the far

photo courtesy of Chris Slattery

end and the livestock does not get enough changes of air. Using the square footage rules mentioned above, we can keep 800 in the two smaller ones and 1,200 in the bigger ones. If you use a mezzanine, of course, you can add that square footage to the total bird capacity.

You need enough linear waterer and feeder space to accommodate enough of the birds so that even timid ones will get enough. In general, feeder space should handle about 20 percent of the birds and waterer about 5 percent. In other words, if you have 100 birds, 5 should be able to get a drink at once. The reason for the difference is that chickens spend far more time at the feeder than they do at the waterer. In general, 6 linear inches per bird is enough. Nest boxes should be 1 per 10 hens. We gather eggs at about 4 p.m. every day. If you're having trouble with the birds soiling or eating eggs, a preliminary gathering around noon can go a long way to protecting eggs from soiling. We also have a simple hay manger for each flock that seems to stimulate production, hold up yolk color in the winter, and give the chickens something to enjoy. Anything to keep them occupied is a good thing.

WINTER HOOP HOUSES

A Typical Layout (poultry only)

Joel's Tip

For the record, I love hoop houses. Plastic should be used for 3 things: hoop houses, water pipe and shade cloths.

Ramp (from Feathernet)
(See page 161)

Standing Nest Box
(See page 459)

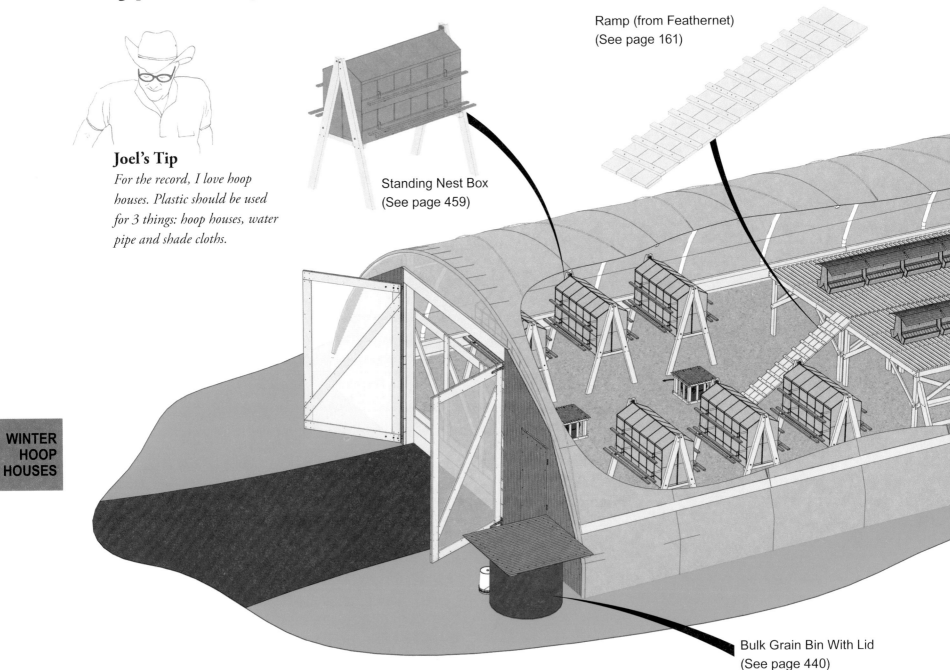

WINTER HOOP HOUSES

Bulk Grain Bin With Lid
(See page 440)

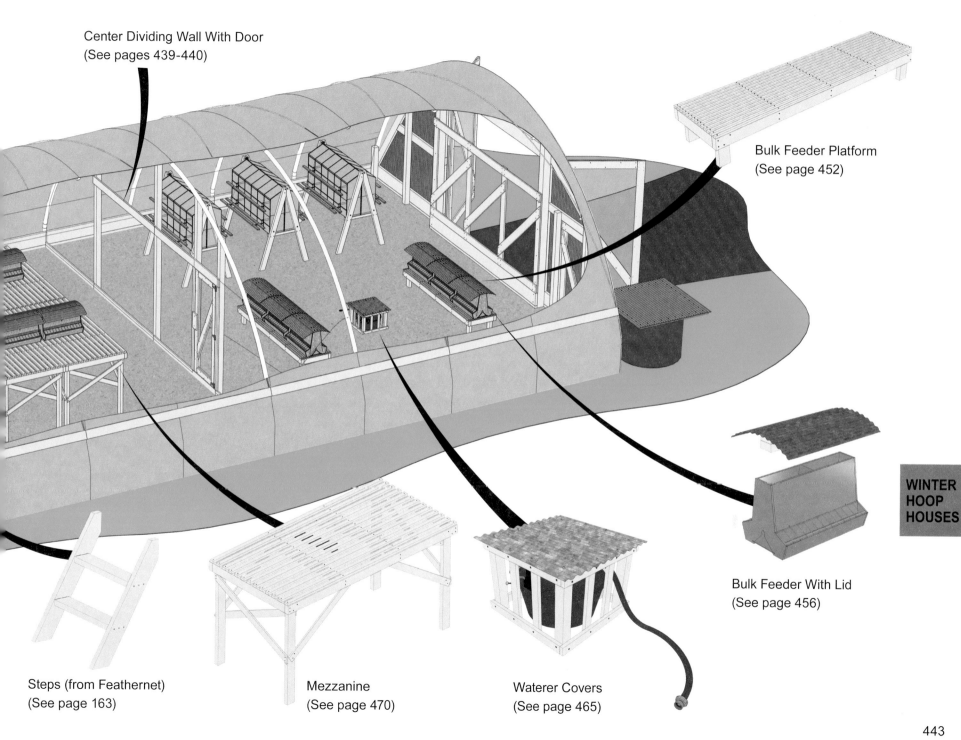

Center Dividing Wall With Door
(See pages 439-440)

Bulk Feeder Platform
(See page 452)

WINTER HOOP HOUSES

Bulk Feeder With Lid
(See page 456)

Steps (from Feathernet)
(See page 163)

Mezzanine
(See page 470)

Waterer Covers
(See page 465)

A Typical Layout (poultry and rabbits aka Raken)

If we are placing rabbit hutches in with the chickens, then we typically do not have mezzanines in that same section. Back to the flying problem. Chickens will inevitably fly from the mezzanines up onto the hutches. We secure the tops of the cages with a chicken wire perimeter, but even still a few flighty hens find their way on top. We often clip a wing on each bird as we bring them in for the winter which helps tremendously with the flying problems. When clipping wings, only clip one wing; not both. This creates imbalance when flying, which is the goal. When you flare out the wing, use sharp scissors to clip the long feathers along the second tier of fluffy ones. This will keep you from going too deep and drawing blood.

Because rabbits need more changes of air than chickens, their height off the floor helps but the key is a layer of metal roofing on top of the hutches. That protects them from the sun. Rabbits are nocturnal, so darkening them down gives them a more natural habitat. Since rabbits have fur, keeping the sun off them also helps them to stay more comfortable.

WINTER HOOP HOUSES

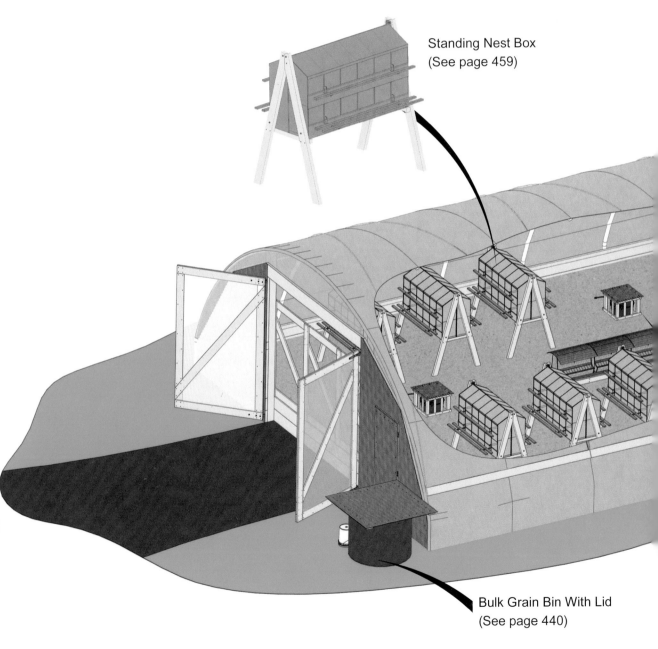

Standing Nest Box
(See page 459)

Bulk Grain Bin With Lid
(See page 440)

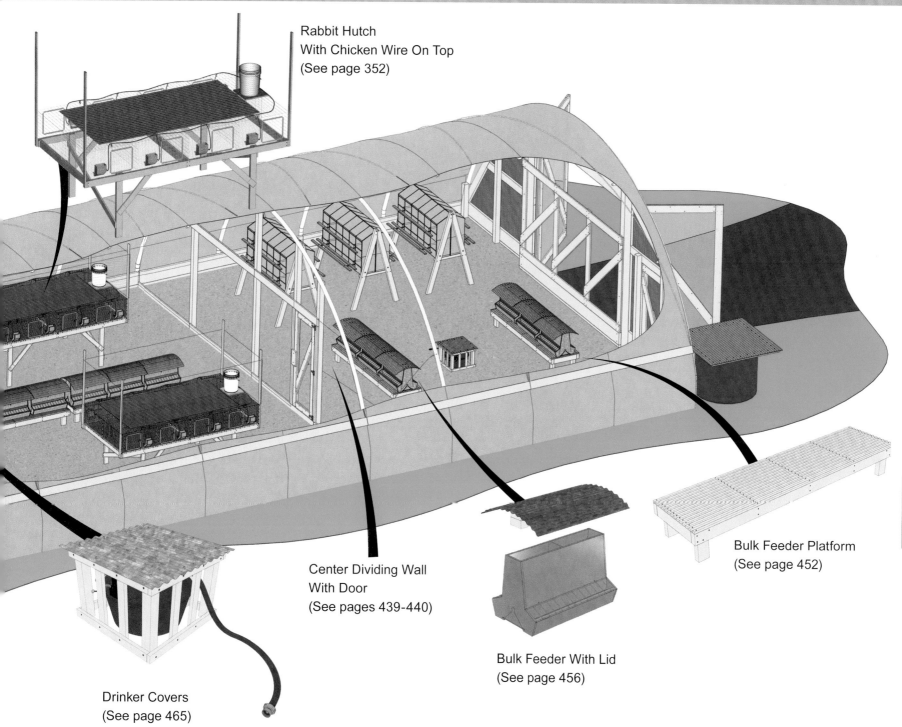

Rabbit Hutch
With Chicken Wire On Top
(See page 352)

WINTER
HOOP
HOUSES

Bulk Feeder Platform
(See page 452)

Center Dividing Wall
With Door
(See pages 439-440)

Bulk Feeder With Lid
(See page 456)

Drinker Covers
(See page 465)

Winter Housing For Pigs

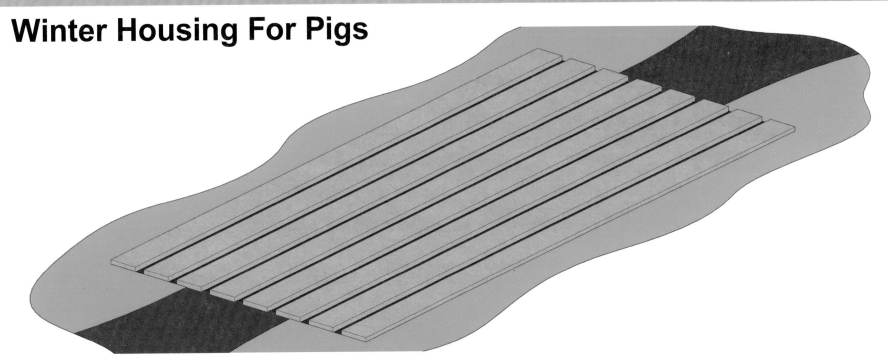

Concrete Floors

Concrete floors are essential, especially when housing large pigs. Given enough time and size, pigs will eventually dig up a dirt floor no matter how deep the bedding. Because we didn't want to lose the hoop house space for growing vegetables, the compromise we settled on was an intermittent floor with rows of concrete followed by alleys of bare dirt. This not only saves on concrete (at the expense of forming and labor) but it allows us to still grow vegetables and therefore maintain the four season functionality of the structure. The gutters also provide a place to drive in T-posts on which to fasten temporary hog panel divisions to keep groupings separate.

The concrete rows are approximately 30" wide with 8-10" spaces in between. We recommend 4" minimum thickness reinforced with $\frac{1}{2}$" (#4) rebar spaced a maximum of 24" apart. If you plan to be driving large equipment (10,000lb plus) over the concrete it should be 5-6" thick and have steel placed closer together. Ultimately your floor is only as strong as the base beneath, so make sure the ground is solid and well compacted. Typical building codes require a minimum 2" of compacted gravel beneath, but this is agricultural so codes may not apply. Consult your local concrete contractors with any questions. This will be by far the most expensive aspect of this infrastructure, so you want to do it right in order for it to last.

We also recommend pouring the floor PRIOR to erecting the hoop structure. Think about ACCESS – the concrete truck needs to be able to reach all the way inside. Having a bunch of hoops in the way will only impede

access. Check with your local concrete plant, but typical American front-load trucks can reach about 20 ft in front of them when head on and about 15 ft when sideways.

Using the building layout and squaring techniques found in the holding pen section (steps 1-9, pages 238-241), we recommend setting your four corners prior to doing anything else.

To save on form lumber you can form and pour a few rows at a time. Basic length x width x thickness volume calculations can help you approximate the yardages needed. Concrete trucks typically hold 10 yards of concrete, so aim to maximize each load while minimizing waste. You will ultimately pay more per yard if you call out several partially full trucks as opposed to a full 10 yard load, so plan accordingly.

Pig Doors & Openings

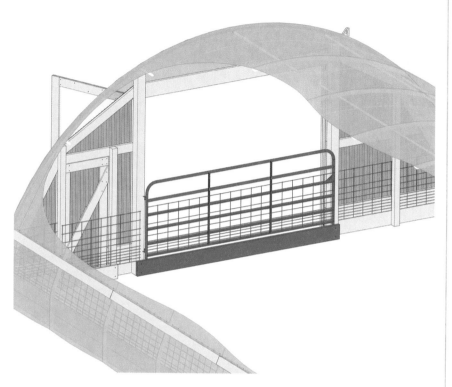

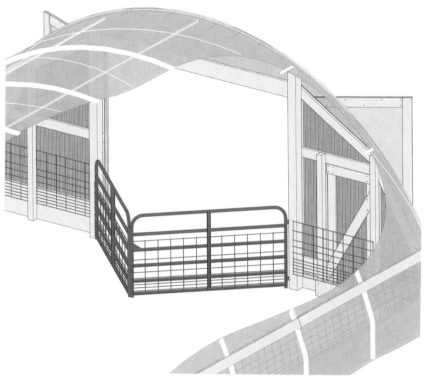

If you plan to only shelter pigs in a hoop house, we recommend filling it with bedding material first. All you really need are heavy metal pipe gates secured across the end openings. All gates should have hog paneling secured on the inside with tie wire to prevent little piglets from slipping between the bars.

Make sure the gate hinges are oriented with one facing downwards to prevent pigs from lifting the gates off the hinges. Also, inserting at least one baseboard across the bottom is helpful in containing the bedding pack, as well as keeping the metal gates out of the corrosive manure.

Another option we frequently use is a "V" shaped configuration. This is great when we split the house down the middle into separate sections as can be seen in future photos. This allows us to access either side independently of the other. It also affords us a little extra room to back an auger buggy or truck up into the opening to fill the bulk feeders.

Also notice the hog-paneling secured to the walls on all sides. We typically hog panel right over any man doors as well to protect them. If you plan to house pigs and chickens together, you will need to combine the use of hog wire, steel gates and chicken wire to contain everyone properly.

WINTER HOOP HOUSES

Pen Divisions

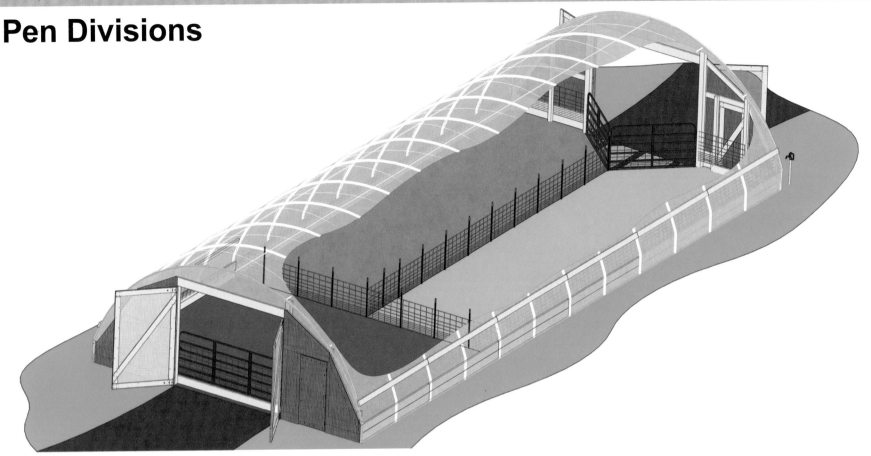

Hoop House Sections

We always divide our hoop houses into separate sections in order to keep each individual group of pigs separate. Combining pig groups inside a shelter is disastrous and leads to fighting and bullying.

Our ideal group size is 35-50 pigs in a group. The reason is that a two-hole waterer and a six-hole self-feeder will service that many. Since you have to check on them every day, economies of scale dictate that the cost of being there is spread over as many pigs as practical, as well as the cost of infrastructure. At about 20 square feet per pig, they have

plenty of room to lounge tightly, which they enjoy, leaving room for the ones who want to run around. Having them on deep bedding, giving them junk hay, cornstalks or other things to tear up and play with go a long way toward reducing bullying and fighting.

In this illustrated scenario we've broken our hoop house into approximate thirds. If you divide it into three equal sections down the length of the shelter, you will be left with a center section and no easy way to fill the feeder in the middle. That's why we split it down the middle as shown. We can keep our feeders for each group close to the entrances

for easy access.

It is imperative that hog paneling be secured along the entire inside perimeter to contain the pigs and prevent damage to the structure. To divide the sections we pound t-posts into the open dirt sections of the floor and secure hog paneling to them. For this reason we recommend putting some thought into how you plan to arrange internal divider fencing PRIOR to pouring your concrete floor. It would be disappointing to have concrete right where you need to drive t-posts. T-posts every 3ft is ideal. The following illustrations show tips on tying hog panel.

Partition Fencing

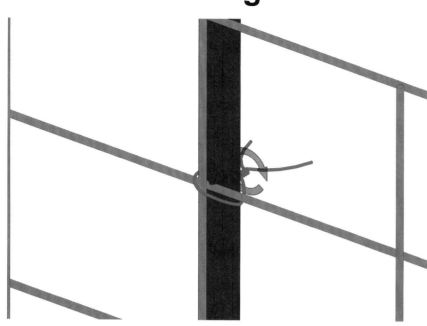

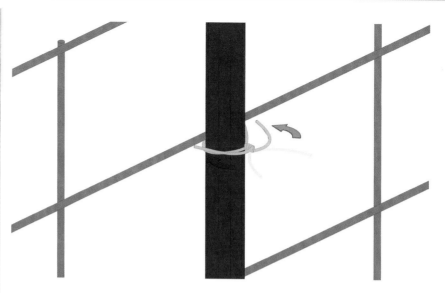

Lastly, be sure to bend the excess wire in a way that nothing can get snagged or cut on it. This will spare animals potential injury and your work clothes from tears.

Properly securing hog paneling to t-posts is critical to successfully containing groups of pigs in this environment. This may seem like a trivial task, but from experience we've found the best way is to use 12-$\frac{1}{2}$ gauge medium tensile wire for most tying because it's cheap, easy to work with, and strong.

Notice that the tie wire crosses over an intersection of a horizontal and vertical section of hog panel. Also notice that there are nubs on the t-post and that we positioned the horizontal wire underneath the nub. This is intentional because the tendency of pigs is to lift up on things. Tying the panel underneath that nub will lock it in place. The pig would now need to uproot the t-post to escape. (This is exactly opposite the normal installation instructions when using T-posts with cattle or sheep, where you want to put the tie ABOVE the nub to keep it from sagging.)

Why Hog Paneling?

The importance of using hog paneling instead of cattle paneling lies in the spacing of the wire. Cattle panels typically consist of 8"x8" squares 48" tall. On the other hand, the largest openings in a hog panel are typically 6"x8" all the way at the top. The spacing at the bottom is much closer and is critical to keeping even the smallest pigs in. An old proverb says that the perfect fence would be "pig-tight, bull-strong, and horse-high." Pig proofing is tough, but with the correct materials it can be done! Both hog paneling and cattle paneling can be readily found at your local farm store or co-op.

WINTER HOOP HOUSES

449

Routing Water Hoses

As mentioned in the pig chapter detailing pig waterers (see pages 310-311), water hoses must be kept up and away from the pigs. We typically loop the hoses through the purlins overhead. We connect "Y" valves when needed, but we will often also use one single float valve/lid assembly and rotate it between the drinkers as needed. It is less to drain the water out of at night, but this only works with constant management and smaller pig groups with less water demand. All hoses route to a frost free hydrant outside. If your hydrant happens to be inside, then you must protect it well from the pigs or they will dig it up and break it off. You can never overestimate a pig's ability to excavate.

A bonus to hanging the hoses close to the ceiling of the hoop house is that they tend to warm up and thaw very quickly from the warmth produced inside. These internal hoses need not be drained unless temperatures are extremely cold, like single digits. Otherwise, although they may freeze at night, they'll thaw as soon as the sun comes up. This is a dance you'll get a feel for over time. Again, while this sounds poor-boy, it's actually much simpler and less fragile than completely frost-free hands-off installations.

Lastly, we always place old wooden pallets underneath of the waterers to keep them up and out of the bedding material. This keeps the water cleaner. If your junk hay builds up around it or the pigs move the bedding around, you can just move the whole installation over to the high spot and start over.

WINTER HOOP HOUSES

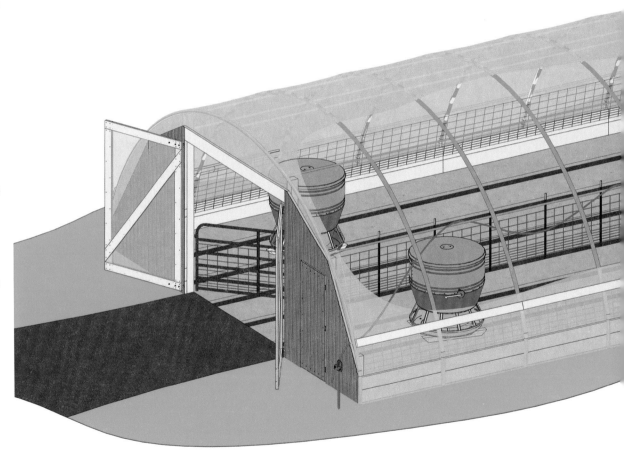

450

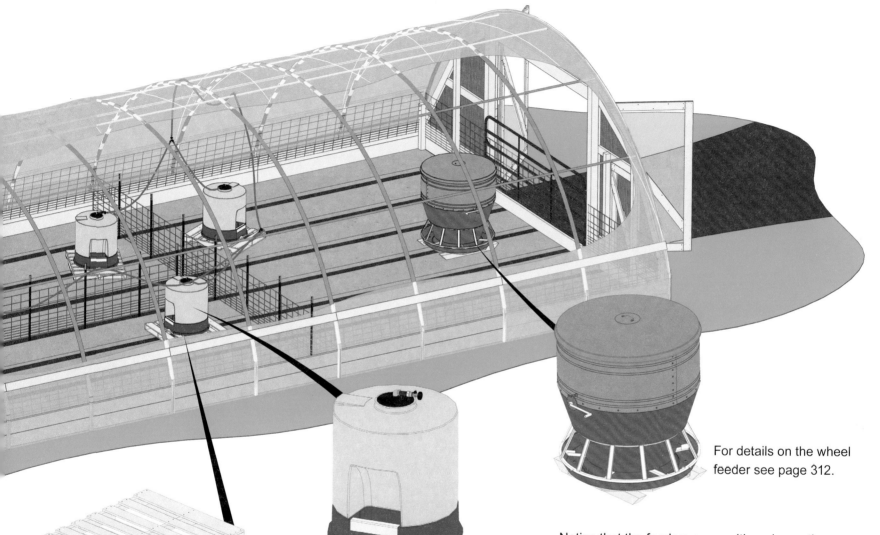

For details on the wheel feeder see page 312.

WINTER
HOOP
HOUSES

For details on pig waterers see page 310.

We always place pallets underneath of the waterers to keep them up and out of the bedding material. This ultimately keeps the water cleaner.

Notice that the feeders are positioned near the entrances, and the waterers are placed on the opposite ends to spread out high impact areas within the pens. It also encourages them to exercise by having to walk between feed and water; exercise encourages health and oxygenation in the muscles. Both muscular activity and oxygen are key to succulence and good taste.

Bulk Feeder Platform

Chickens like to scratch. Deep bedding is critical for their health in a housing situation, but it can present problems in feeders and waterers. When the chickens scratch aggressively, the bedding can get up into feeders and counteract the hygiene that the deep bedding offers.

In order to keep that from happening, these simple slatted platforms prohibit chickens scratching in the bedding immediately proximate to the feeder. It makes a huge difference in keeping the feed clean.

The slats allow the chicken manure to go through onto the bedding. Every week or so we can move the platform and the feeder to another location, which keeps the impaction zones spread around the floor. It's a simple solution to a significant problem.

The platform design can be easily scaled to meet the space requirements on the feeders you are using.

Tools

❑ Hand Saw

❑ Circular Saw

❑ Power Drill

❑ Driver Bits (for screws)

❑ Speed Square

❑ Marking Pencil

❑ Tape Measure

Wood Cutlist

Wood scraps are highlighted in red.
Do not discard until project is completed.

QTY	SIZE
☐ **14x**	1x1x8ft
☐ **4x**	2x4x8ft

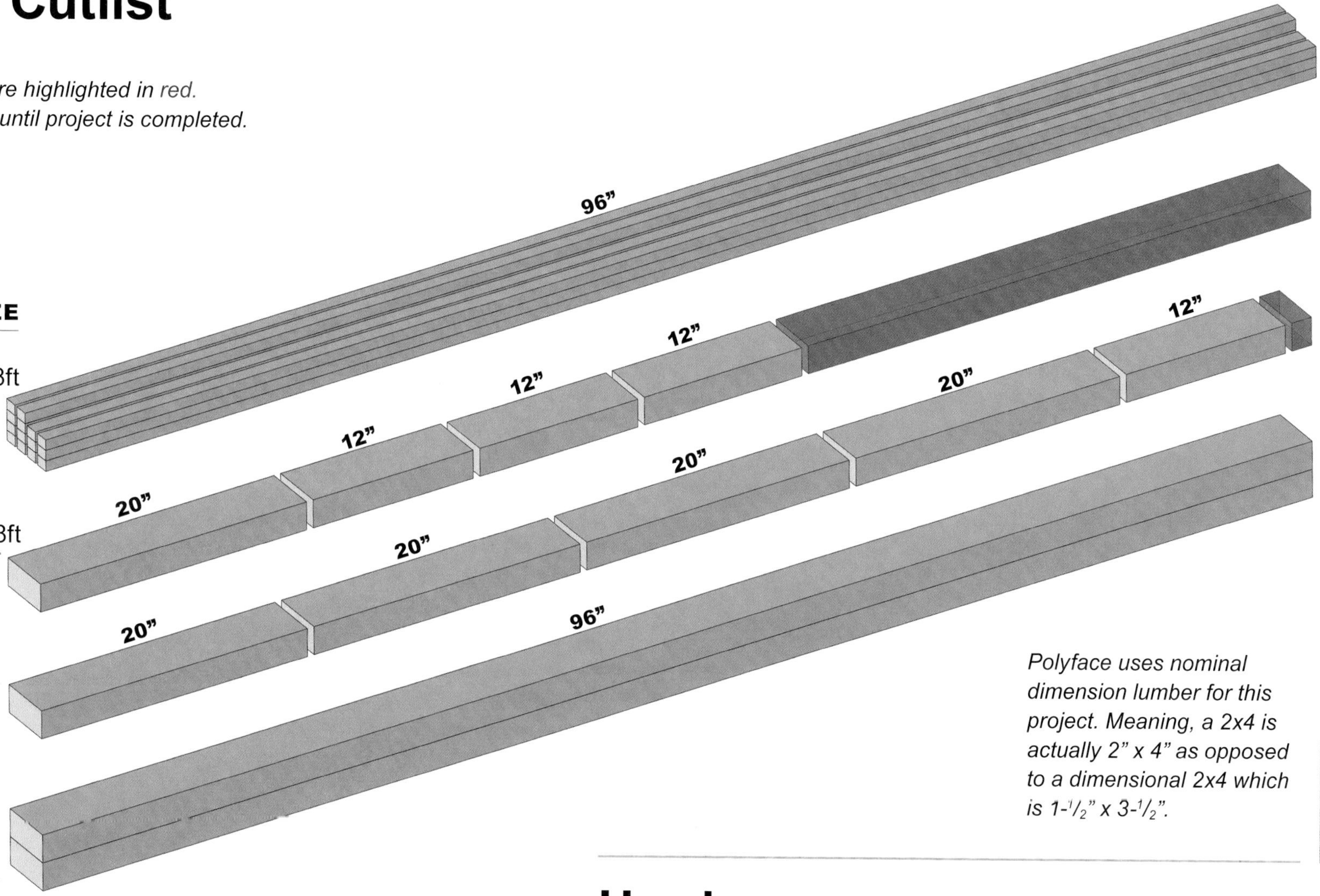

96"

12" 12" 12" 20" 12"

20" 12" 20" 20"

20" 96"

Polyface uses nominal dimension lumber for this project. Meaning, a 2x4 is actually 2" x 4" as opposed to a dimensional 2x4 which is 1-1/2" x 3-1/2".

WINTER HOOP HOUSES

Hardware

DECK SCREWS

☐ **70x**	2"	
☐ **36x**	3"	

1

1x 2x4x96"

2

4x 3" Deck Screws
2x 2x4x20"

WINTER
HOOP
HOUSES

Square and screw the two end supports first.

3

6x 3" Deck Screws
3x 2x4x20"

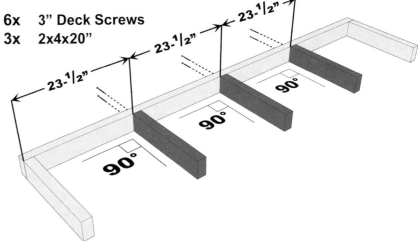

23-¹⁄₂"   23-¹⁄₂"   23-¹⁄₂"

90°   90°   90°

Space the center supports evenly. Square them and screw them in place.

4

10x 3" Deck Screws
1x 2x4x96"

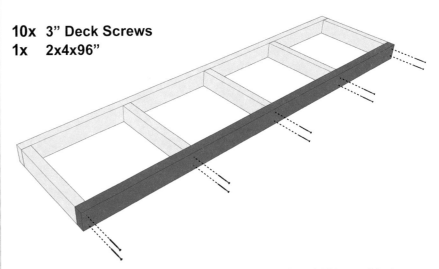

Complete the frame structure by screwing the other 96" long side in place.

5

4x   3" Deck Screws
1x   2x4x12"

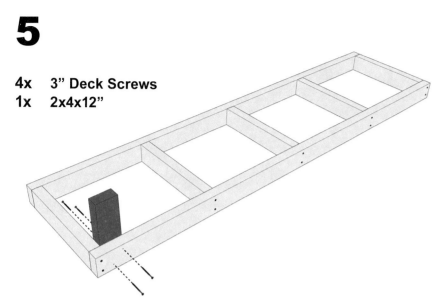

Mount the first leg with screws from both directions to ensure a sturdy connection.

7

5x   2" Deck Screws
1x   1x1x96"

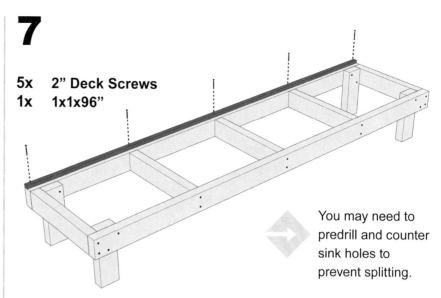

You may need to predrill and counter sink holes to prevent splitting.

Place the first 1x1" slat flush with the end and fasten with screws.

6

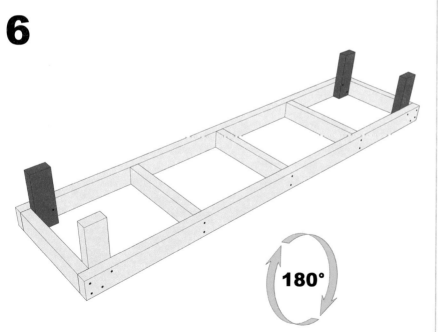

180°

Repeat step 5 for other 3 legs and turn right side up.

8

5x   2" Deck Screws
1x   1x1x96"

Cut a few 3/4" spacer blocks and use them to create a uniform spacing between the slats.

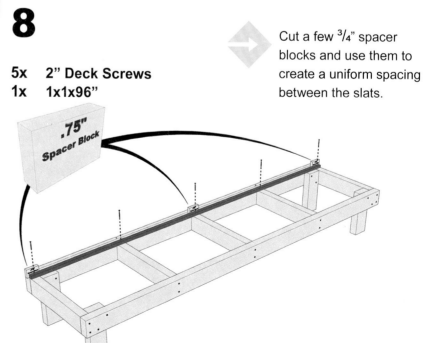

.75"
Spacer Block

WINTER HOOP HOUSES

9

60x 2" Deck Screws
12x 1x1x96"

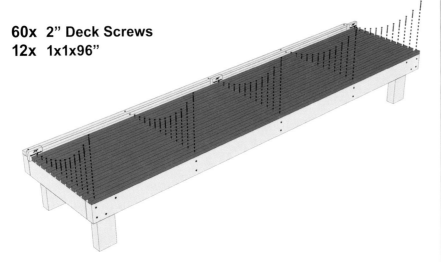

Repeat step 8 using the spacer blocks to attach the rest of the 1x1 slats.

10

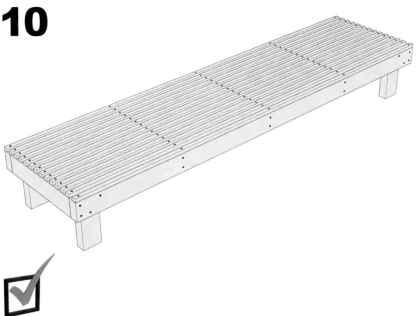

The bulk feeder platform is complete and holds 3 bulk feeders.

WINTER HOOP HOUSES

Bulk Feeder Lid

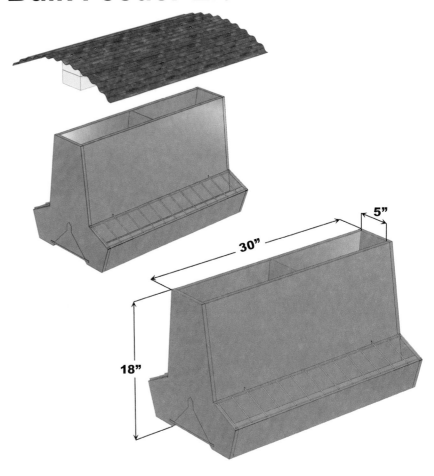

This is a typical bulk feeder that Polyface uses for chickens. These are out of old CAFO houses and are acquired at auctions. They are versatile because we use them in Eggmobiles during the summer months and in hoop houses during the winter months. They hold 100 pounds of feed and have enough linear space to handle 20 chickens at a time. That's plenty for a flock of 100 chickens.

Some of the feeders we find do not have any sort of lid on top. If that is the case, we typically fashion our own using some scrap lumber and roofing metal. This prevents chickens from roosting on top and soiling the feed. If your feeder differs in size from this one then you will need to resize your lids accordingly.

Tools

- ❑ Hand Saw
- ❑ 8" Tin Snips
- ❑ Power Drill
- ❑ Driver Bits (for screws)
- ❑ Speed Square
- ❑ Marking Pencil
- ❑ Tape Measure

Hardware

- ❑ **8x**   3" Deck Screws

- ❑ **8x**   1-½" Roofing Screws

Cutlist

QTY	SIZE
❑ ≈24"	2x4x2ft

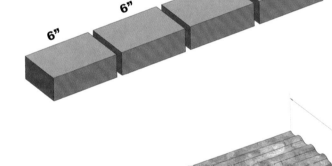

❑ 1x	Corrugated Metal Roofing

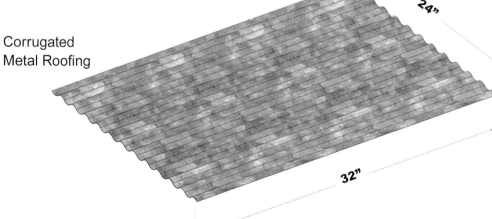

WINTER HOOP HOUSES

1

4x   3" Deck Screws
2x   2x4x6"

Screw two wood blocks together. Wood should be sized to fit into the opening at the top of the feeder you are using (see step 5).

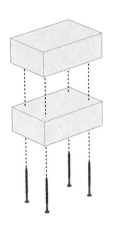

2

Repeat step 1 to create a second block.

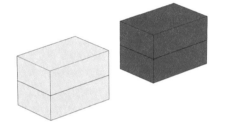

3

8x   1-½" Roofing Screws
1x   2x4x6"

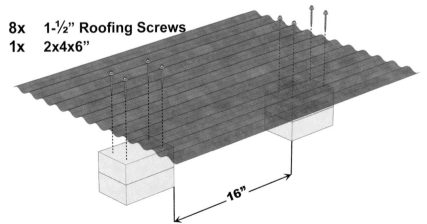

16"

Your spacing may vary. The goal is to center the metal roofing over the top of the structure.

4

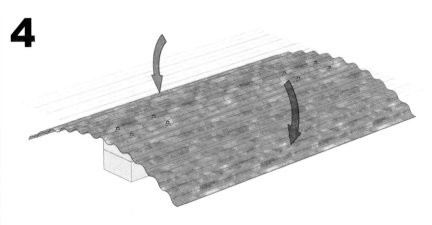

Bend metal to create a pitch in the lid.

5

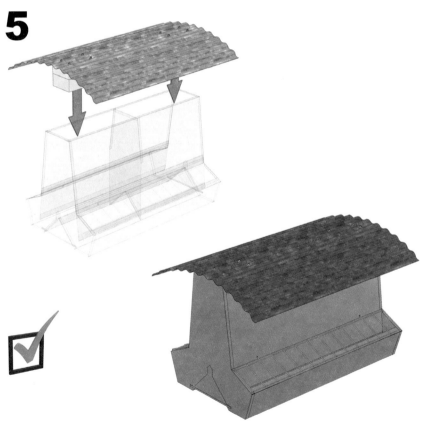

The bulk feeder lid is complete!

Standing Nest Boxes

photo courtesy of Jessa Howdyshell

The most significant management technique to ensure clean eggs is to reduce loitering around the nest boxes. Elevation is the key. If a nest box is low enough for a hen to see in as she walks around, she'll be enticed to enter and cause mischief.

Nests should always be above eye level for that reason. We mount the boxes on legs to get them up and away from the birds. That way the only birds up there are the ones laying eggs. Loitering occurs elsewhere.

This simple "nest boxes on stilts" arrangement keeps everything mobile. We can move them around wherever the configuration works best. In the summer, when we grow vegetables in these hoop houses, we move them all to one side and let tomatoes or vines grow up over them. Obviously you'd never use nest boxes specifically for a trellis, but since they're there, they work just fine. Cucumbers in

nest boxes, anyone?

When we gather eggs in the late afternoon, we fold up the perches on front, which then exclude the chickens from the nest boxes overnight. We open them in the morning. Our goal is 90% clean eggs. If you must wash more than 10%, something is wrong with your management.

Having the nest boxes high facilitates gathering because you don't have to bend over as far.

Some folks might wonder why we don't use roll-away nests. The reason is that chicken behavior is always the same: the hen enters the nest, scootches down, and then begins placing pieces of nesting material around. To deny her that basic instinctual expression seems wrong to me. So we use nests that honor that basic nest-building desire. It's part of the chickeness of the chicken.

Tools

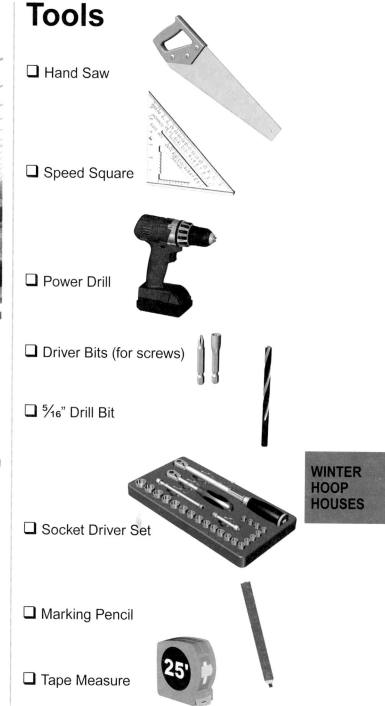

☐ Hand Saw

☐ Speed Square

☐ Power Drill

☐ Driver Bits (for screws)

☐ ⁵⁄₁₆" Drill Bit

WINTER HOOP HOUSES

☐ Socket Driver Set

☐ Marking Pencil

☐ Tape Measure

Hardware

☐ **10x**   3" Deck Screws

☐ **18x**   6-20 x ⅝" Self-tapping Screws*

☐ **2x**   ⁵⁄₁₆-18 x 1" Hex Bolts

☐ **8x**   ⁵⁄₁₆-18 x 2-½" Hex Bolts

☐ **34x**   ⁵⁄₁₆" Washers

☐ **10x**   ⁵⁄₁₆-18 Hex Nuts

Self-tapping screws are only needed if your box does not have a back on it. These are used to secure the plywood to the backside.

Wood Cutlist

Wood scraps are highlighted in red.
Do not discard until project is completed.

WINTER HOOP HOUSES

QTY	SIZE
☐ **4x**	2x4x8ft
☐ **1x**	¼" Plywood**

**Only needed if nest box does not have a back. Other material such as sheet metal, lexan, etc works great too.*

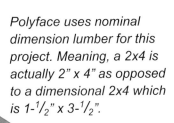

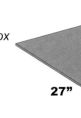

Polyface uses nominal dimension lumber for this project. Meaning, a 2x4 is actually 2" x 4" as opposed to a dimensional 2x4 which is 1-½" x 3-½".

26"

66"

66"

60"

27"

Supplies

☐ **2x**   Nest Boxes

> If your boxes are missing the mounting tabs in the top left and right corners then you will need to get creative as to how to hang and attach them. The steps outlined in this chapter will need to be altered if that is the case.

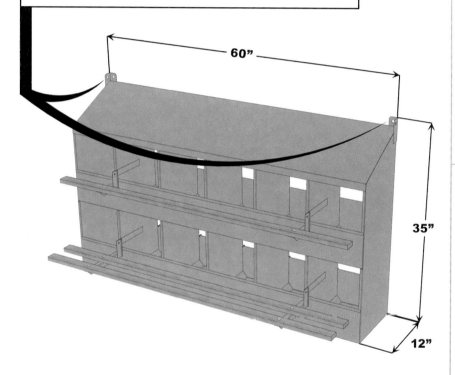

Typical nest boxes used by Polyface come from old CAFO houses and are acquired at auctions. They are typically galvanized steel and have removable pans inside of each box for cleaning purposes. This box happens to be a 12 box configuration, but we also use 10 box setups as well – both work well.

1

1x   ¼x27x60 Plywood
18x   6-20 x ⅝" Self-tapping Screws

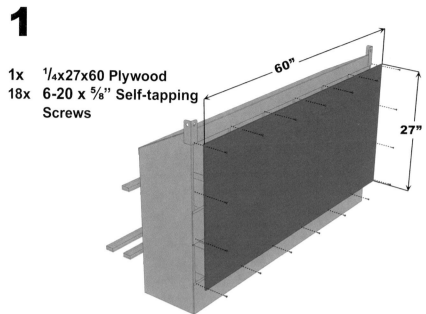

This step is only necessary if your nest boxes need a back.

2

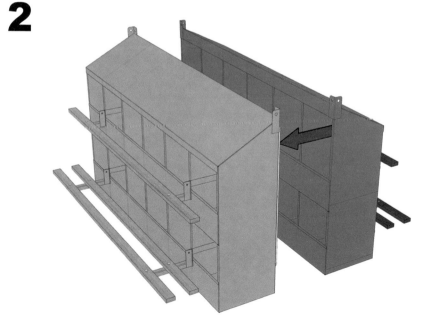

Only one of the boxes needs to have a back on it.

WINTER HOOP HOUSES

3

2x $\frac{5}{16}$-18 x 1" Hex Bolts
≈18x $\frac{5}{16}$" Washers
2x $\frac{5}{16}$-18 Hex Nuts

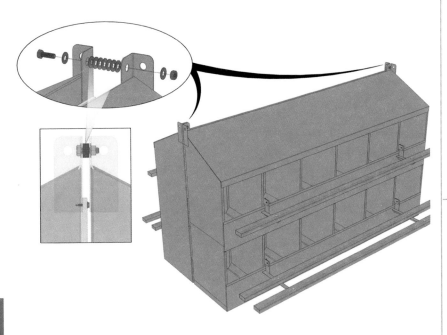

The number of washers may vary depending on the configuration of your boxes and the gap present.

4

2x   2x4x66

1 $\frac{3}{4}$"

6"

Chamfer ends as shown.

5

1x   3" Deck Screw

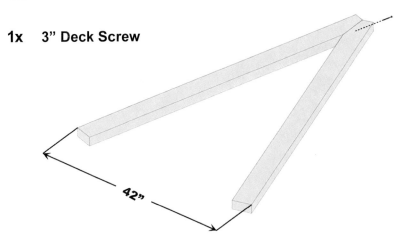

42"

The screw is to temporarily hold the frame together until it is bolted to the nest boxes.

6

4x   3" Deck Screws
1x   2x2x26

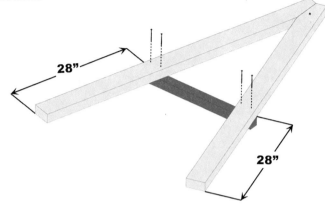

Measure and screw the A-frame support in place as shown.

8

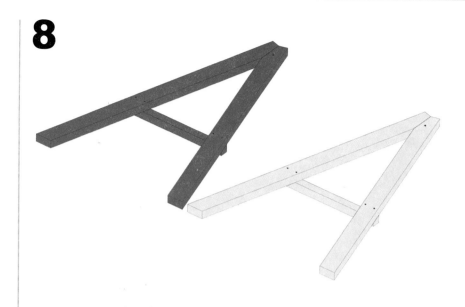

Repeat steps 4-7 to create a second A-frame.

7

Using a straight edge or chalk line, mark from corner to corner on the legs as shown. Using this method for cutting the feet will ensure that they sit flat on the ground.

9

Mount the A-frame as shown. The 2x2 should nest underneath the boxes and the top of the "A" should line up with the holes of the mounting tab above. If they do not, you may need to tweak the location of the 2x2 brace.

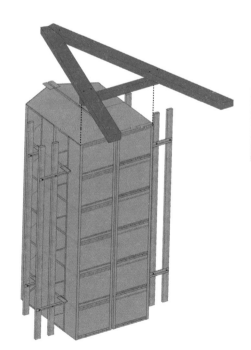

WINTER HOOP HOUSES

10

- 4x $\frac{5}{16}$-18 x 2-$\frac{1}{2}$" Hex Bolts
- 8x $\frac{5}{16}$" Washers
- 4x $\frac{5}{16}$-18 Hex Nuts

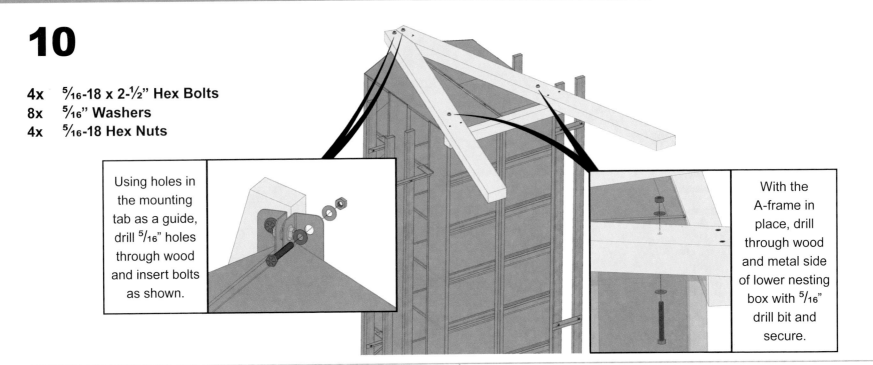

Using holes in the mounting tab as a guide, drill $\frac{5}{16}$" holes through wood and insert bolts as shown.

With the A-frame in place, drill through wood and metal side of lower nesting box with $\frac{5}{16}$" drill bit and secure.

11

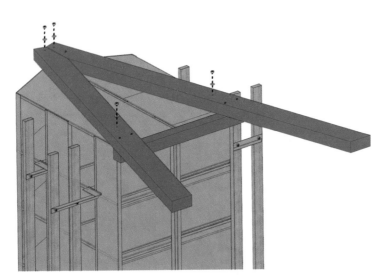

Repeat steps 9-10 on other side.

12

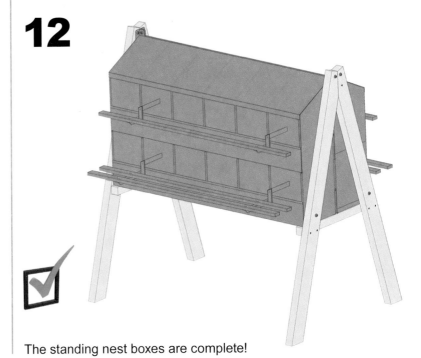

The standing nest boxes are complete!

Waterer Cover

photo courtesy of Jessa Howdyshell

Tools

- ❑ Hand Saw
- ❑ 8" Tin Snips
- ❑ Power Drill
- ❑ Driver Bits (for screws)
- ❑ Speed Square
- ❑ Marking Pencil
- ❑ Tape Measure

We use simple rubber tubs to water the chickens in the summer on pasture. They don't get in it to soil it and it's almost indestructible. In the pasture setting, they have much more to keep them occupied so they never perch on the edge of it, where they could poop into it.

But in the hoop house, their tendency to perch on the edge and the level of dust is much higher. To add a level of protection, we made this slatted box with cover. The chickens can drink easily through the slats but they can't soil the water.

We could use a different type of waterer, of course, but why? Multi-function is cheaper. The only thing we have to add to the set-up that's different than summer time is this protective cover. That keeps things simple. These tubs come in with the birds and go out with the birds in the spring. When not in use during the summer, we simply stack these protective covers in a corner.

WINTER HOOP HOUSES

465

Hardware

☐ **16x**   1-⅝" Deck Screws

☐ **16x**   2" Deck Screws

☐ **12x**   1" Roofing Screws

☐ **1x**   6 Gallon Water Drinker Tub*

See page 531 for details on the drinker tub assembly.

Polyface prefers using a 6 gallon flexible, rubber water tub 18" in diameter and 9" deep for this application. This size provides ample water for the livestock while still being small enough to relocate during pasture moves.

Cutlist

➡ **Before beginning this project:** If using a water tub other than this size, you will need to make this structure larger. See page 469 for details pertaining to modification of this design.

QTY	SIZE
☐ **1x**	Corrugated Metal Roofing

24"   24"

Wood scraps are highlighted in red. Do not discard until project is completed.

This project is built with dimensional lumber purchased from a building supply store. Meaning, a 2x4 is actually 1-1/2" x 3-1/2".

☐ **1x**   2x2x8ft

15"	15"	15"	15"	

☐ **3x**   1x6x10ft (cut into thirds lengthwise)

15"	15"	15"	15"	15"	15"	
21"	21"	21"	21"	15"	
21"	21"	21"	21"	15"	

1

1x   2x2x15"

2

1x   2" Deck Screw
1x   1x2x21"

Square and fasten the bottom
rail with a single screw as shown.
(You may need to predrill
holes to prevent
splitting).

3

1x   2" Deck Screw
1x   1x2x21"

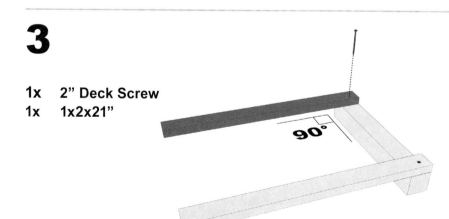

4

2x   1-⅝" Deck Screws
1x   1x2x15"

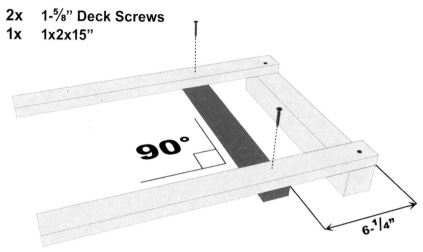

6-¼"

5

2x   1-⅝" Deck Screws
1x   1x2x15"

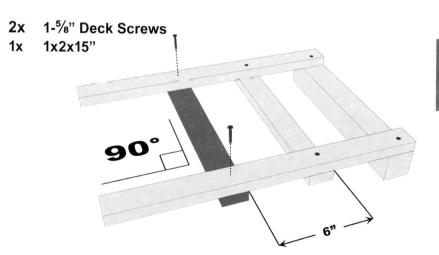

6"

Space vertical slats as shown. See the modification advice at the end of
this section prior to deviating from the dimensions described.

**WINTER
HOOP
HOUSES**

467

6

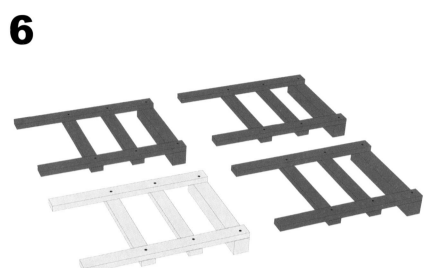

Repeat steps 1-5 to create 3 more identical sides.

8

2x   2" Deck Screws

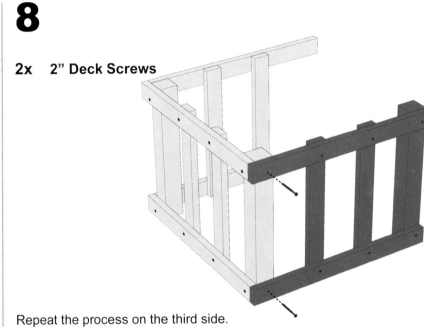

Repeat the process on the third side.

7

2x   2" Deck Screws

Square and fasten the two sides together.

9

4x   2" Deck Screws

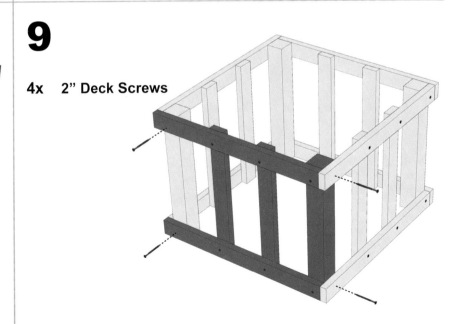

Finish attaching the fourth and final side.

10

≈12x   1" Roofing Screws
1x   24x24" Corrugated Roofing Panel

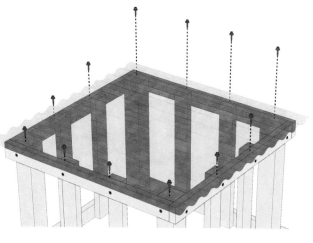

Center the roofing metal over the frame and secure with roofing screws.

11

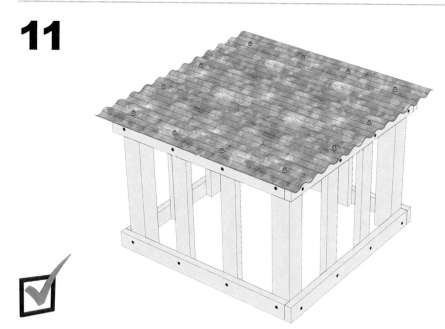

The waterer cover is complete!

12

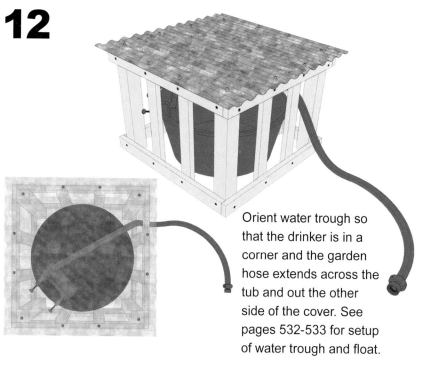

Orient water trough so that the drinker is in a corner and the garden hose extends across the tub and out the other side of the cover. See pages 532-533 for setup of water trough and float.

Modify The Design

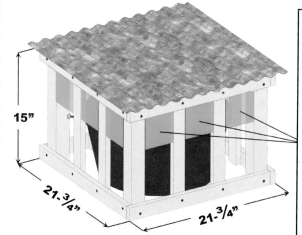

15"

21-³/₄"

21-³/₄"

If you must make this structure larger to accommodate a larger water tub, be aware of the spacing between vertical rails as well as the overall height above the rim of the water tub.

The green squares are approximately 4x4" and represent the space chickens have to get their heads into the drinker. It is critical that this space between the wood rails is no larger or else chickens will be able to squeeze through and will get trapped in the water. (We've had it happen).

WINTER HOOP HOUSES

469

Mobile Mezzanine

WINTER HOOP HOUSES

Sometimes the simplest things can make the biggest differences. By now you know here at Polyface we're addicted to multiple use infrastructure and fully occupying space. With that in mind, we wanted to put young piggies in the hoop houses with the chickens but knew we couldn't let the pigs have access to the much more fragile chicken waterers and feeders.

A 4ft x 8ft frame made from 2x4s and 1x2" slats provides us with a 4ft high platform. When we put multiple units together, it creates a modular and mobile mezzanine floor. We can tailor the quantity and orientation of the platforms in any number of ways to suit our needs for a particular season.

This also adds floor space to the hoop house by utilizing an additional layer. It's like adding a second floor. That means we can put many more chickens in the space. The chickens tend to spend lots of rest time up on the slats and of course that's where we place their feeders and waterers. The slats are assembled just like the floor in the Eggmobile; close enough to comfortably walk but far enough apart to let manure fall through: ideally 1-$\frac{3}{4}$" spacing between 2" wide slats.

For clean out, we pick these up with the front end loader and take them outside. If you didn't want pigs underneath, this mezzanine would also work with sheep or even small calves. It's a simple way to add a species and/ or create a lot more floor space to increase production in the hoop house.

For summer storage, we stack a second layer upside down; this way they take up half the space they do when in use. In the summer, they can serve as poor-boy trellises for climbing vegetables. The mobile mezzanine greatly increases the versatility and productivity of the hoop house.

Tools

- ☐ Hand Saw
- ☐ Power Drill
- ☐ Driver Bits (for screws)
- ☐ Speed Square
- ☐ Marking Pencil
- ☐ Tape Measure
- ☐ Circular Saw

Wood Cutlist

Wood scraps are highlighted in red.
Do not discard until project is completed.

QTY	SIZE
☐ 13x	1x2x8ft
☐ 4x	1x3x8ft
☐ 6x	2x4x8ft

96"

46"

50"

96"

44"

48"

Polyface uses nominal dimension lumber for this project. Meaning, a 2x4 is actually 2" x 4" as opposed to a dimensional 2x4 which is 1-1/2" x 3-1/2".

WINTER HOOP HOUSES

Hardware

DECK SCREWS

☐ 84x	2"
☐ 32x	3"

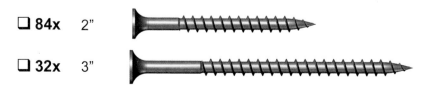

1

1x   2x4x96"

2

4x   3" Deck Screws
2x   2x4x44"

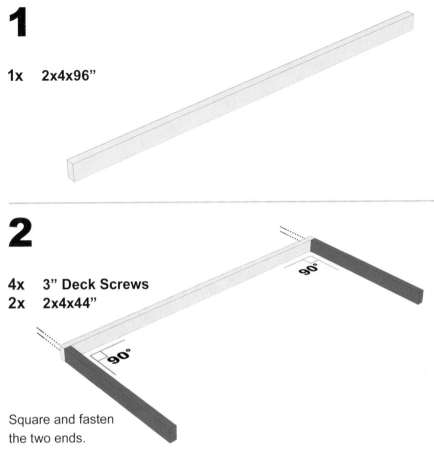

Square and fasten
the two ends.

3

4x   3" Deck Screws
2x   2x4x44"

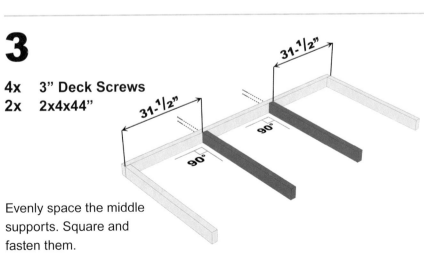

Evenly space the middle
supports. Square and
fasten them.

4

8x   3" Deck Screws
1x   2x4x96"

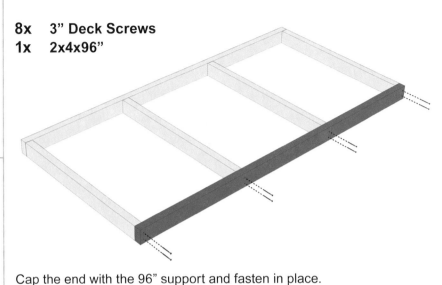

Cap the end with the 96" support and fasten in place.

5

2x   3" Deck Screws
2x   2x4x48"

Install two legs and
fasten with a single
screw. (The reason
for only one screw is
so that the leg can be
maneuvered easily while
squaring and adding
braces. Additional
fasteners will be added
in step 11.)

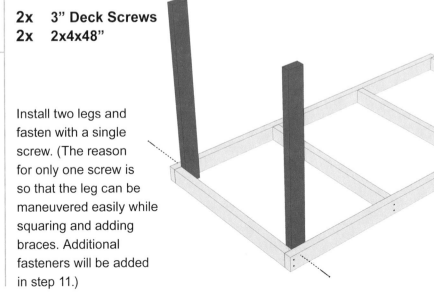

6

4x   2" Deck Screws
1x   1x3x50"

Square the leg in the direction shown and fasten the diagonal brace.

8

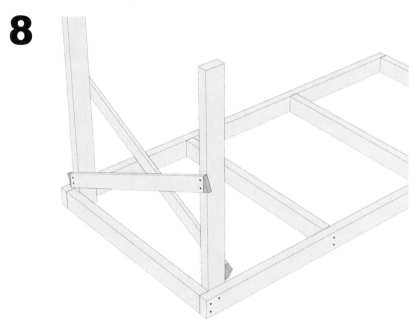

Trim edges (shown in green) in four places with a saw.

7

4x   2" Deck Screws
1x   1x3x50"

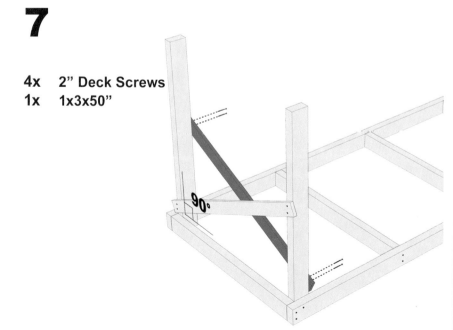

Repeat step 6 for the other leg.

9

4x   2" Deck Screws
1x   1x3x46"

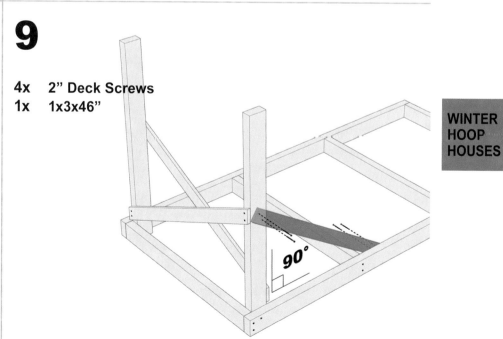

Square the leg in the direction shown and fasten the diagonal brace.

WINTER HOOP HOUSES

10

4x   2" Deck Screws
1x   1x3x46"

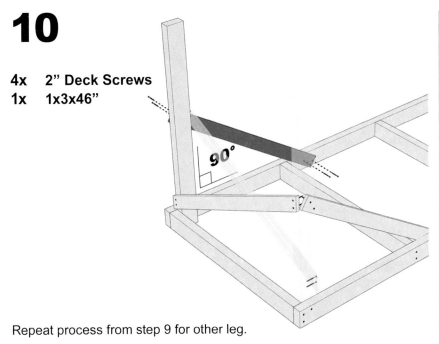

Repeat process from step 9 for other leg.

11

6x   3" Deck Screws

WINTER
HOOP
HOUSES

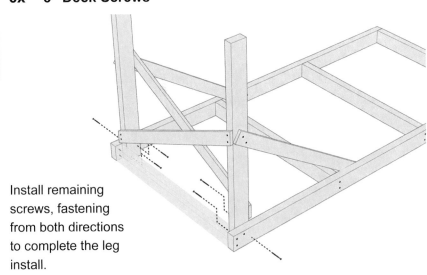

Install remaining
screws, fastening
from both directions
to complete the leg
install.

12

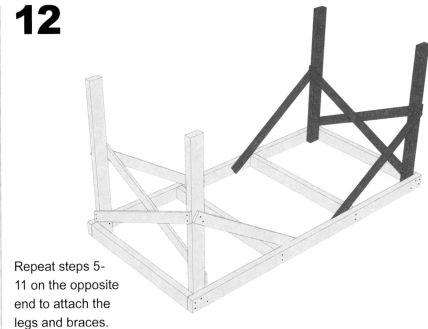

Repeat steps 5-
11 on the opposite
end to attach the
legs and braces.

13

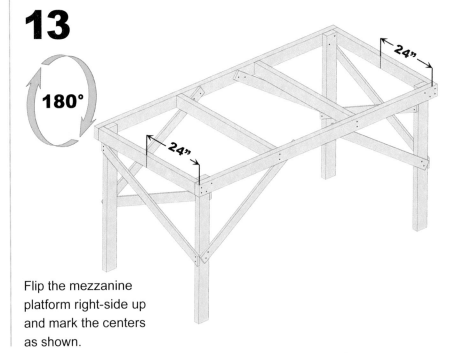

Flip the mezzanine
platform right-side up
and mark the centers
as shown.

14

4x   2" Deck Screws
1x   1x2x96"

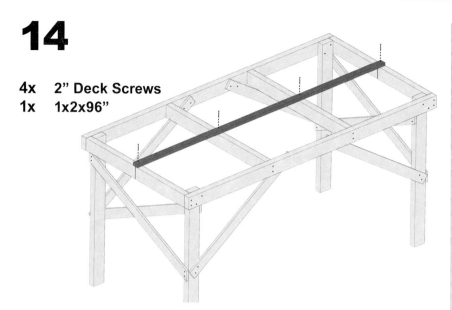

Center the first 1x2 slat on the marks and fasten. (You may need to predrill and counter sink holes to prevent splitting.)

15

4x   2" Deck Screws
1x   1x2x96"

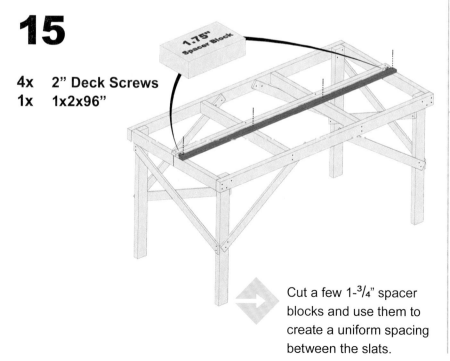

Cut a few 1-³/₄" spacer blocks and use them to create a uniform spacing between the slats.

16

20x 2" Deck Screws
5x   1x2x96"

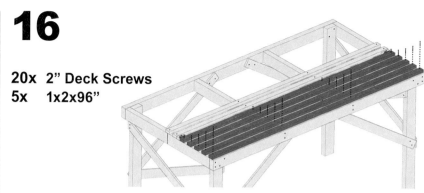

Repeat step 15 using the spacer blocks to attach the 1x2 slats.

17

24x 2" Deck Screws
6x   1x2x96"

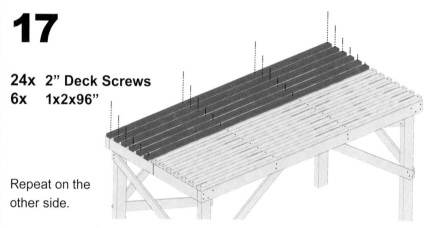

Repeat on the other side.

18

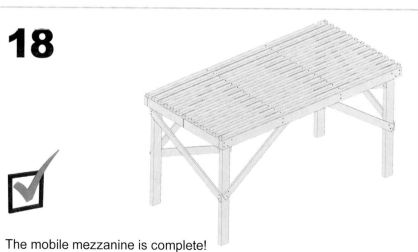

The mobile mezzanine is complete!

WINTER HOOP HOUSES

BULK FEED BUGGY

photo courtesy of Jessa Howdyshell

Farm efficiency hinges on minimizing trips to the field with heavy equipment, minimizing material toting, and decreasing chore time. Feeding chickens out in the field certainly qualifies as problematic for all these efficiencies.

The bulk feed buggy offers a way to have 2-4 weeks' worth of feed storage on site. It's accessible right where the feeders are and can even be hooked to the portable shelter to eliminate any extra move time. Trains can be in your pasture; they don't have to be on tracks. I've seen too many folks load 5 gallons buckets of feed onto an ATV, cart them out to the field, unload them; this is all extra toting and handling. Inventorying your feed right next to the feeders saves both vehicular and physical energy and time.

Normally we want an axle that can handle 2 tons. With about 900 pounds of tongue weight, that gives plenty of wiggle room in case you overfill. The bulk feed buggy is really nothing more than a box tight enough to hold feed.

The front and back utilize a garner board set up. Back in the old days when people used to handle grain by hand, they would shovel it into tight rooms in the barn. Because grain flows easily, they needed a way to put on retainer boards as they filled the room. And they needed an easy way to remove these boards as they shoveled

the room out.

Precisely cut boards for the width of the room slid down a slot made by two furring strips. Our bulk feed buggy uses this basic idea, except instead of precise dimensions and furring strips, we have a space board that lets us shove these boards in from the side. Having these garner boards on both front end and back end reduces distance to the feed when you're standing outside. Why have to step in if you can help it?

The roof is flat and either completely removable or has a strip that's removable. Held down with bungees, the removable roof makes it easy to auger feed into the buggy. Always remember to fill it with feed, starting in the front,

and always empty it starting in the back in order to maintain correct balance.

Like the Eggmobile, since it's always parked on a bit of a slant, as long as the roofing pieces are long enough to go completely side to side, it doesn't matter which way they're slanted; the water runs off either direction.

If you're handy at all, you can make these buggies for a couple hundred dollars. If you can find used gravity wagons for scrap metal prices, they're certainly an option. But normally those sell for $1,000 or more and they're more complicated and heavier. The simple homemade feed buggy is a poor-boy way to create efficiencies without spending much money.

Tools

☐ Power Drill

☐ ½" HSS Drill Bit

☐ Drill Bit for Torx Screws*

☐ Driver Bits (for screws)

☐ Marking Pencil

☐ Tape Measure

☐ Hammer

☐ Speed Square

☐ 4ft Level

*Torx floor screws require a pilot hole to be drilled. Consult screw supplier for drill bit size.

☐ Welder

☐ Circular Saw

☐ Grinding Wheel

☐ Cut-Off Wheel

☐ Angle Grinder

☐ Socket Driver Set

Hardware

☐ **2x** Double-Sided Spring Hooks

These hooks are used to hold the lid on to the buggy. Bungee cords also work very well.

☐ **2x** 2-½" Barrel Bolts With Fasteners

☐ **6x** Brass Utility Hinges With Fasteners

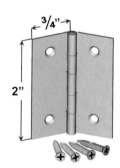

3/4"

2"

☐ **4x** Fencing Staples

MATERIALS HANDLING

☐ **≈16x** 1-½" Torx Floor Screws*

☐ **225x** 2" Deck Screws

☐ **32x** 3" Deck Screws

☐ **56x** 20D Galvanized Nails*

*It is recommended that these nails be either spiralled or ribbed for superior holding power

☐ **≈60x** 1" Roofing Screws

☐ **1x** Adjustable Clevis Hitch Part 18078 (see note B)

☐ **1x** Bolt-On A-Frame Channel-Up Part 15349-52 (see note B)

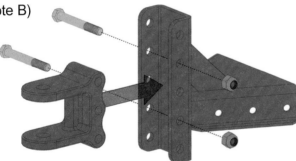

MATERIALS HANDLING

A savvy person could come up with a "poor man's" hitch that would do the job. However, in our effort to simplify this build for a novice and provide a final product that will be safe and durable, we have opted to showcase an off the shelf option that can handle the abuse. A bonus of this option is the flexibility to change the hitch type (ball, clevis, pintle, etc) according to one's needs. It also allows adjustability in height which is beneficial, especially when hooking a Gobbledygo behind it.

Notes:

A.) The trailer jack we prefer is one with a tube mount configuration. The small steel tube can be easily welded to a chassis and has proven to be a very durable design. In 2020, this jack could be found at Tractor Supply stores for less than $60. (Part # 11630699)

B.) In 2020, these specific trailer parts can be found online from Croft® Trailer Supply. You will need to purchase the appropriate hardware to mount the hitch. Refer to installation manual for details.

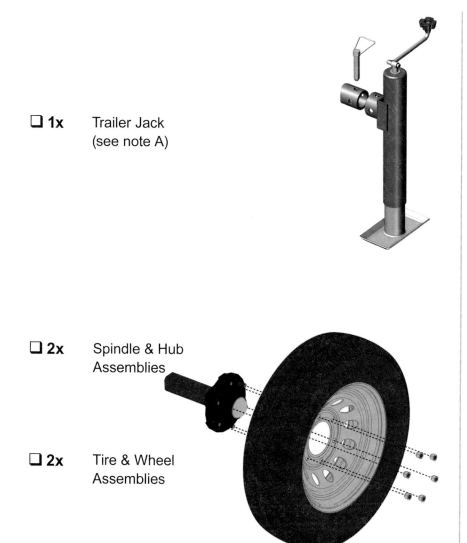

☐ **1x**   Trailer Jack
(see note A)

☐ **2x**   Spindle & Hub
Assemblies

☐ **2x**   Tire & Wheel
Assemblies

Spindle, hub, and tire/wheel assemblies come in all shapes and sizes. You should look for something in the 5-7K capacity range. If you are buying everything separately be aware that wheel and bolt patterns vary. Make sure that the combination you purchase will pair up. Another option is to purchase a pre-existing axle and chop it to narrow the track width to size. The key to keeping your costs down on this project is to try and use what you have available.

On this project we have chosen to illustrate a chassis made of steel C-channel. While we have many buggies in operation with undercarriages made from heavy oak beams, we realize that this is not cost effective for those without access to timber and a lumber mill of their own. For the masses, we have found that it is in fact cheaper and lighter to build the chassis from C-Channel. While this is indeed a durable design, we reiterate that we do not want to stifle your imagination in this area. We've got buggies made from old truck frames, old trailers, etc. If you are an innovator, then use what you've got and glean from this chapter what you can!

MATERIALS HANDLING

Wood Cutlist

Wood scraps are highlighted in red.
Do not discard until project is completed.

QTY	SIZE
☐ **15x**	2x4x8ft
☐ **2x**	1x4x8ft
☐ **5x**	1x3x8ft
☐ **7x**	1x6x8ft
☐ **1x**	2x6x8ft

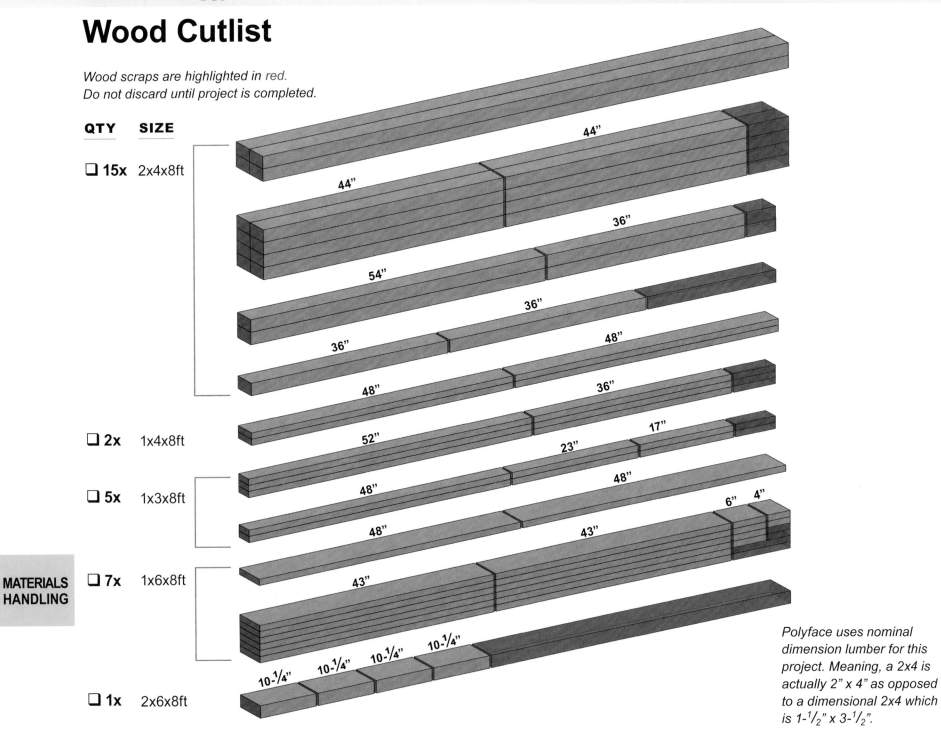

44"
44"
36"
54"
36"
36"
48"
48"
36"
52"
23"
17"
48"
48"
48"
43"
6"
4"
43"
10-¼"
10-¼"
10-¼"
10-¼"
10-¼"

MATERIALS HANDLING

Polyface uses nominal dimension lumber for this project. Meaning, a 2x4 is actually 2" x 4" as opposed to a dimensional 2x4 which is 1-¹/₂" x 3-¹/₂".

QTY	SIZE
☐ **1x**	4x8x½" Plywood Sheet*

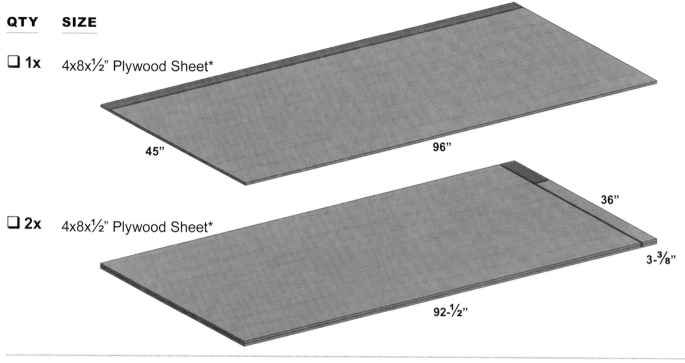

45" 96"

While we typically use plywood or sheathing that's treated for exposure to the elements, we do not like to use it in situations like this because it is directly touching feed. Humidity and temperature changes cause chemicals in treated wood to leach. While we have no data to support this, we still prefer NOT to take any chances.

QTY	SIZE
☐ **2x**	4x8x½" Plywood Sheet*

36"

3-³⁄₈"

92-½"

Metal Cutlist

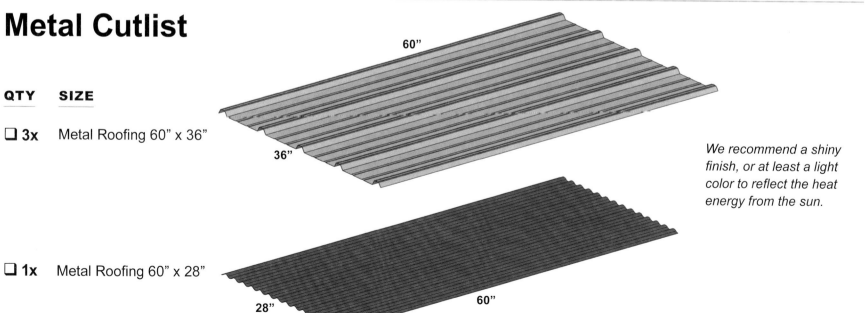

60"

QTY	SIZE
☐ **3x**	Metal Roofing 60" x 36"

36"

We recommend a shiny finish, or at least a light color to reflect the heat energy from the sun.

MATERIALS HANDLING

QTY	SIZE
☐ **1x**	Metal Roofing 60" x 28"

28" 60"

Metal Cutlist

Measurements in red are approximates.
Cut each piece to fit.

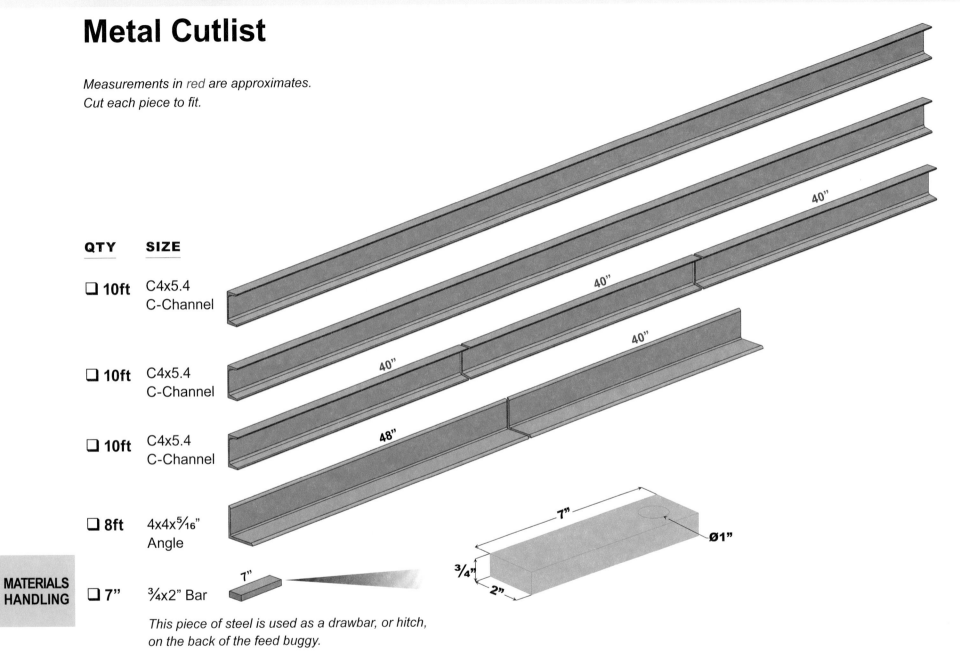

QTY	SIZE
☐ 10ft	C4x5.4 C-Channel
☐ 10ft	C4x5.4 C-Channel
☐ 10ft	C4x5.4 C-Channel
☐ 8ft	4x4x⁵⁄₁₆" Angle
☐ 7"	¾x2" Bar

This piece of steel is used as a drawbar, or hitch, on the back of the feed buggy.

1

1x   C-Channel 10ft

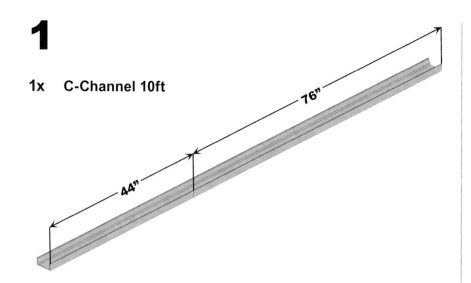

Mark all sides of the c-channel. This will be the location of the notch and bend in the next steps.

2

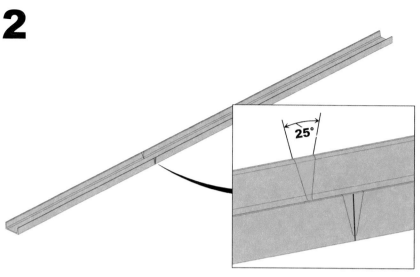

Cut the notches in the shape of a "V" at the angle prescribed.

3

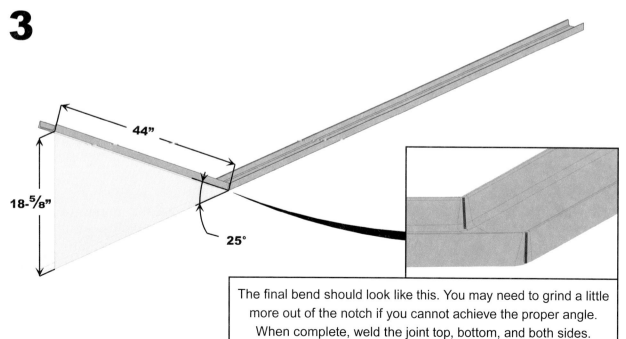

The final bend should look like this. You may need to grind a little more out of the notch if you cannot achieve the proper angle. When complete, weld the joint top, bottom, and both sides.

With the notch cut, the next step is to create the bend. One way to determine whether or not your angle is correct is to do some simple trigonometry using inputs that we already know. Since we know that we want a 25° angle and that the length of the hypotenuse of our triangle is 44", we can thereby calculate the opposite side of the triangle (vertical measurement) using the following equation:

Sin(25°)*Hypotenuse=Opposite

0.42262*44"=18.6" ≈ 18-⁵⁄₈"

MATERIALS HANDLING

4

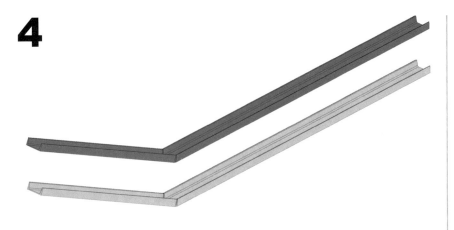

Repeat steps 1-3 to create an identical second side.

5

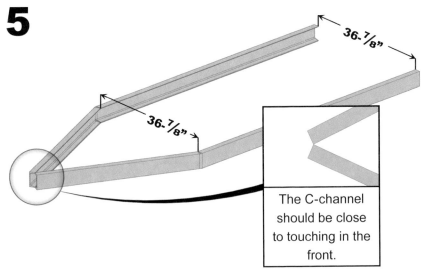

36-⁷⁄₈"

36-⁷⁄₈"

The C-channel should be close to touching in the front.

The front and back dimensions may vary slightly from shown. What is important, is that they equal each other. That ensures that the two c-channel frame rails are parallel. Record this measurement for use in step 6.

6

1x   C-Channel 40"

Cut notches into the ends of the c-channel so that it will fit inside of another member perpendicular to it. Remove the pieces shown in green.

a)

1-³⁄₈"

b)

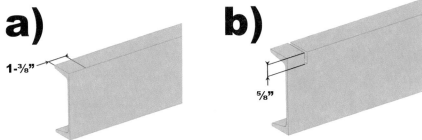

⁵⁄₈"

c)

d)

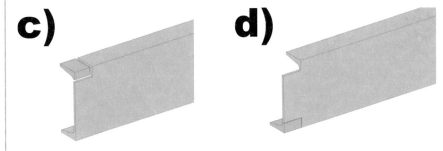

e)

Dimension from step 5

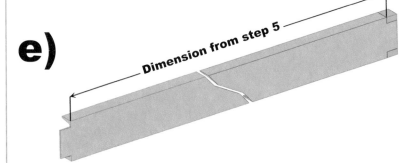

7

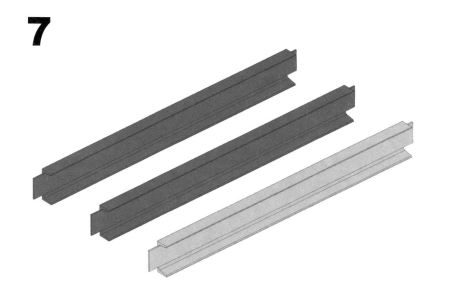

Repeat step 6 to create a total of 3 braces.

8

3x   C-Channel From Step 6

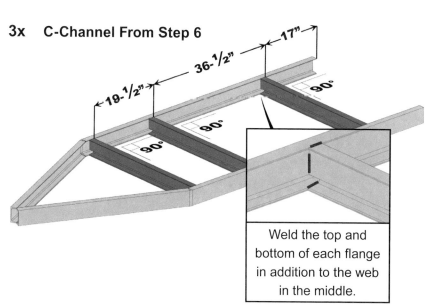

18-¹/₂"   36-¹/₂"   17"

90°   90°   90°

Weld the top and bottom of each flange in addition to the web in the middle.

9

1x   4x4x⁵⁄₁₆" Angle, 40" Long

40"

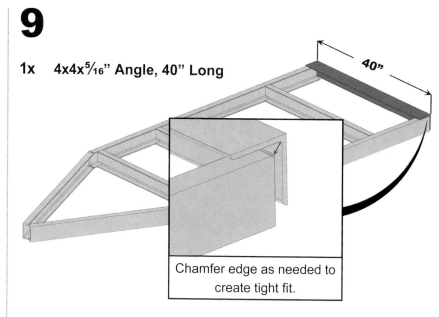

Chamfer edge as needed to create tight fit.

Cut length of angle iron to fit. Weld in place.

10

1x   ¾x2x7" Drawbar

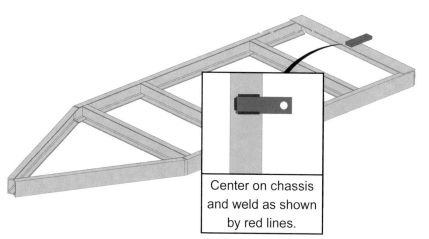

Center on chassis and weld as shown by red lines.

MATERIALS HANDLING

11

1x   **Bolt-on A-Frame Channel-up Assembly**

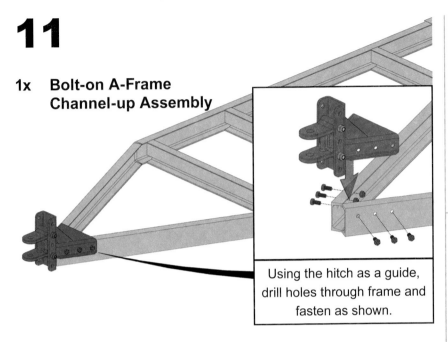

Using the hitch as a guide, drill holes through frame and fasten as shown.

12

1x   **Weld-on Mount For Trailer Jack**

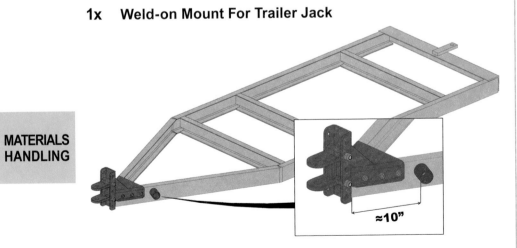

≈10"

Choose a location relatively close to the hitch to mount your jack. Weld mount to the frame.

13

1x   **4x4x⁵⁄₁₆" Angle, 48" Long**

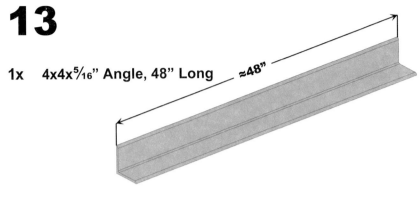

≈48"

14

2x   **Spindle & Hub Assembly**

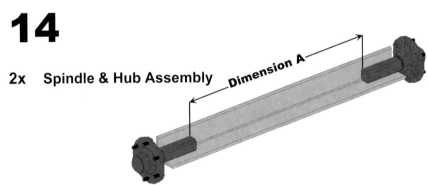

Dimension A

Since tires and hub/spindle configurations vary in size, you will need to figure out "Dimension A" on your own. An inside dimension between the tires of 48" will provide approximately 1-½" of clearance on either side from the walls of the buggy. Record "Dimension A" to position spindles on angle iron.

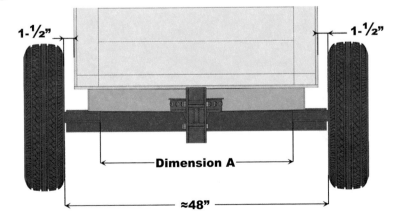

1-½"   1-½"

Dimension A

≈48"

15

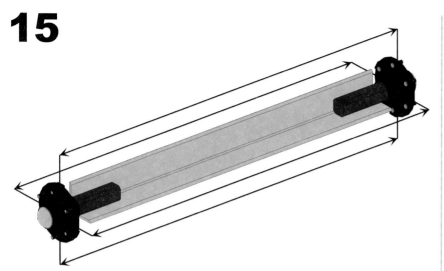

Carefully check measurements from the face of each hub in several directions to ensure that they are parallel with each other. Tack weld and double check measurements.

16

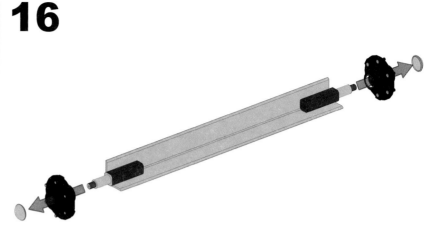

Remove hubs and bearings prior to finish welding. (Excessive heat can damage the bearing components).

17

Find the center of the chassis and make sure the axle is perpendicular to it. The center of the axle should also be 35-½" from the rear.

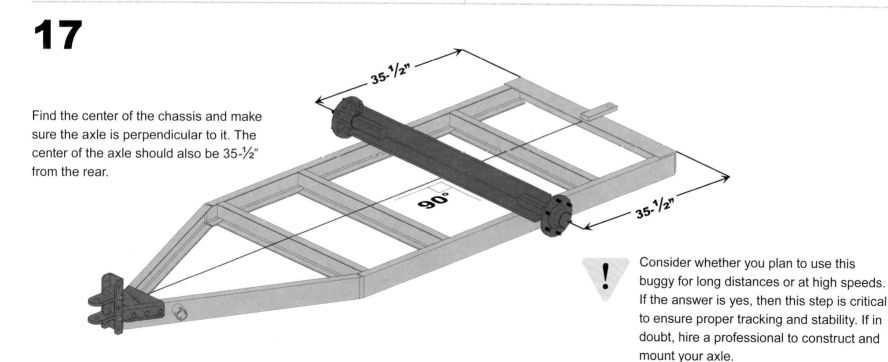

35-½"

90°

35-½"

! Consider whether you plan to use this buggy for long distances or at high speeds. If the answer is yes, then this step is critical to ensure proper tracking and stability. If in doubt, hire a professional to construct and mount your axle.

MATERIALS HANDLING

18

2x   Tire & Wheel Assembly
1x   Trailer Jack

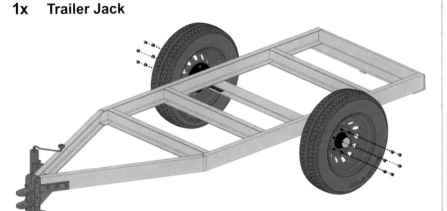

Flip chassis over using a tractor or loader. Install tires/wheels and jack.

20

1x   45x96x½" Plywood Sheet

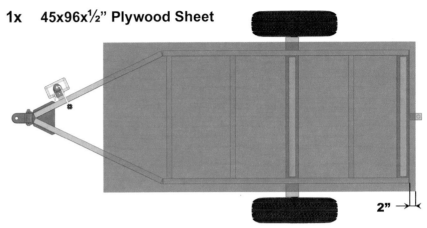

2"

Position the plywood approximately 2" from the back and centered on the chassis.

19

2x   2x4x36"

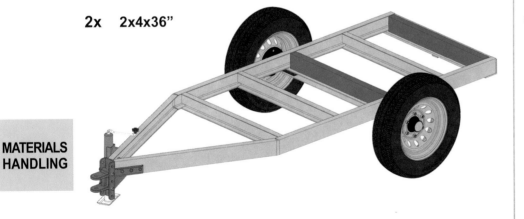

These pieces of wood need to be the same height as the c-channel (4"). They will be fastened to the plywood decking in a later step. Cut them if needed to fit flush.

21

≈16x   1-½" Torx Floor Screws

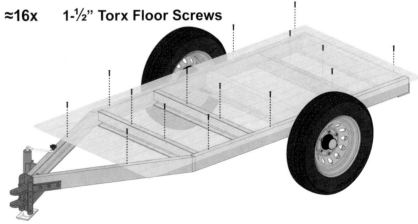

Pre-drill one hole at a time, starting with the corners. Once the corners are fastened, fill in between.

22

6x 2" Deck Screws

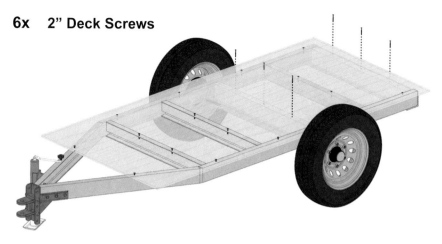

Be sure that the 2x4x26" blocking from step 19 is still in it's proper position, and fasten with deck screws.

23

2x 2x4x96"

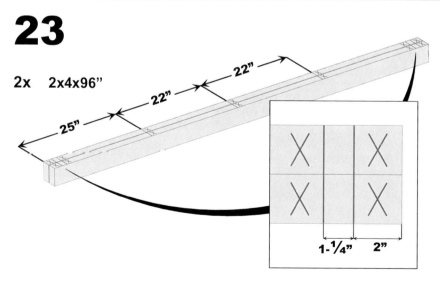

The 1-¼" dimension assumes that you are using 1" thick garner boards per the cutlist provided in this chapter. Should you decide to use different sized lumber, then this dimension should be approximately 1/4" larger than the thickness of the garner boards you plan to use.

24

7x   2x4x44"
28x 20D Galvanized Nails

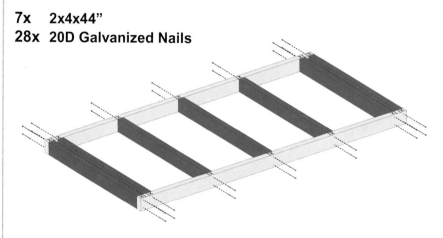

Align studs with the marks on the top and bottom plates. Fasten everything in place.

25

1x   48x92-½x½" Plywood Sheet
36x 2" Deck Screws

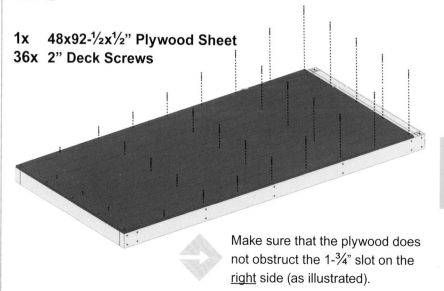

Make sure that the plywood does not obstruct the 1-¾" slot on the <u>right</u> side (as illustrated).

Align the plywood with the studded frame and fasten in place.

**MATERIALS
HANDLING**

26

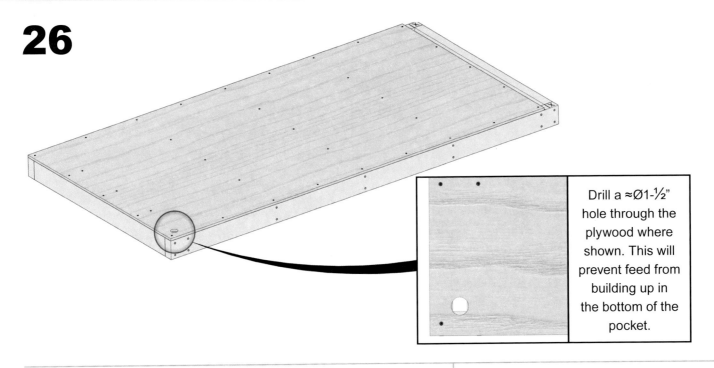

Drill a ≈Ø1-½" hole through the plywood where shown. This will prevent feed from building up in the bottom of the pocket.

Build A Door For The Garner Boards

27

1x **36x3-⅜x½" Plywood Sheet**

36"

3-⅜"

Garner boards do not move when there is weight/grain pushing against them. However, when there is no weight holding them in place, vibrations during transit rattle these boards loose. One option is to take all of the boards out and stow them. The other option, shown here, is to make a simple door device to hold these boards in place. This is purely an optional feature, but it is nice to have.

28

3x **Brass Utility Hinge With Fasteners**

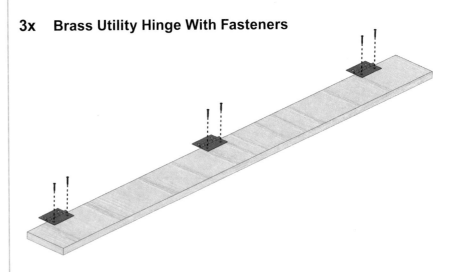

Space the three hinges evenly apart.

29

1x   2-½" Barrel Bolt with Fasteners

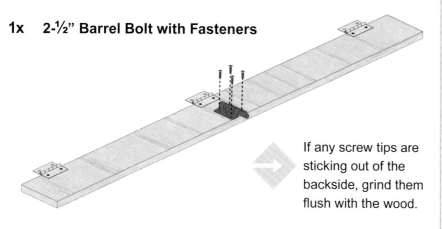

If any screw tips are sticking out of the backside, grind them flush with the wood.

Mount the barrel bolt approximately where shown.

30

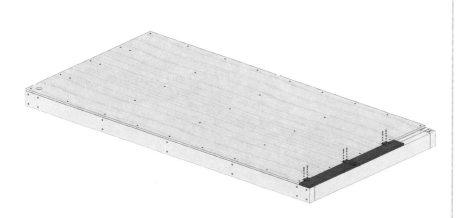

Install the door as shown. Door must not stick out past the wall.

31

1x   1x6x4"
4x   2" Deck Screws

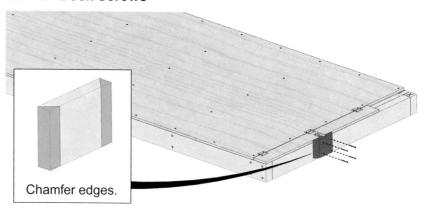

Chamfer edges.

Align 1x6x4" block with the barrel bolt and fasten it in place.

32

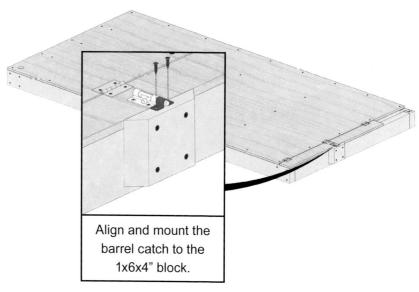

Align and mount the barrel catch to the 1x6x4" block.

MATERIALS HANDLING

33

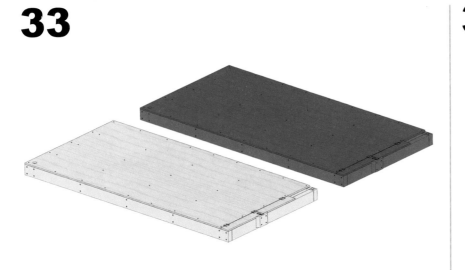

Repeat steps 23-32 to create a second identical side

34

10x 2" Deck Screws

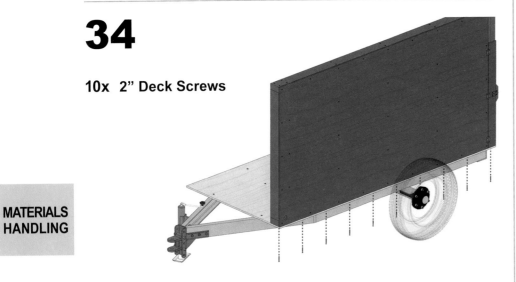

Carefully note the orientation of the wall panel—the restraint door should be in the back of the buggy. Align it flush with the plywood deck and fasten screws from the bottom of the deck through the bottom plate of the wall.

35

Repeat step 34 on the other side. Note that on the other side, the restraint door should now be in the front of the buggy.

36

2x 2x4x44"
16x 3" Deck Screws

Fasten top braces to the front and back making sure that the side walls are square.

37

2x   2x4x36"
10x 2" Deck Screws

Cut 2x4s to fit and fasten from the bottom as shown.

39

16x 3" Deck Screws

Mount cross braces inside as shown. Check sides again for square before securing braces.

38

2x   2x4x54"

a)

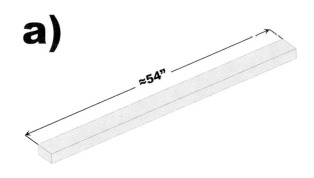

≈54"

Cut and fit cross braces to be used in step 39 to stabilize the box structure.

b)

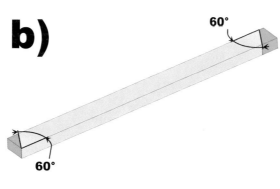

60°

60°

Miter angles on each end as shown. Check fit and trim as necessary.

c)

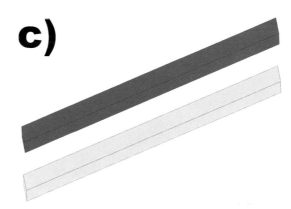

Repeat process to create a second cross brace.

MATERIALS HANDLING

40

4x   1x4x48"
32x   2" Deck Screws

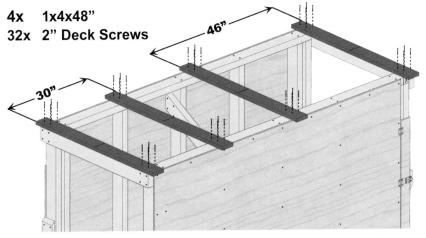

Fasten roof supports as shown. Mark a centerline down the middle to aid in the positioning of the purlins in the following steps.

42

3x   1x3x52"
12x   2" Deck Screws

Fasten purlins as shown.

41

3x   1x3x36"
12x   2" Deck Screws

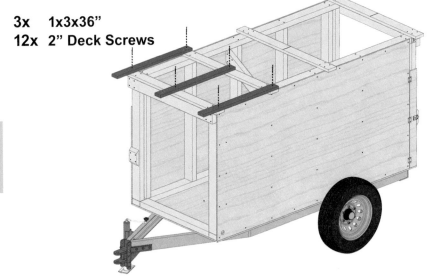

Fasten purlins as shown.

43

1x   60" Roofing Panel
15x   1" Roofing Screws

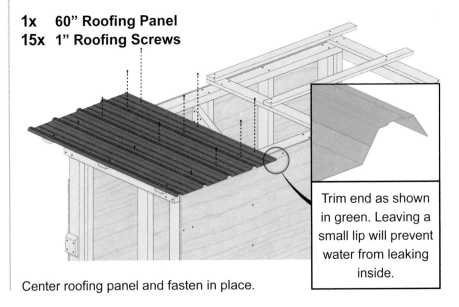

Trim end as shown in green. Leaving a small lip will prevent water from leaking inside.

Center roofing panel and fasten in place.

44

1x 60" Roofing Panel
12x 1" Roofing Screws

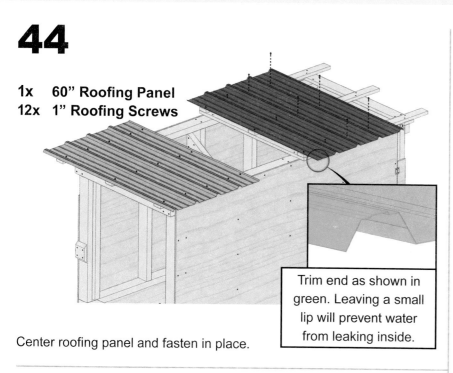

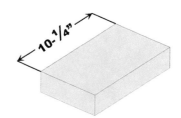

Trim end as shown in green. Leaving a small lip will prevent water from leaking inside.

Center roofing panel and fasten in place.

45

1x 60" Roofing Panel
9x 1" Roofing Screws

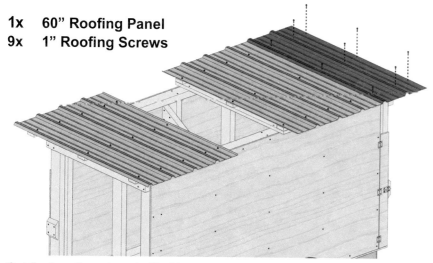

Cut final roofing piece to fit. Check underneath and grind off any screw points that penetrated through the wood to prevent injuries later on.

Build The Optional Garner Board Shelves

46

4x 2x6x10-¼"

This is a non-essential luxury for this buggy, but it does save time and hassle by stowing these garner boards (when not in use) in an organized and secure fashion.

a)

10-¼"

b)

60°

Trim and remove piece shown in green.

c)

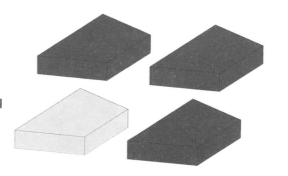

Make 3 more identical pieces.

MATERIALS HANDLING

47

1x   **1x6x6"**
3x   **2" Deck Screws**

Fasten 1x6 bottom to the side as shown. Pre-drilling may be necessary to prevent wood from splitting.

49

1x   **1x6x48"**
6x   **2" Deck Screws**

Fasten 1x6 front to the sides as shown. Pre-drilling may be necessary to prevent wood from splitting.

48

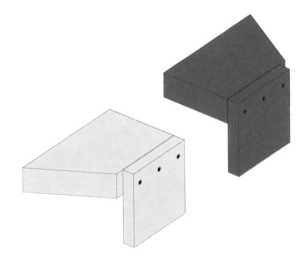

Repeat process to create an opposite hand version.

50

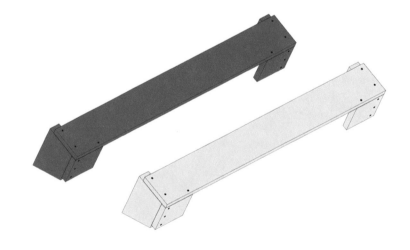

Repeat steps 47-49 to create a second identical shelf.

51

6x   2" Deck Screws

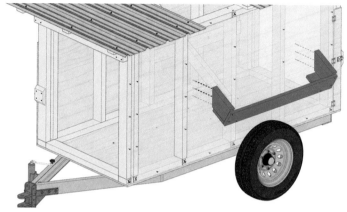

Mount shelf approximately where shown. Choose a height that is convenient for the user. Mount from the inside.

52

6x   2" Deck Screws

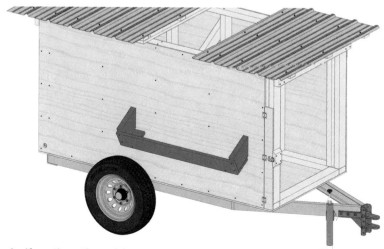

Mount shelf on the other side.

53

6x   1x6x43" Garner Boards

54

6x   1x6x43" Garner Boards

Simply unlatch and open the door and slide in the desired number of boards you need. Any unused boards can now be stored on the shelves. When finished, close and latch the door shut.

MATERIALS HANDLING

Build The Lid

55

2x   1x3x48"

1-½"

1-½"

Start by marking the following locations.

56

2x   1x3x23"
8x   2" Deck Screws

90°

90°

90°

Square frame and fasten together.

57

2x   1x3x17"
8x   2" Deck Screws

180°

Fill in gaps and secure as shown.

58

2x   Fencing Staples

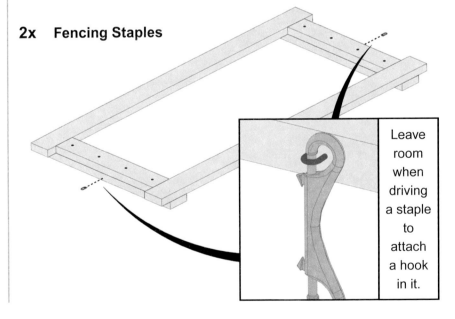

Leave room when driving a staple to attach a hook in it.

59

1x   60" Roofing Panel
22x 1" Roofing Screws

Center the roofing panel on the frame
and fasten. Check to see if any screws
penetrated through the bottom and grind
them flush to prevent injuries.

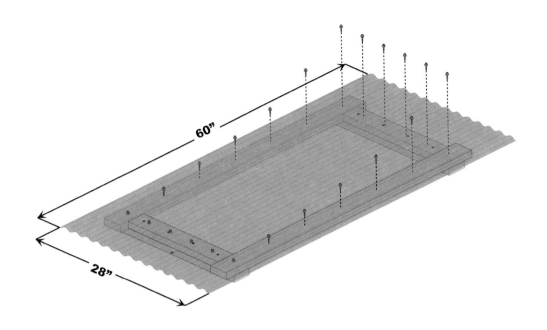

60

2x   Fencing Staples
2x   Double-Sided Spring Hooks

Use the double-sided spring hooks
to secure the lid onto the buggy.
(Another option is to use bungee
straps. In that case, you will need
to mount the fencing staples farther
down in a location that provides
adequate tension on the straps.)

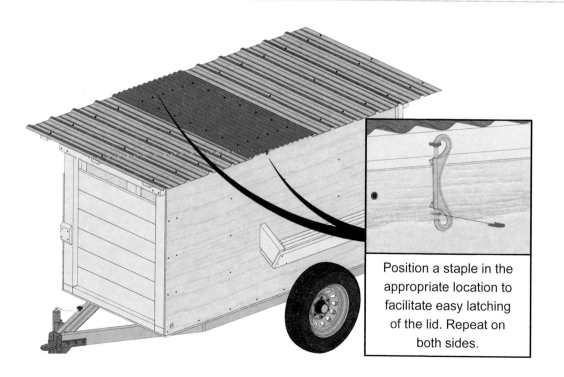

Position a staple in the
appropriate location to
facilitate easy latching
of the lid. Repeat on
both sides.

**MATERIALS
HANDLING**

61

The Bulk Feed Buggy is complete!

Joel's Tip

Always empty the feed buggy from the rear first. Doing the opposite may put too much weight behind the axle causing the tongue to lift off of the ground.

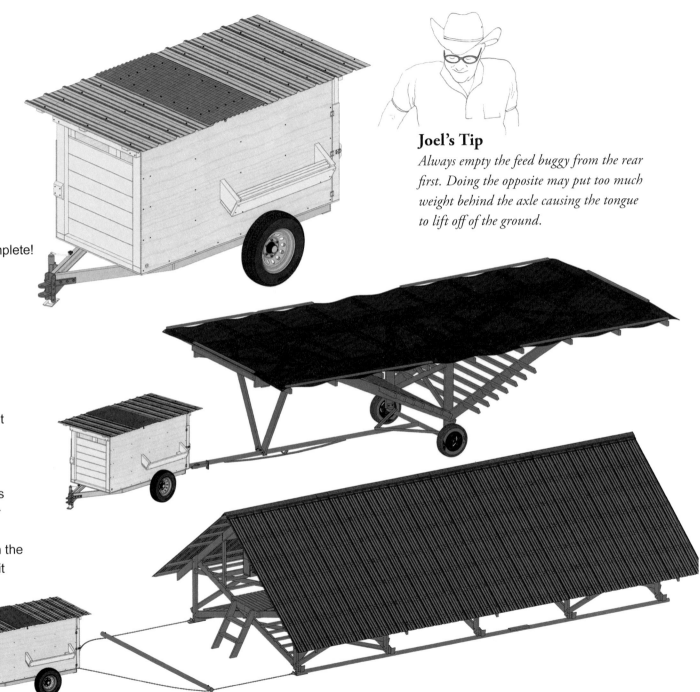

When equipped with the rear drawbar, this Bulk Feed Buggy can attach to the front of either the Gobbledygo or the Millennium Feathernet. Hooking these structures together consolidates chores and makes relocation easier and more efficient. You can either refill the feed buggy in the field or disconnect it and fill it elsewhere.

MATERIALS HANDLING

HAY WAGON CHASSIS

photo courtesy of Jessa Howdyshell

There are as many variations to building a hay wagon deck as there are ways to design a door. We will touch upon a few and then provide step-by step instructions on our favorite method.

All hay wagon decks have a few things in common:

1.) They all have a pair of structural supporting beams that span the length of the running gear.

2.) They all have a sturdy weather resistant deck surface.

Beams vary in length, size, and material. Size of the beam is determined by perceived payload, and the length of the deck. The longer the deck, the longer the span of the beam, hence the bigger it needs to be.

Because of our access to a bandsaw mill and plentiful hardwoods, wood is our material of choice. It is recommended that you use a strong, rot resistant species like white oak for these beams.

Our typical deck size is 16ft, although we do have some as long as 20ft. We prefer the shorter decks because of our topography. We hay some narrow fields and navigate hilly terrain. Our hay storage is close to the fields, so overall carrying capacity of an individual wagon is not as critical as it would be if we were ferrying hay a mile or two. If you live in the plain states, or transport hay over longer distances, a larger capacity wagon may suit you better. Bear in mind though that the longer the wagon, the farther you have to move bales to stack them, assuming you're stacking bales on the wagon behind the baler. For those of you who've never loaded a wagon, you always start from the back and work your way forward (We've included tips on stacking at the end of this chapter.)

With a 16ft long deck, you can probably get away with a 4x8" hardwood beam, but we typically beef ours up more and go with 4x10" white oak.

Depending on your resources, you may find it more economical to use steel in place of wood for the supporting structure of the deck. In this case, C-channel really is the way to go. Square/ rectangular tubing is typically more expensive, heavier, and more rigid by nature. C-Channel will allow a little more "flex" as your wagon twists and contorts over uneven terrain, while still supporting the payload it is carrying.

MATERIALS HANDLING

Decking Options

Whichever beam size and material you choose, you will need to be aware of the clearance between the deck and the top of the tires. You may need to install a taller beam than actually necessary, just to have adequate clearance.

With a beam structure chosen, it is now time to consider your decking options. The planks on your deck can run in one of two ways: Long-ways or short-ways.

In order for deck planks to run long ways (parallel to the beams) there needs to be an intermediary supporting structure on which to fasten the planks. A common building method is to space 4x4 timbers (or steel C-Channel) perpendicular to the beams. With a spacing of 2-3ft, the thickness of the deck plank material can be 1" to 1-¼" thick. Wagons always have thickened edges called "rails" to help keep the hay from sliding off sideways.

A benefit to this method brings us back to the tire clearance issue. With this style of construction, the deck is elevated by the height of the beam AND the 4x4 supports. When clearance is an issue, this construction method will help you gain needed clearance.

A downside to this deck style (and why we don't typically use them) is with traction. After some use, and many bales being dragged or slid across the decking, it becomes slippery. When covered in hay chaff it can be hard for bale handlers to keep their footing. The direction of the deck boards also leads to the stacked hay slipping forward when going down a hill. The wagon jostles along, and lengthwise planking on a down-grade can get exciting for both stacker and stack.

Decking Longways C-channel Frame

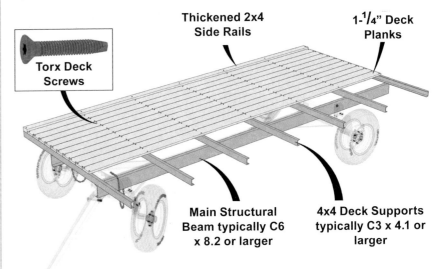

Torx Deck Screws

Thickened 2x4 Side Rails

1-¼" Deck Planks

Main Structural Beam typically C6 x 8.2 or larger

4x4 Deck Supports typically C3 x 4.1 or larger

Decking Longways Wooden Frame

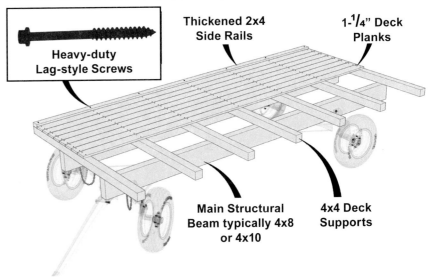

Heavy-duty Lag-style Screws

Thickened 2x4 Side Rails

1-¼" Deck Planks

Main Structural Beam typically 4x8 or 4x10

4x4 Deck Supports

Decking Shortways

Our preferred method of construction results in deck boards that are fastened short-ways, or perpendicular to the main beams. When building a deck this way, we always use 2" rough-sawn oak for our deck planks. Because of the orientation, the joints between the planks act as cleats, and give both bale handlers and bales traction. We find our hay shifts much less with this style of deck.

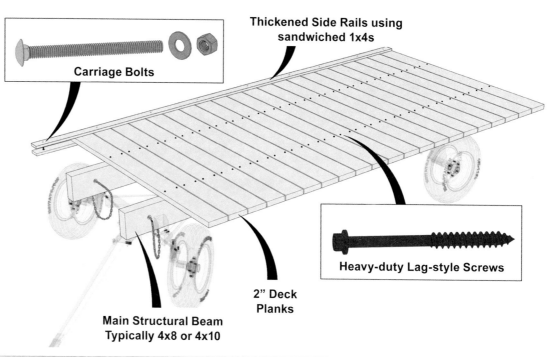

Carriage Bolts

Thickened Side Rails using sandwiched 1x4s

Heavy-duty Lag-style Screws

2" Deck Planks

Main Structural Beam Typically 4x8 or 4x10

Make sure the bolster stakes (4 brackets protruding off the top of the running gears) are intact, since these brackets are what connect the bed to the undercarriage. Replacing these and mounting hardware can easily cost $100 or more, if you can find them. Of course, you can always replace them with homemade pieces.

You can often find serviceable chassis at farm auctions. At Polyface, we've never bought a brand-new running gear. A good deck often quadruples the price of a hay wagon. In fact, some people with an on-farm sawmill make a side hustle out of buying chassis and putting decks on them.

MATERIALS HANDLING

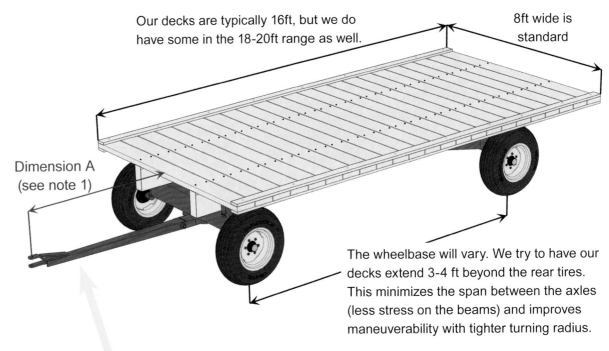

Our decks are typically 16ft, but we do have some in the 18-20ft range as well.

8ft wide is standard

Dimension A (see note 1)

The wheelbase will vary. We try to have our decks extend 3-4 ft beyond the rear tires. This minimizes the span between the axles (less stress on the beams) and improves maneuverability with tighter turning radius.

Joel's Tip
Choosing a Running Gear:

We tend to be a little more scrupulous when choosing a running gear to be used as a hay wagon than, say, a Shademobile. For starters, hay wagons carry a lot more weight. They also carry precious cargo. Yes, quality hay is precious, but I am talking about human beings. A component failure on a wagon full of hay with two bale handlers on it could be disastrous. Our wagons also double as shuttles for all of our farm tours, and get heavy use year-round for all sorts of farm tasks. In short, a chassis needs strength, integrity, and proper maintenance. Page 208 explains what we typically look at when buying a running gear. Other than keeping tires inflated, maintenance requires lubricating wheel bearings and greasing the few joints on the steering end. Pretty straight forward.

Also, check the weight ratings on the running gear. A wagon with a 16ft deck can hold upwards of 150 bales, so it needs to be rated for that amount of weight. We typically use 5-8 ton gears for our wagons.

MATERIALS HANDLING

Dimension B (see note 1)

photo courtesy of Chris Slattery

Note 1:
The most critical element is the distance from the tongue to the front of the deck (Dimension A) and 60" is pretty standard on this measurement. Whatever your distance is, you want it to be consistent across all of your wagons so that when you hook to your baler, you maintain a constant distance between the bale chute and wagon (Dimension B).

Tools

- ☐ Welder

- ☐ Circular Saw

- ☐ Socket Driver Set

- ☐ Grinding Wheel

- ☐ Cut-Off Wheel

- ☐ Angle Grinder

- ☐ Power Drill

- ☐ Driver Bits (for screws)

- ☐ Marking Pencil

- ☐ Tape Measure

- ☐ Speed Square

- ☐ $^3/_8$" HSS Drill Bit

- ☐ $^1/_2$" HSS Drill Bit

- ☐ $^5/_8$" HSS Drill Bit

MATERIALS HANDLING

Hardware

☐ **10x**　2" Deck Screws

☐ **92x**　4" TimberLok Hex-Head Heavy-Duty Wood Screws**

BOLTS

☐ **6x**　³⁄₈-16 x 3-½" Carriage Bolts

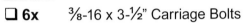

☐ **20x**　³⁄₈-16 x 4-½" Carriage Bolts

☐ **2x**　⅝-11 x 5" Grade 8 Hex Bolts*

☐ **6x**　½-13 x 5-½" Grade 8 Hex Bolts*

*Bolt diameter, length, and quantity may vary depending on the size of your beams and the size/quantity of the holes pre-drilled into the bolster stakes on your running gear.
**Quantity may vary depending on the width and number of planks used for the deck surface.

☐ **≈8ft**　³⁄₁₆" High Test Chain

☐ **2x**　³⁄₁₆" Quick Links

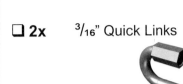

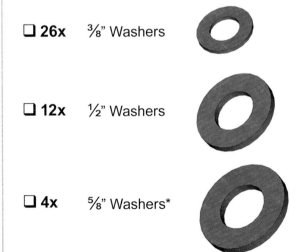

You will need a drill bit large enough to pass this chain through a hole.

☐ **26x**　³⁄₈-16 Hex Nuts

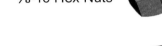

☐ **6x**　½"-13 Hex Nuts

☐ **2x**　⅝-11 Grade 8 Hex Nuts*

☐ **26x**　³⁄₈" Washers

☐ **12x**　½" Washers

☐ **4x**　⅝" Washers*

MATERIALS HANDLING

Wood Cutlist

Wood scraps are highlighted in red.
Do not discard until project is completed.

We highly recommend using rot resistant
hardwoods like white oak for this project.

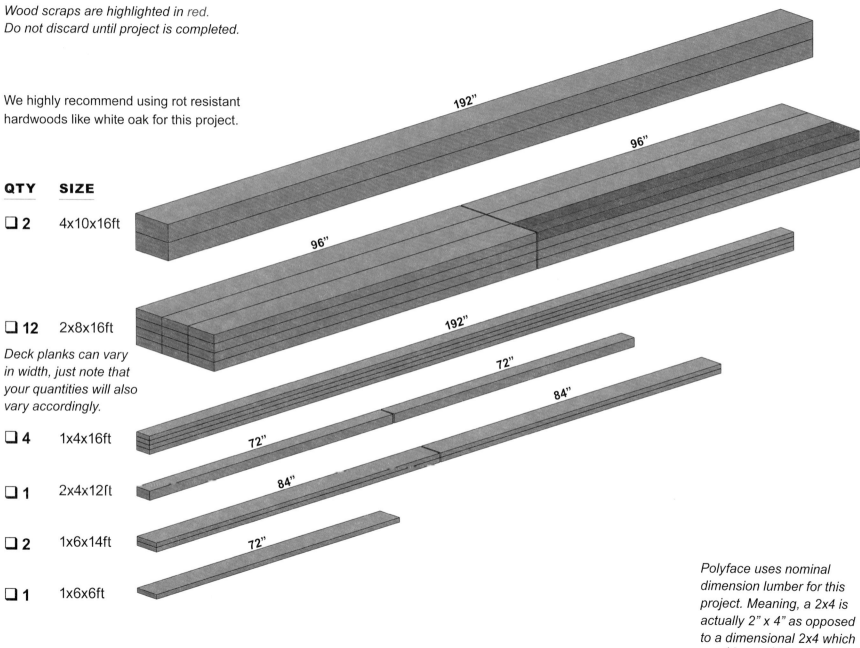

QTY	SIZE
☐ 2	4x10x16ft
☐ 12	2x8x16ft

*Deck planks can vary
in width, just note that
your quantities will also
vary accordingly.*

QTY	SIZE
☐ 4	1x4x16ft
☐ 1	2x4x12ft
☐ 2	1x6x14ft
☐ 1	1x6x6ft

**MATERIALS
HANDLING**

*Polyface uses nominal
dimension lumber for this
project. Meaning, a 2x4 is
actually 2" x 4" as opposed
to a dimensional 2x4 which
is 1-$\frac{1}{2}$" x 3-$\frac{1}{2}$".*

Metal Cutlist

! *Measurements in red are approximates.*
Cut each piece to fit.

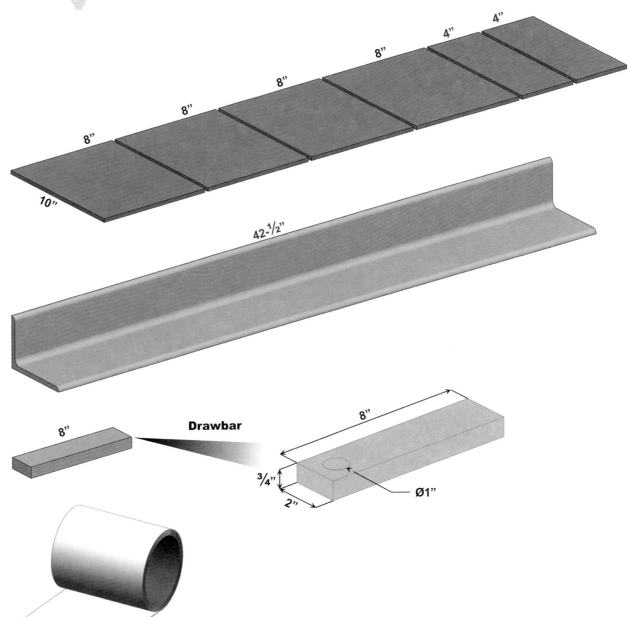

QTY	SIZE
☐ ≈4ft	¼x10" Plate

Items highlighted in yellow are optional materials used to build a hitch. Refer to step 26 and procure only if needed.

☐ ≈4ft	5x3-½x⁵⁄₁₆" Angle

☐ ≈8"	¾x2" Bar

☐ ≈32ft	Aluminum Drip Edge Flashing 6" Wide Roll

This is optional. Refer to step 4 for more information.

Drawbar

10"

8" 8" 8" 8" 8" 4" 4"

42-½"

8"

8" ¾" 2" Ø1"

6"

1

1x   4x10x192" Oak Beam
1x   ⅝-11 x 5" Grade 8 Hex Bolt
1x   ⅝-11 Grade 8 Hex Nut
2x   ⅝-11 Washers

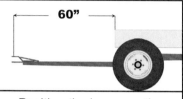

Position the beam on the running gear so that you achieve the proper clearance (60") to the tongue.

Using the hole(s) in the bolster stake as a guide, drill through beam and fasten with bolt hardware.

2

≈4ft ³⁄₁₆" Chain
1x   ³⁄₁₆" Connecting Link

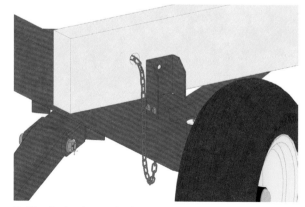

Drill a hole through the beam large enough to slide the chain through.

It is important not to bolt both ends of the beam to the running gear. This gear is designed to undulate with the terrain, so bolting the front end and back inhibits the intended movement and can lead to failures. We find it best to bolt the back end and attach the front end loosely with chains.

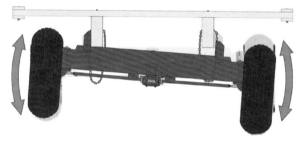

Correct slack

Too much slack

The chain length will vary between running gears. The goal is to leave enough slack to allow the gear to flex. However, too much slack and the wood beam could lift up and over the bolster stake. Adjust accordingly and secure the chain to itself using a threaded quick link.

MATERIALS HANDLING

3

Repeat process on
the other beam.

4 Optional

≈32ft   6" Wide Aluminum Drip Edge Flashing

While this step is optional, it is an old trick to help
preserve the wood beams. The idea is to cover the
tops of the beams with flashing before installing the
deck planks. Use some readily available aluminum,
tin, or galvanized roof flashing and bend it over the
top and an inch or so down the sides. This flashing
prevents water from collecting on the top surface and
soaking into the wood. (It's always the top surface that
rots first.) Any water that contacts the sides or bottom
runs off and dries. Once fitted, it can be tacked in
place with construction or roofing adhesive.

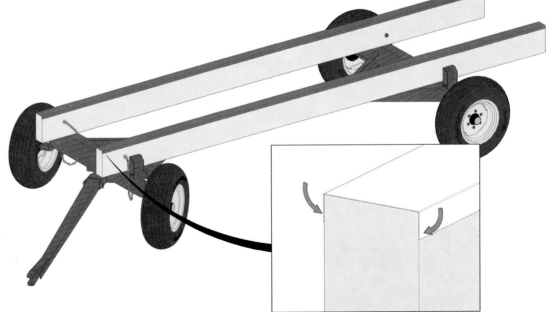

5

1x 2x8x96" Oak Plank
4x 4" TimberLok Wood Screws

Center your first plank along the front and fasten in place.

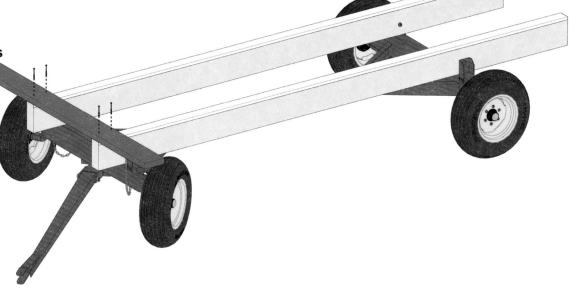

6

≈22x 2x8x96" Oak Planks
88x 4" TimberLok Wood Screws

Repeat process across the rest of the deck. If your deck planks are green (still full of moisture) you can probably butt the joints up tight. If the wood is seasoned (dry) then we recommend $\frac{1}{2}$" gaps between the planks. These gaps allow chaff and debris to fall through the floor which helps keep it cleaner while stacking.

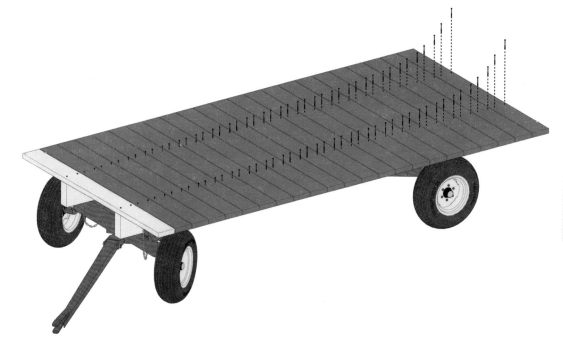

MATERIALS HANDLING

Install The Side Rails

7

1x   1x4x192"

We typically run a single 1x4 along the outer edge and bolt it every few feet with carriage bolts. It helps hold bales on the wagon, but it still allows the deck boards to become uneven. Over time, it can become a trip hazard.

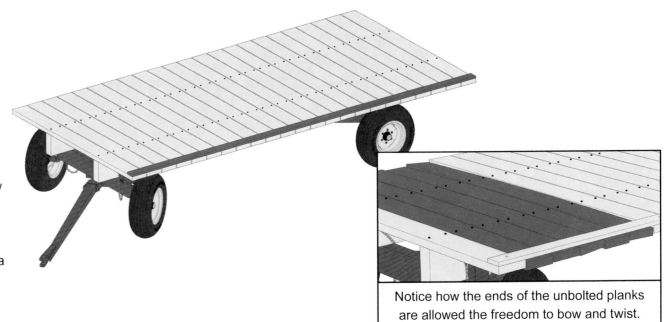

Notice how the ends of the unbolted planks are allowed the freedom to bow and twist.

8

1x   1x4x192"
≈9x   ⅜-16 x 4-½" Carriage Bolts
≈9x   ⅜" Washers
≈9x   ⅜- 16 Hex Nuts

**MATERIALS
HANDLING**

By placing a second 1x4 underneath and sandwiching the ends of the planks together, it helps prevent warping from occuring. Simply clamp pieces together and drill holes every 18-24" and fasten with carriage bolt hardware.

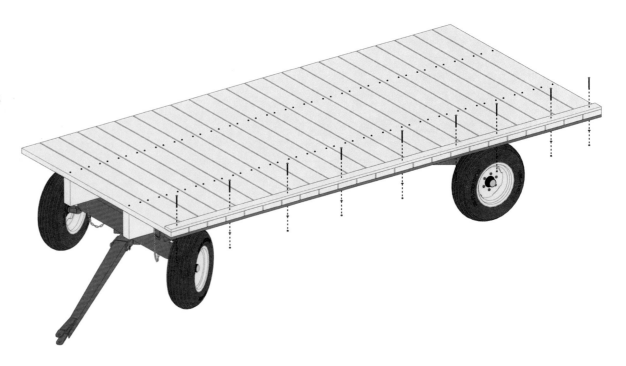

9

Install the side rails on the other side.

Make Tailgate Brackets

10

1x   ¼x10x8" Plate

a)

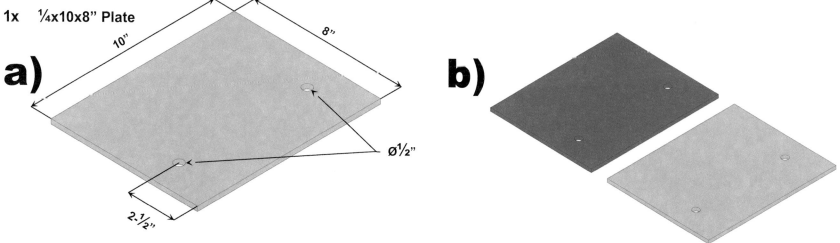

Drill holes approximately where shown.

b)

Repeat process to create a second plate.

11

2x   ½-3 x 5-½" Grade 8 Hex Bolts
2x   ½-13 Hex Nuts
4x   ½" Washers

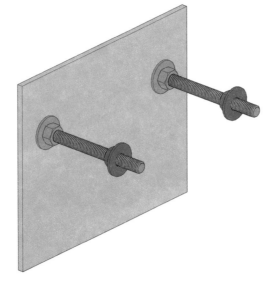

Fasten bolt with nuts as shown.

13

1x   ¼x10x8" Plate From Step 10
2x   ½-13 Hex Nuts
2x   ½" Washers

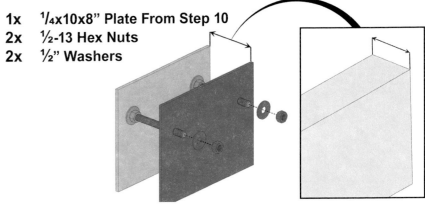

The distance between the plates will ultimately be determined by the width of the beams. The objective is to make this bracket a hair wider than the beam so it slides on easily. Adjust the nuts to achieve desired clearance.

12

2x   ½-13 Hex Nuts
2x   ½" Washers

Thread a second set of nuts part way onto bolts. These nuts are temporarily used to hold the plates in their proper positions during fabrication.

**MATERIALS
HANDLING**

14

1x   ¼x10x4" Plate

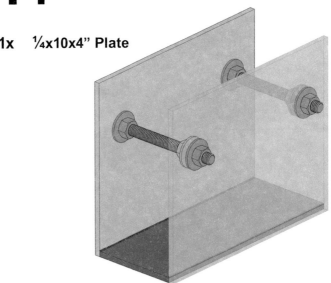

Cut the end plate to fit in between the sides and weld in place.

15

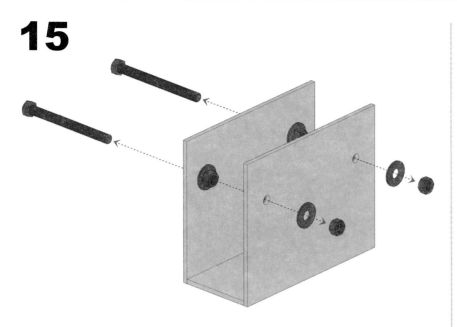

Remove all bolt hardware and test fit on the back of wagon.

16

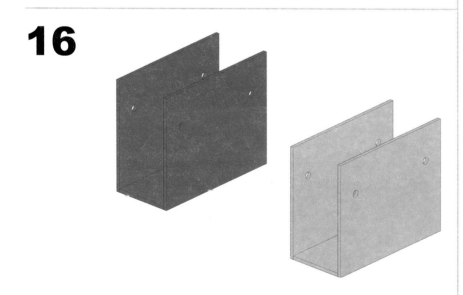

Repeat steps 10-15 for the second bracket.

17

2x ½-13 x 5-½" Grade 8 Hex Bolts
2x ½-13 Hex Nuts
4x ½" Washers

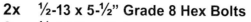

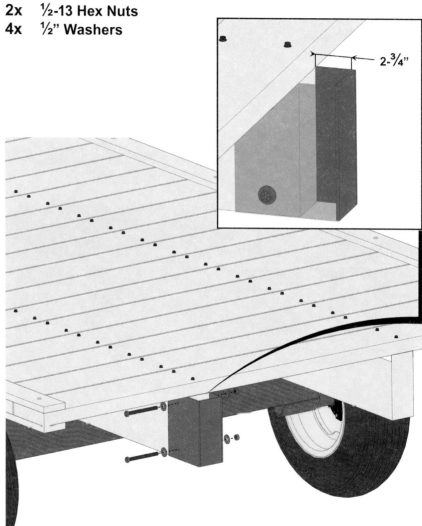

Bracket should be approximately 2-³/₄" inches from the rear end of the beam. This space creates a pocket for the tailgate to sit. With the bracket in place, use the holes as a guide and drill through the wood beam. Secure with bolt hardware. Repeat process for the other bracket.

MATERIALS HANDLING

Build The Tailgate

18

2x   2x4x72"

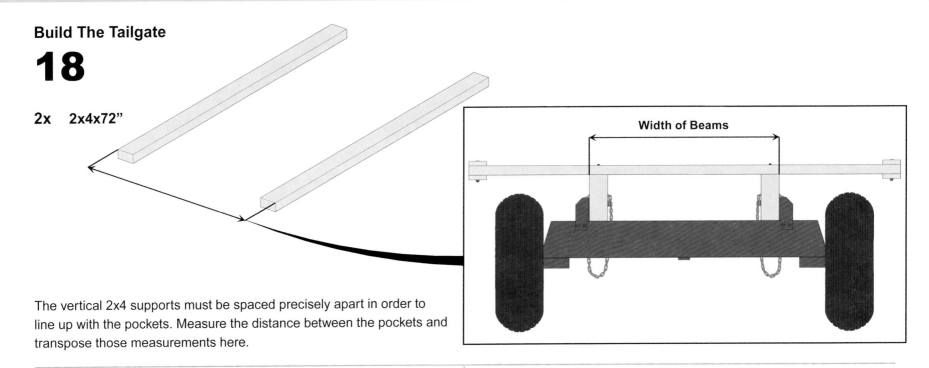

Width of Beams

The vertical 2x4 supports must be spaced precisely apart in order to line up with the pockets. Measure the distance between the pockets and transpose those measurements here.

19

1x   1x6x84"
2x   2" Wood Screws

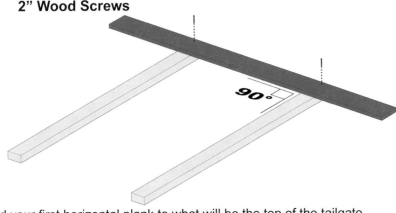

90°

Add your first horizontal plank to what will be the top of the tailgate. Center it so that it over hangs equally off of both sides. Square it and tack it in place with screws.

20

3x   1x6x84"
6x   2" Wood Screws

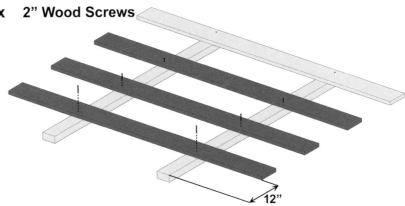

12"

Screw on three more planks. The bottom plank should be 12" from the bottom. The spacing between the other planks is not crucial, just make them look even.

MATERIALS HANDLING

21

1x   1x6x72"
2x   2" Wood Screws

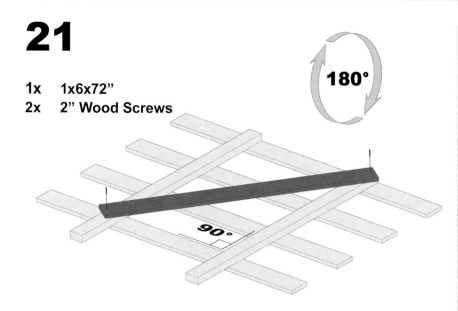

Flip the tailgate over and check that it is square. Secure the diagonal brace in place with screws.

23

2x   ⅜-6 x 4-½" Carriage Bolts
2x   ⅜" Washers
2x   ⅜-16 Hex Nuts

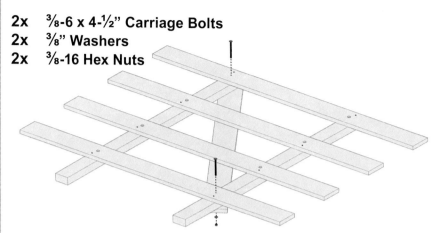

Longer bolts are required in these two locations because of the added thickness of the diagonal brace underneath.

22

6x   ⅜-16 x 3-½" Carriage Bolts
6x   ⅜" Washers
6x   ⅜-16 Hex Nuts

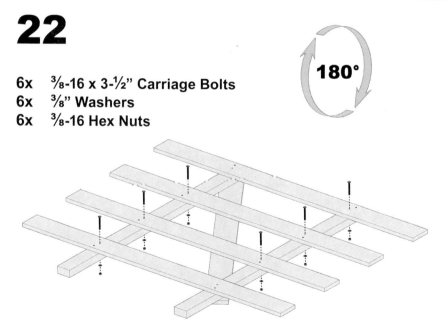

Flip the gate back over and drill Ø⅜" holes through all of the joints. Fasten bolts in the 6 locations shown.

24

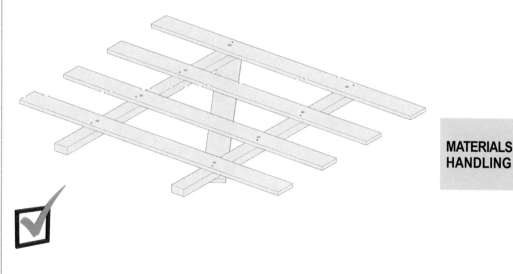

The tailgate is complete!

MATERIALS HANDLING

25

Test fit gate. Sand the ends of the 2x4 uprights as needed to ensure a snug fit.

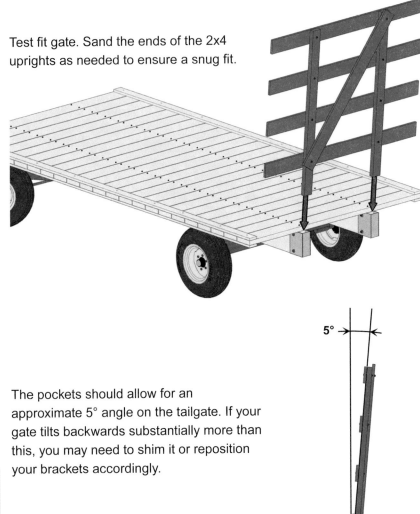

The pockets should allow for an approximate 5° angle on the tailgate. If your gate tilts backwards substantially more than this, you may need to shim it or reposition your brackets accordingly.

MATERIALS HANDLING

Weld An Optional Rear Hitch

26 Optional

a)

If you plan on hooking multiple wagons onto one another like a train, then you will need to assess the positioning of your rear hitch. Some hitches extend rearward and are adjustable. Others (like what is shown) are welded to the rear of the chassis. In a situation like this, it may be necessary to fabricate your own hitch.

b)

2x ½-13 x 5-½" Grade 8 Hex Bolts
2x ½-13 Hex Nuts
4x ½" Washers
1x 5x3-½x⁵⁄₁₆ Angle, 42-½" Long
1x Drawbar

There are many ways to fabricate your own hitch, however in this example, we've simply utilized the existing tailgate brackets and welded a drawbar between them on some angle iron. If following this approach, we recommend beefing up or adding extra grade 8 bolts to handle the increased loading on the brackets.

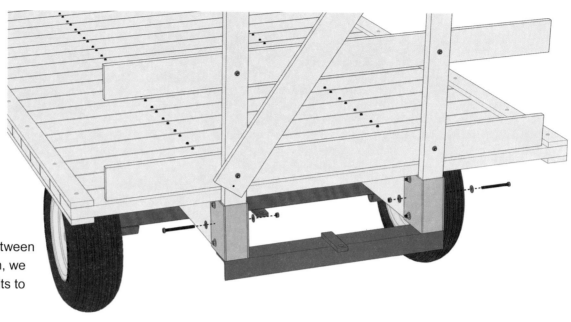

27

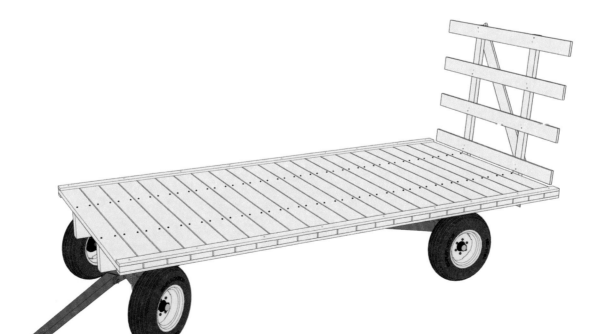

The Hay Wagon is complete!

MATERIALS HANDLING

both photos courtesy of Chris Slattery

The Art of Loading a Hay Wagon

For some this may seem too trivial to take up space in this book, but with round baler prevalence, many folks are unfamiliar with the art of stacking a wagon. In fact, the majority of our apprentices have never stacked a wagon before coming to Polyface. Most country children of my generation stacked hay wagons as a matter of course, but today it's a rare activity.

Today's young people miss out on an important part of growing up. In my day, teenagers weren't in the gym all summer, they were on wagons. The best training for high school athletics was bucking bales all summer. Something about sitting on an empty wagon at the end of a hot day, slurping down a slice of ice-cold watermelon looking at all the work you accomplished that day builds character. Alright, I will get off my soap box now.

Learning how to properly stack a wagon is paramount for safe and secure transport of bales from your fields to their final destination. Each layer is staggered and joints overlap to lock everything in place.

We often compete to see how high we can stack and how many bales we can fit on a wagon, but for on farm transport we won't go higher than 8 rows, and if we are hauling over road, we keep it 6 high max.

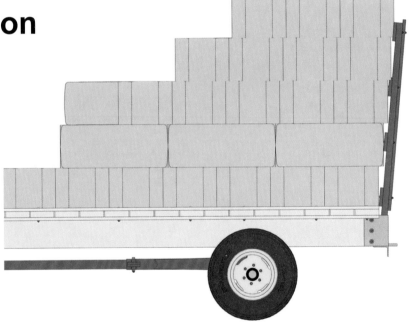

Wagons are always stacked starting in the back and working toward the front. Our tailgates have approximately a 5° angle backwards, and this allows us to stagger our bales slightly toward the back. This helps lock everything in. As our stack grows taller, we often stairstep our rows in order to get those last bales to the top.

MATERIALS HANDLING

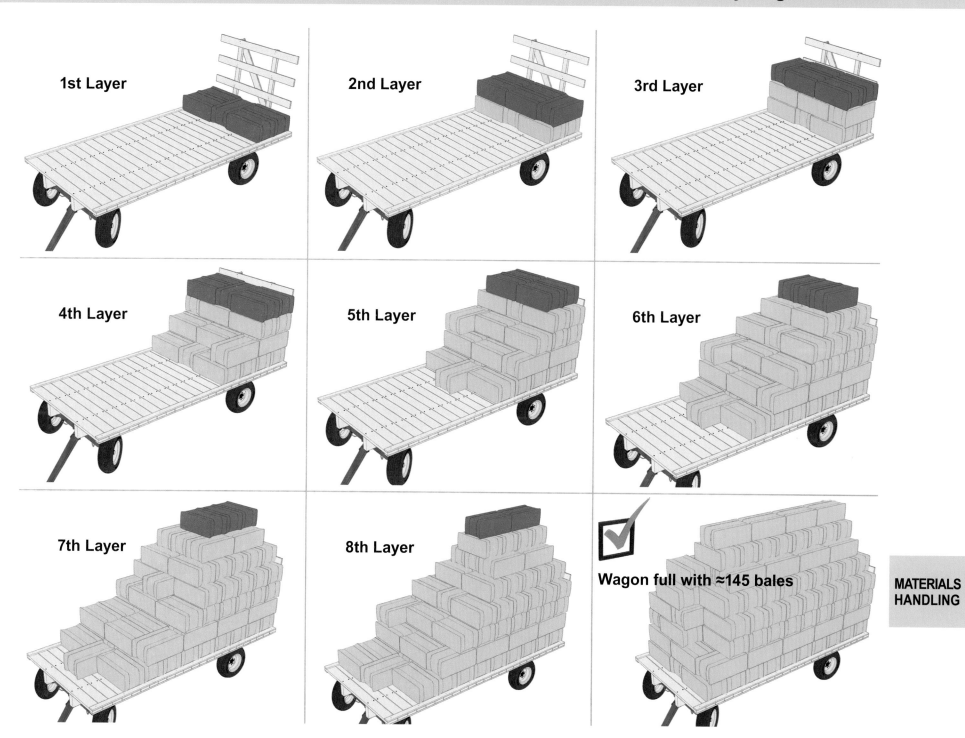

1st Layer

2nd Layer

3rd Layer

4th Layer

5th Layer

6th Layer

7th Layer

8th Layer

Wagon full with ≈145 bales

MATERIALS HANDLING

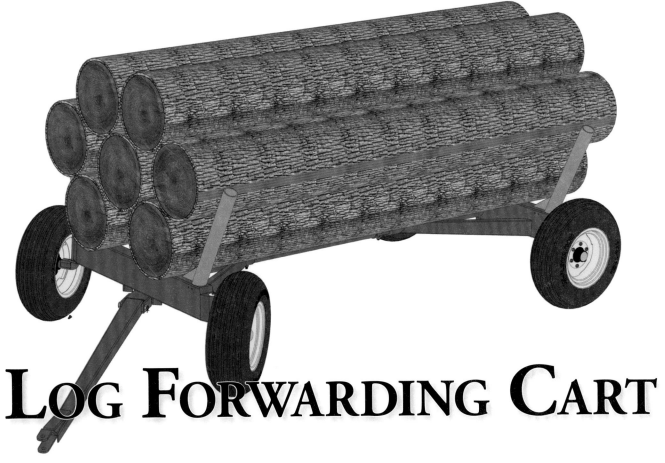

LOG FORWARDING CART

Self-sufficiency in lumber, on a farm, is what business folks call an unfair advantage. All sorts of projects that would break the bank otherwise suddenly become affordable and doable if you mill your own logs into lumber.

We initially thought we'd use our portable mill and take it to the logs, but that proved problematic. It takes time to set up and its length made it unhandy to snake around tight places in the woods. The main problem, though, involves momentary access. We found that the saw mill is best utilized as a filler. Have an hour? Go mill a couple of logs.

That makes the up-in-the-woods setup a problem. You can't just walk out the back door and mill for an hour if it's half an hour away up in the woods. We found it much more handy and efficient to park it near the house where any time of day we could walk out and run it for a little bit.

That reality forced us to figure out an efficient way to transport the logs from the woods to the mill. We started out with hay wagons, but they are not built heavy enough to handle logs. Inevitably, we overfilled them and broke

something. Not fun. We have a dump truck, and that works fine for small logs, but it's dangerous with big 16 foot logs to lift them up that high.

The logging industry uses forwarding carts for this purpose. They're specially made, heavy duty carts with pipe wings extending out from a simple steel frame, called the deck. In Austria nearly every farmer has one of these with a built-in knuckleboom loader on the back end.

Ours isn't that sophisticated, but it does have the other elements. We made it out of a heavy duty wagon chassis. Make sure you put

photo courtesy of Chris Slattery

gussets on the wings; butt welding won't be strong enough by itself. Highly maneuverable in the woods, it's light enough to easily pick up and move around to position perfectly before you start loading.

Because it has heavy duty tires and frame, it'll handle several tons of logs without needing any strap down. Unlike a hay wagon, it doesn't have any wood so it can handle weathering well. We have two that we can hook together. Of course, you can put some logs in a heavy duty truck for ballast and tow one of these forwarding carts home behind that as well.

Many forwarding carts are trailers rather than wagons, which makes them easier to maneuver in tight spots. But we generally unhitch to load with the tractor forks, and unhitching a wagon is a lot easier than unhitching a trailer. Hooking back up is also easier. So we're happy with the wagon as opposed to a trailer.

Tools

- ❏ Grinding Wheel

- ❏ Cut-Off Wheel

- ❏ Angle Grinder

- ❏ Tape Measure

- ❏ Marker

- ❏ Welder

Cutlist

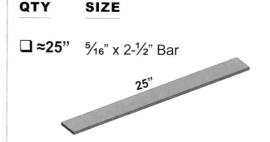

QTY	SIZE	
❏ ≈25"	5/16" x 2-1/2" Bar	

25"

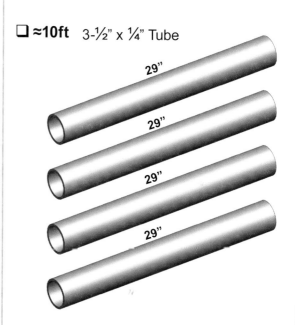

| ❏ ≈10ft | 3-1/2" x 1/4" Tube | |

29"

29"

29"

29"

MATERIALS HANDLING

1

Choose a heavier duty running gear for this application. Gears rated for 10k+ lbs are ideal, as these carts get a lot of abuse. Running gears that come off of silage wagons or gravity wagons are excellent choices. See page 208 for additional tips on choosing a running gear.

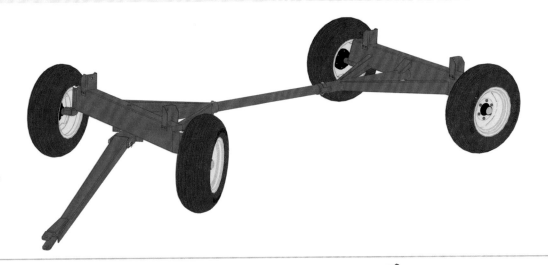

2

Remove bolster stakes from running gear. Save them in the event you procure a running gear later that is missing some or all of them.

3

MATERIALS HANDLING

Adjust the wheelbase of your running gear to accommodate the length of logs that you will be hauling. We regularly haul 12ft logs, so the wheelbase on our wagons is typically 10ft.

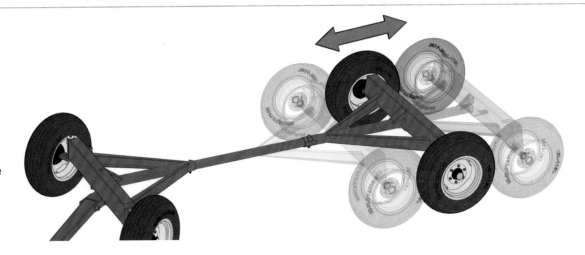

4

1x   3-½" x ¼" Tube, 29" Long

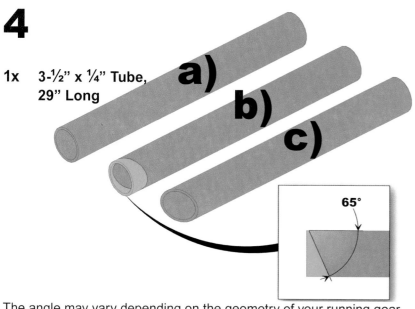

65°

The angle may vary depending on the geometry of your running gear. Adjust as needed.

5

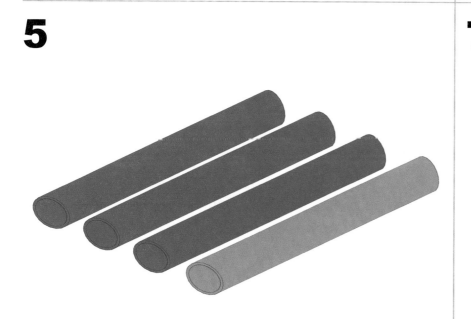

Repeat process to create a total of four upstands.

6

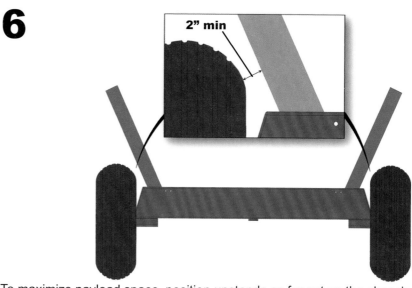

2" min

To maximize payload space, position upstands as far out on the chassis frame as possible. Be sure to maintain adequate clearances from tires. You may have to temporarily remove tires for welding.

7

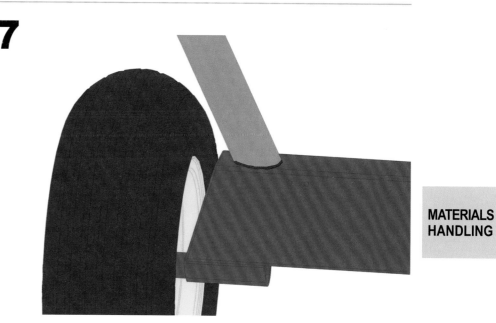

MATERIALS HANDLING

Weld tubes (shown in red) all the way around to the chassis.

8

1x   **⁵⁄₁₆" x2-¹⁄₂" Bar, 25" Long**

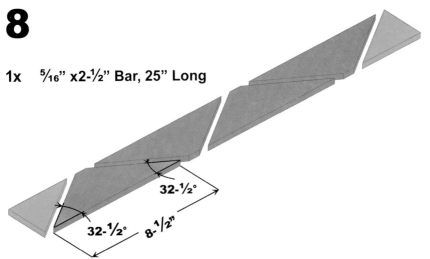

32-¹⁄₂°

32-¹⁄₂°   8-¹⁄₂"

Above is the best arrangement to minimize waste when cutting gussets. The angles may vary depending on how exact you mitered the upstands in step 4. If in doubt, make a template from cardboard and test fit prior to cutting steel.

10

Weld (shown in red) on front and back of gusset.

9

Test fit gusset in location shown.

11

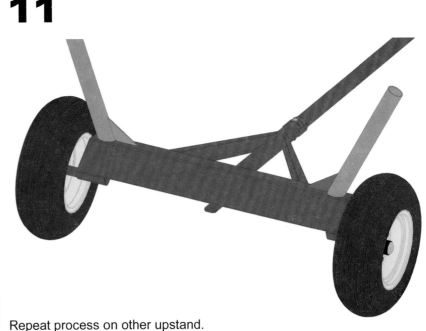

Repeat process on other upstand.

12

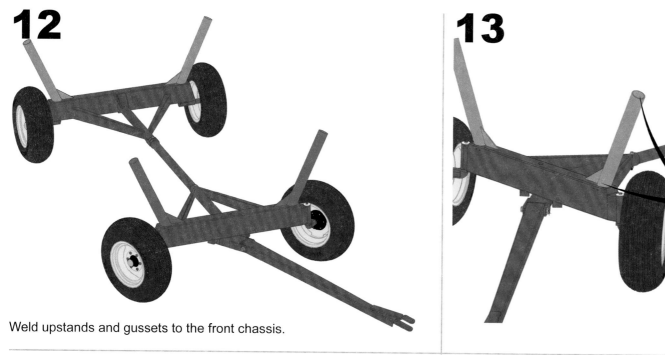

Weld upstands and gussets to the front chassis.

13

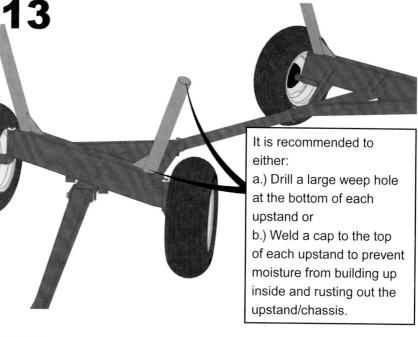

It is recommended to either:

a.) Drill a large weep hole at the bottom of each upstand or

b.) Weld a cap to the top of each upstand to prevent moisture from building up inside and rusting out the upstand/chassis.

14

The Log Forwarding Cart is complete!

MATERIALS HANDLING

Plasson® Broiler Drinker

Many chicken watering systems exist. We've tried many over the years and prefer the Plasson® low profile broiler drinker. The reason is that you can see in an instant whether it is working, enabling an efficient walk-by check at chore time. If the trough has water, all is well. The problem with nipples is that you can't tell if they're working. The tiny little cup systems are better, but they typically mount rigidly on the shelter side; if you're on a steep hill, it's hard to make them level so the water doesn't run out. These hanging waterers always seek level.

How It Works

This drinker uses a spring to resist the force of gravity and open the valve to release water. As the trough in the bell fills with water it gets heavier and heavier until it compresses the spring and forces the valve to contact the rubber washer and create a seal, cutting off the flow of water. As water is removed from the trough, the bell becomes lighter until the spring is able to break the seal and open the valve again.

Valve Opened

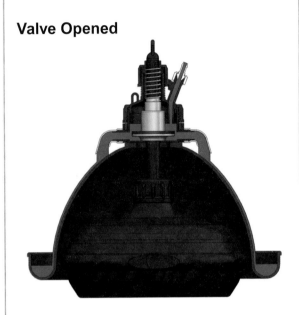

Valve Closed

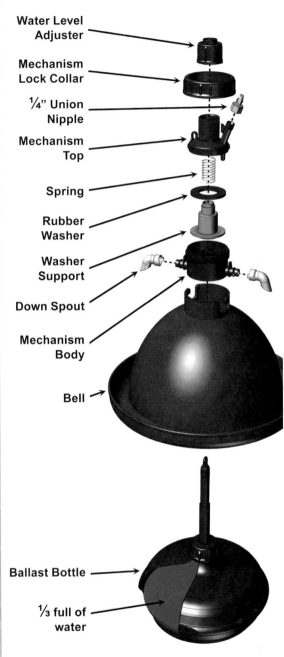

Water Level Adjuster

Mechanism Lock Collar

¼" Union Nipple

Mechanism Top

Spring

Rubber Washer

Washer Support

Down Spout

Mechanism Body

Bell

Ballast Bottle

⅓ full of water

Adjust the Water Level

Raise Water Level

The water level adjuster threads onto the mechanism top and its function is to increase/decrease the compression on the spring. By threading it clockwise (1), more force will be applied to the spring (2) resulting in a larger amount of water filling the trough (3) before closing the valve.

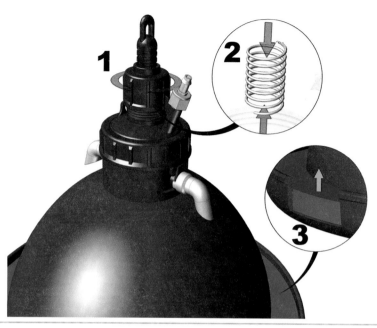

Lower Water Level

The opposite is also true. By threading the water level adjuster counter clockwise (1), it reduces the compression on the spring (2), permitting the valve to close with less water in the trough (3).

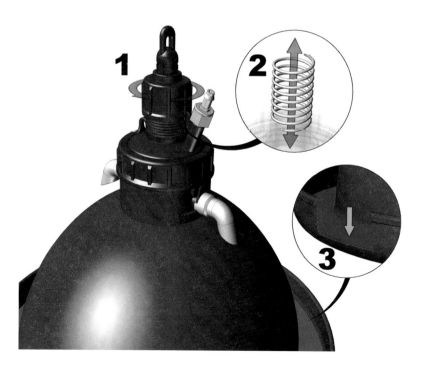

Joel's Tip

We strive to achieve a water level somewhere in the middle so as to supply ample water for the birds, while not spilling excessively as they bump the drinker around. Observe the water level and adjust accordingly.

APPENDIX

Plasson® Broiler Drinker Troubleshooting

What we like about these drinkers is that they are simple. If you have a problem it's generally one of two things:

1.) No water is flowing or
2.) The water will not shut off.

Here are some explanations regarding these scenarios and their solutions.

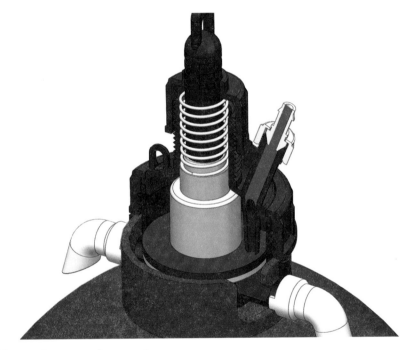

No water coming out

When no water flows into the bell, it indicates a blockage somewhere between the bucket and mechanism body. It is common for an insect to drown in the bucket and get lodged in the hose. Try blowing on the orifice exit to dislodge debris.

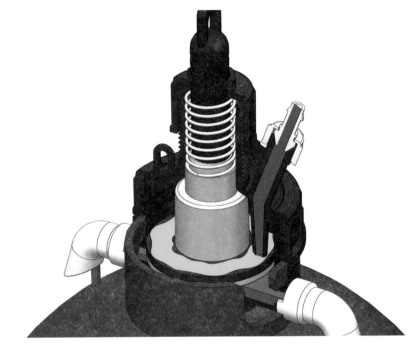

Water flow never ceases

The telltale sign that there is an issue is when only one bucket in an entire row of shelters is empty. The most common issue is a buildup of debris around the rubber gasket, causing it to not form a proper seal.

Subsequently the water continues to flow until the bucket is empty. This can be corrected by removing and cleaning the washer. If the washer is damaged, it can be flipped over to utilize the other side.

Another common mistake is setting the height of the drinker too low to the ground, in which case the bell hangs up on the grass and is unable to move downward far enough to close the valve. This can be corrected by raising the drinker height or cutting the grass ahead of the shelter. Proper height for the waterer rim is about a half inch below the chickens' beak. Seeing them stretch a bit for water is just fine.

Watering Setup

For Eggmobile, Gobbledygo, Millennium Feathernet, and Winter Hoop House Accommodations.

While the Plasson® drinkers are typically used for the broiler shelters exclusively, the following setup is used across a number of our enterprises: Eggmobile, Millennium Feathernet, Gobbledygo. We even utilize these same drinkers in our winter hoop house accommodations in combination with the waterer cover discussed on page 465. The simplicity and versatility of this setup makes it a real winner!

This is almost a bullet proof set-up, meaning it seldom fails. These indestructible rubber pans last for years. Since we move these every day or two, we rinse out any soiling inside routinely enough to keep things clean. One of these can easily service 500 chickens or 200 turkeys. When we go above those numbers in a flock, we simply split the feeder hose with a Y and add a second pan.

Supplies

❑ **1x**   6 Gallon Water Tub

Polyface prefers using a 6 gallon water tub for these applications. The tubs are made of a flexible rubber material and are typically 18" in diameter and 9" deep. This size has proven to provide ample water for the livestock while still being small enough to easily relocate during pasture moves.

❑ **1x**   Float Valve

We typically use these cheap but functional stock tank float valves made by Dare. They work great for low flow applications like this.

❑ **≈4ft**   Scrap Garden Hose

❑ **≈4ft**   12-$\frac{1}{2}$ Ga Medium Tensile Wire

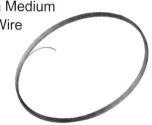

❑ **1x**   Male Hose End

❑ **1x**   Female Hose End

Pay attention to what size hose you have: $\frac{1}{2}$", $\frac{5}{8}$" or $\frac{3}{4}$" are standard sizes.
While we have shown cheap plastic fittings in our illustrations, if you can afford it, it really is worth spending the extra money to buy high quality brass fittings connected with durable hose clamps.

APPENDIX

531

1

1x Male Hose End
1x Female Hose End
≈4ft Scrap Garden Hose

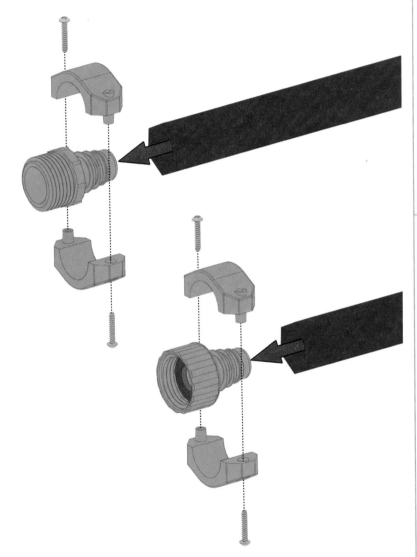

Affix the male and female couplings to either end of the garden hose.

2

1x 6 Gallon Water Tub
1x Float Valve

These floats mount on the top rim of a tank and are designed to maximize water level in the reservoir. The tank needs to be on fairly level ground. If situated on a slope, the float always needs to be positioned on the lowest side of the tank, or else you risk overflowing the tank and wasting water.

3

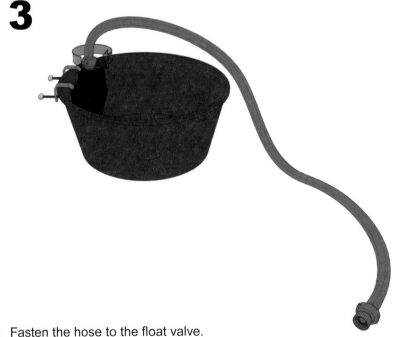

Fasten the hose to the float valve.

4

≈4ft 12-¹/₂ Ga Medium Tensile Wire

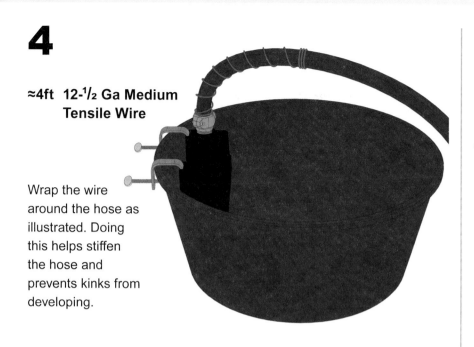

Wrap the wire around the hose as illustrated. Doing this helps stiffen the hose and prevents kinks from developing.

Joel's Tip

Another nuance with this style of float that is important to understand, is the position of the hose relative to the tank and float. This valve relies on a float to close the valve and stop water flow. These rubber tubs are more flexible than rigid cattle troughs, allowing the float valve to move around relative to the water; the float valve has a tendency to want to push upwards and out of the water. By positioning the hose so that it is routed over and across the tub, the weight of the hose helps hold the valve down in the water. We often use a block of wood to give the clamps something to grab.

Wrong way! Positioning the hose as shown will cause the float to lift up, potentially allowing water to run without shutting off. Always orient the hose across the top of the drinker trough.

5

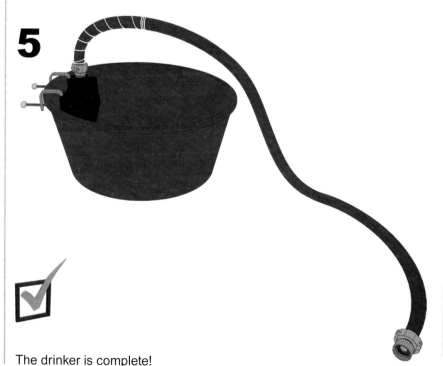

The drinker is complete!

APPENDIX

Shade Cloth Folding Technique

Consider it due diligence to routinely fold and store shade cloths when not in use. Doing so will prolong the life of your investment and is well worth the time and effort. With a strategy and a little practice, this can become an easy task that even one person can tackle on a calm, non-windy day.

The trick to setting up a shade cloth lies in how you fold it to start. By following our step-by-steps below and folding the shade in on itself, it makes setup a walk in the park.

When stowing these shades away for the winter, it is good practice to tag and label the size and/or identify which structure the shade cloth belongs to. This spares the guesswork during the spring hustle.

APPENDIX

1

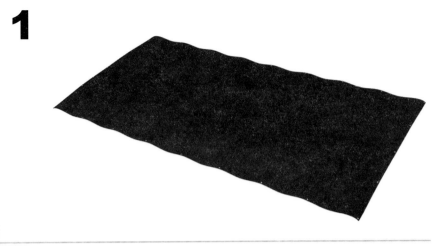

2

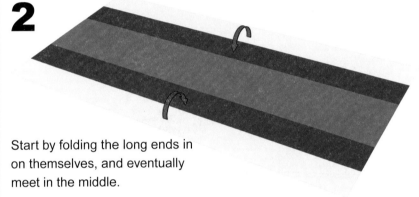

Start by folding the long ends in on themselves, and eventually meet in the middle.

3

4

5

Continue the same process along the short edges until you eventually meet in the middle and are left with a manageable sized unit.

6

7

8

9

Don't forget to label it so you know which structure it belongs to!

APPENDIX

Shade Cloth Unfolding Technique

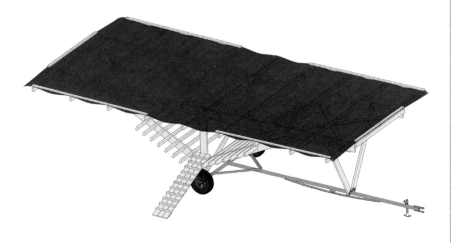

2

Unfold the short edge along the supporting member.

1

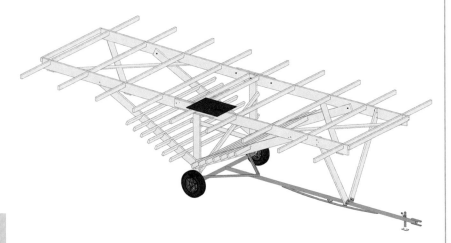

Start with the folded up unit and position it centrally along one of the main supports running lengthwise on the structure.

3

4

5

Finally, start unfolding the long edges and adjust it until it is in position.

6

7

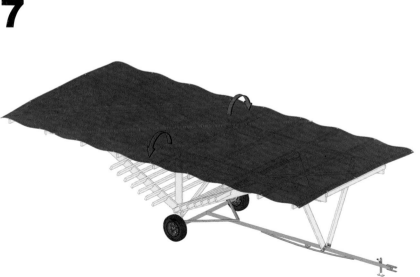

For further instructions on securing the shade cloth to your structure, reference the Shademobile on page 227 or the Gobbledygo on page 429.

APPENDIX

Fencing Techniques

Controlling livestock is foundational to owning them. Too many farms invest too much in fencing. It needs to be adequate, but not overly so. With that in mind, we will dig into our poor boy basic fencing techniques. We make three distinctions: boundary, permanent internal, and temporary (portable). We believe boundary should be permanent and physical. If boundary fence requires spark, I can't sleep at night; that's too risky. Our preference is woven wire as opposed to 9-strand high tensile.

Internal permanent can be single or multistrand electric, depending on class of livestock. Except for corrals or special training/receiving paddocks, we see no reason for interior physical fence except to help drain your bank account. If your animals are not respecting an electric fence, here are the reasons:

1.)   Incorrect height.
2.)   Lack of visibility (vegetation encroachment, rust).
3.)   Weak spark (weak energizer, too small a ground, shorting out, poor conductivity).
4.)   Too loose.
5.)   Lack of training (the animals did not get adequately shocked when first encountering the wire).
6.)   Inconsistency (intermittent spark maintenance).

Temporary portable fencing comes in all stripes and we don't have a big bias on type. The technology to weave great polywire material is now good enough that most brands seem to work fine. For reels, we use the $5 orange plastic extension cord reels from Lowe's or Home Depot. You can buy expensive reels but when you crack plastic, replacing a $5 reel is easier than a $50 one.

Permanent internal fencing defines topographical areas and the fields you're going to use. For this fence we use a single strand of 12-$\frac{1}{2}$ gauge aluminum high tensile wire. If we ran sheep or goats, we'd use three strands. We have no stationary permanent energizer installation, preferring complete mobility. Yes, the energizer moves with the livestock. These are the fences we will delve into in this section. These basic principles apply to all internal single or multistrand permanent electric fences, regardless of animal type.

We use inexpensive gate handles one step up from the cheapest variety. Again, we've used expensive ones and find that when a deer runs through it, cost doesn't matter. You might as well use something cheaper so when it breaks it doesn't break the bank. These gates are light and cheap and conveniently offer wide options. Our typical access gate is 30 feet.

The system we'll describe here only costs

photo courtesy of Jessa Howdyshell

a few cents per foot. We save enormous costs by cutting and manufacturing our own posts. While superb lightweight fiberglass and vinyl post systems are available from commercial suppliers, with a bountiful supply of rot resistant trees and our own bandsaw mill, we opt to make our own out of wood. Normal species are black locust, cedar, or Osage orange. Because we use a single strand of aluminum wire, posts only need enough strength and diameter to hold a staple for the insulators. They can also be short: we cut them about 5 ft. Because aluminum is lighter than steel, we can space them 15 yards apart if the terrain permits. We use slightly beefier posts for corners and dispense entirely with braces. Count the savings you will have in posts alone, and it more than makes up for the premium paid for the aluminum wire. Furthermore, aluminum never rusts, so it stays bright and visible for both livestock and wildlife; and it conducts much better than steel.

Another perk with this fencing system is the modular aspect of construction. As you will see on the pages that follow, our insulator end-loop assemblies (which are the most time-consuming aspect of this project) can all be made ahead of time. When we head to a farm to build fence, we have all of our posts precut and sharpened, all of our insulators ready to go, and wire on hand. I typically lead the way placing posts where they need to be. A two-person crew sets the posts

with the tractor mounted post pounder. Another person staples insulators to line-posts and runs wire. As soon as I set out the posts, I usually install the white knobs on all end points. This totals a 4-person crew, which can install half a mile of fence per hour.

It goes in easy and comes out easy. If we lose a grazing contract on a leased property, we simply roll up our wire, pull out our posts, and leave. It's fast and minimizes losses and therefore risk.

The End-loop Assemblies

As with any electric fencing setup, you need a way to insulate the electricity from the fence posts to prevent it from grounding out. We achieve this with the use of porcelain donut insulators. They are a smidgen more expensive than plastic, but they are more durable. We attach our donut insulators to our fence posts using what we call an "end-loop assembly," made out of the same aluminum wire used for the fence. End-loop assemblies will vary in size, depending on the diameter of the post you are wrapping them around. To further confuse you, we use two different types of end-loop assemblies.

Bear with me, it will make more sense as you read on. For now, just remember the formation of the end-loop is exactly the same between the two; the only difference is the extra loop on the insulator.

Standard End-loop Assembly:
Has a basic donut insulator threaded onto it.

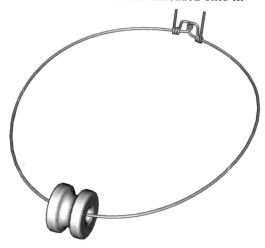

Gate End-loop Assembly:
Identical to the end-loop above, with the addition of a wire loop formed around the donut insulator. This loop acts as an attachment point for a gate handle to hook to.

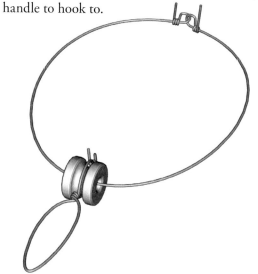

APPENDIX

539

Standard End-loop Assembly

1

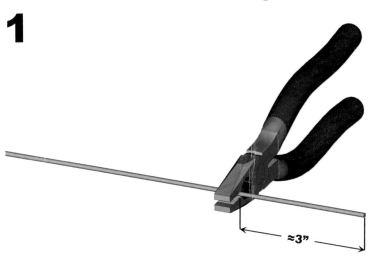

≈3"

Start with a ≈40" long piece of aluminum wire. Grab the wire with pliers roughly 3" from one of the ends.

2

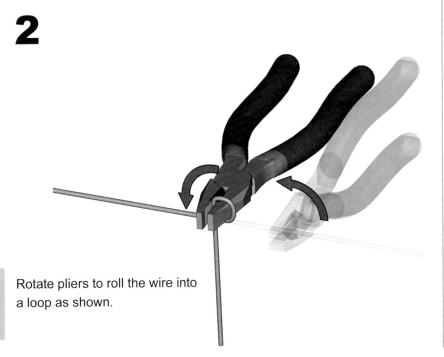

Rotate pliers to roll the wire into a loop as shown.

3

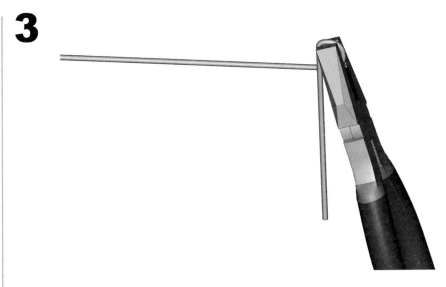

Now grab the loop with the pliers like so.

4

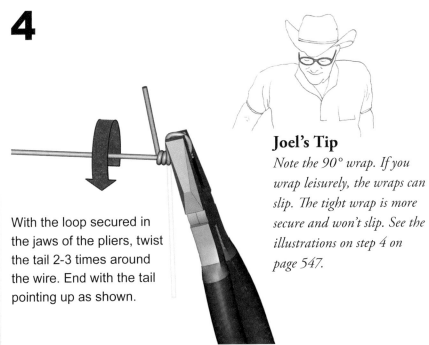

With the loop secured in the jaws of the pliers, twist the tail 2-3 times around the wire. End with the tail pointing up as shown.

Joel's Tip

Note the 90° wrap. If you wrap leisurely, the wraps can slip. The tight wrap is more secure and won't slip. See the illustrations on step 4 on page 547.

5

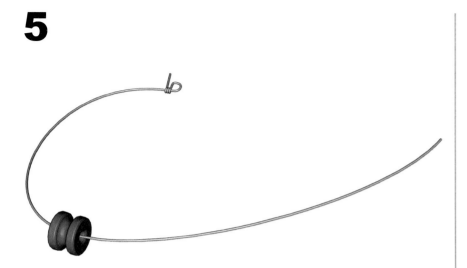

Slide an insulator onto the untied end.

6

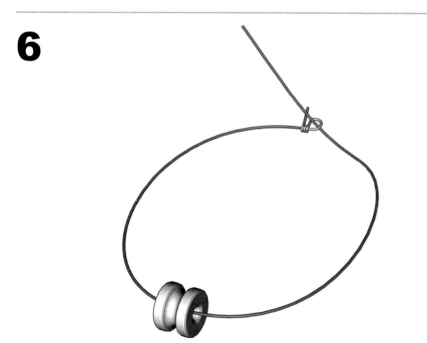

Thread the untied end through the loop.

7

Repeating steps 1-4, tie a knot in the other end.

8

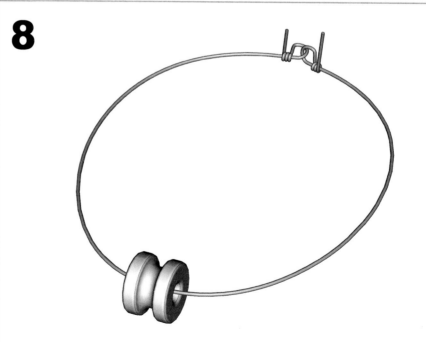

The standard end-loop is complete!

APPENDIX

Gate End-loop Assembly

1

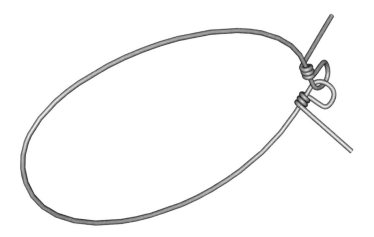

Start with ≈24" long piece of aluminum wire, and create a loop using the same techniques as before.

2

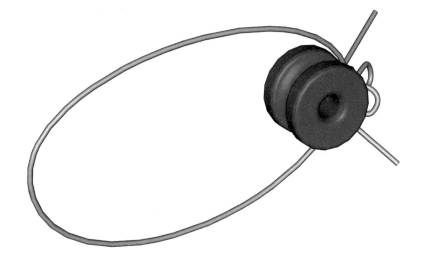

Place an insulator inside the loop as shown.

3

Pinch and twist the wire snugly around the insulator approximately 1-½ to 2 turns.

4

Cut the tails (highlighted in green) to prevent any accidental shorts.

5

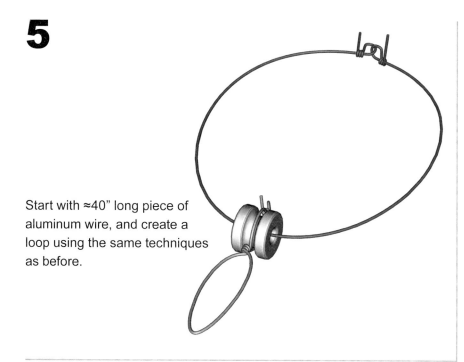

Start with ≈40" long piece of aluminum wire, and create a loop using the same techniques as before.

6

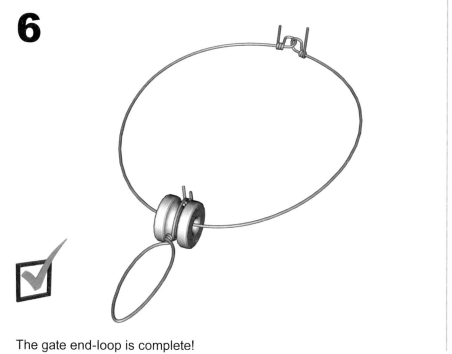

The gate end-loop is complete!

Now that you know how to construct the two basic end-loop assemblies, it's time to calculate how many you will need out in the field. As mentioned before, we really like this setup because it is a form of modular construction. On a rainy day, we can easily fill boxes of these ready made end-loop assemblies. With them all premade, it makes our in-field setup time that much quicker.

When we tally how many we will need, we start with the number of gate openings. For every gate handle you plan on using, you will need one gate end-loop assembly and three standard end-loop assemblies.

photo courtesy of Jessa Howdyshell

APPENDIX

Install End-loop Assemblies

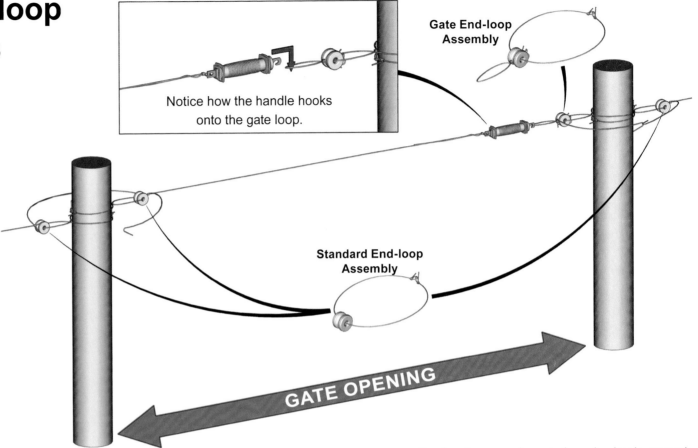

Notice how the handle hooks onto the gate loop.

Gate End-loop Assembly

Standard End-loop Assembly

GATE OPENING

You will also need a standard end-loop assembly for every other start or stop in the fence that you plan to have. This is very simple to figure. I typically make some quick sketches on paper and then count out my quantities from there. If you're wondering about these assemblies sagging and making the handle hard to hook up without getting shocked, remember that our energizer is always nearby so it's easy to turn off the spark when we need access.

For you sophisticated young folk, Google maps lets you zoom into your pastures and draw out your fencing. Armed with this information you will also know precisely how many fence posts you need, how much wire, and even how large your paddock subdivisions will be. Pretty cool!

As far as where to position these fences and gates, that's honestly an entire book topic in and of itself. I've written about it extensively in *Salad Bar Beef* and plenty of other fencing layout design systems are there for you to enjoy. If you're troubled about positioning, a rule of thumb is to make everything questionable temporary; if you don't move it in 3 years, upgrade to permanent. In any case,

the fencing we advocate here is simple enough that it can be repositioned quickly, cheaply, and easily. You can always add additional gate locations later on down the road. Even pulling a stretch of posts and relocating your fence isn't a huge undertaking. You can't say that about one of those 6 wire barbed wire fence rows. Over the years we've accommodated all sorts of on-going developments on rented properties. Change the yard, add a garden, fence a parking lot, build a pond and fence it out. All of these changes are easy with this simple fencing arrangement.

1

Ok, so now you've tallied how many end-loop assemblies you need and have them all premade, ready to go. Let's head out to the field and show you how to install them.

Slip end-loop assembly over the top of the post. For cattle we aim for about 30" above the ground.

2

Orient the insulator in the direction the wire will be routed and gently pull on it as shown.

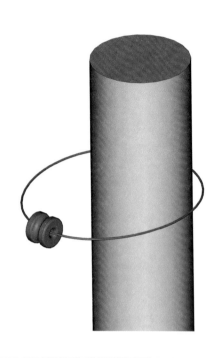

3

Pinch the loop around the post and give it 1-½ to 2 twists. Be careful not to over tighten the loop around the post because it will over-stress the knot and break. The objective here is to make it just tight enough to stay in place and not slide down the post. No staple necessary.

4

The first install is complete!

5

Depending on your configuration, you can add several end-loops to the same post; one for each fence wire that will be attached to it. If you are running with pigs or sheep, you would simply add a second set of end-loops at the desired elevation.

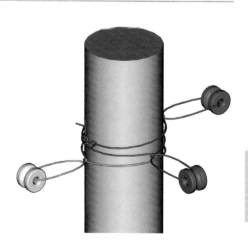

APPENDIX

Install Fencing Wire

Joel's Tip

With all of your end-loops installed, it is time to route the aluminum fence wire. We order our wire on whatever spool size offers the cheapest price. Sometimes a 4,000 foot spool is cheaper than a one mile; I have no idea why. Take a broom handle or stick and run it through the center of the reel as a handle. You can then walk the fence line, unrolling the wire as you carry it along.

2

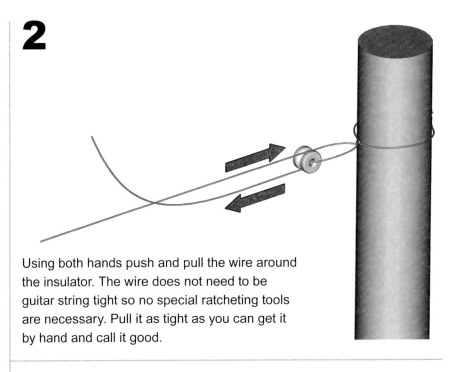

Using both hands push and pull the wire around the insulator. The wire does not need to be guitar string tight so no special ratcheting tools are necessary. Pull it as tight as you can get it by hand and call it good.

1

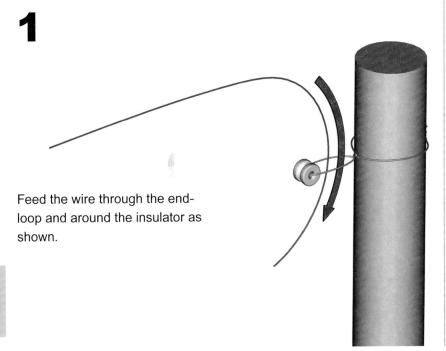

Feed the wire through the end-loop and around the insulator as shown.

3

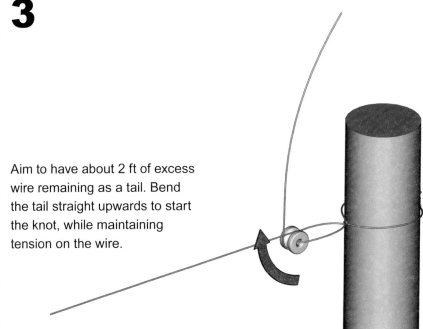

Aim to have about 2 ft of excess wire remaining as a tail. Bend the tail straight upwards to start the knot, while maintaining tension on the wire.

APPENDIX

4

Wrap the wire 2-3 times to secure it. Note the correct way to wrap in the illustrations.

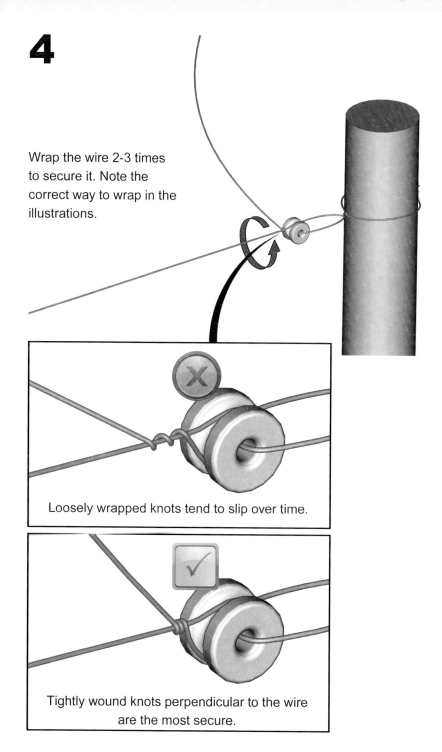

Loosely wrapped knots tend to slip over time.

Tightly wound knots perpendicular to the wire are the most secure.

5

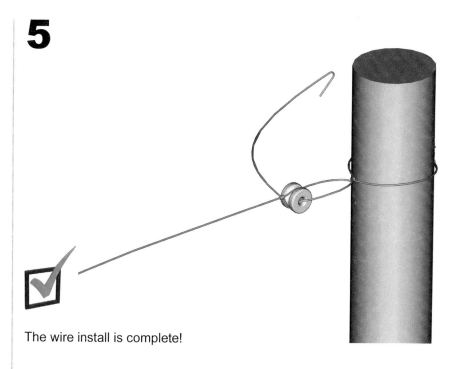

The wire install is complete!

photo courtesy of Chris Slattery

APPENDIX

Route Electricity

This is where that 2ft tail you left yourself comes in handy. Not only is it nice to have extra material in case you have a break and need to untie and let some wire out of one end, but it also helps you jump the electricity to other wires. We typically take our tails and bend a hook on the ends.

The following illustrations show just how easy it is to route electricity and turn sections of fence on and off. No expensive gate switches or complicated wiring. We keep it simple. Again, steel wire eventually corrodes and rusts; aluminum holds up for decades. We've never had a problem getting enough contact to assure good conductivity.

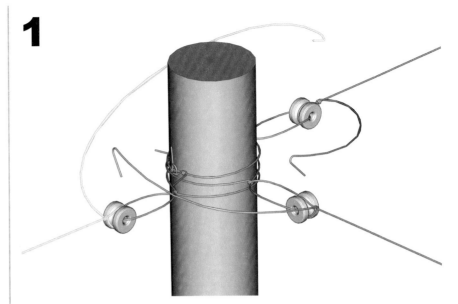

1

With no jumpers connected, the electrified wire (highlighted in yellow) ends at the junction.

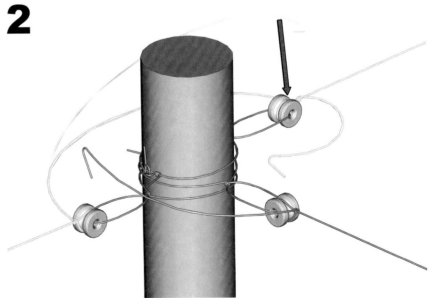

2

You can see how we connected the jumper to the opposite side and it electrified that wire.

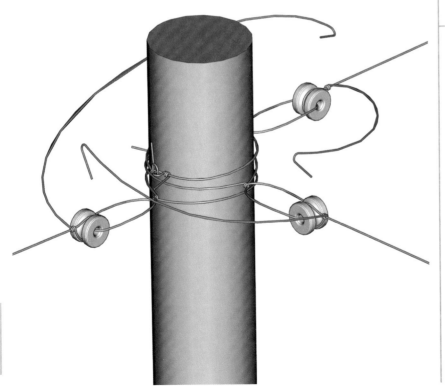

3

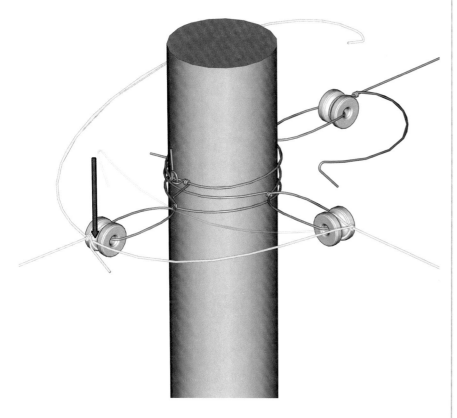

With properly arranged jumpers, you can isolate any combination of wires you would like.

4

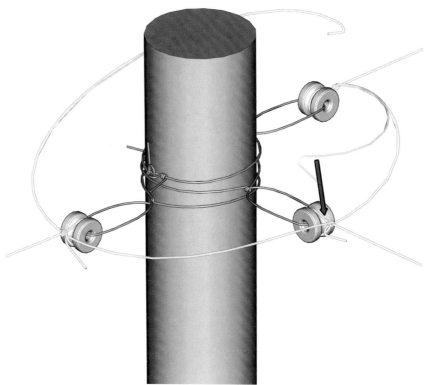

All wires are hot.

Cut A Notch

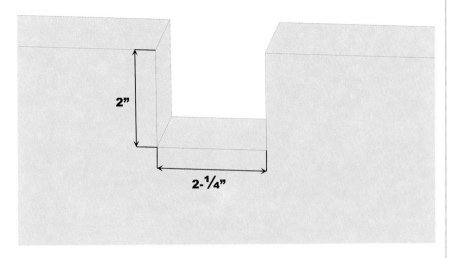

Notches are used on a couple of projects in this book, so we felt it would be valuable to step you through one way to create notches. There are certainly other ways to accomplish this (and some may be better), but we felt this was the easiest way while using minimal tools.

Joel's Tip

Determine the width and depth of your notch. When using nominal dimension lumber (i.e. a 2x4 is actually 2" x 4"), our standard notch size will be as shown above. If using lumber of different sizes, the rule of thumb is to oversize the notch by at least ¼" to prevent any binding from occurring as the lumber warps and shrinks.

1

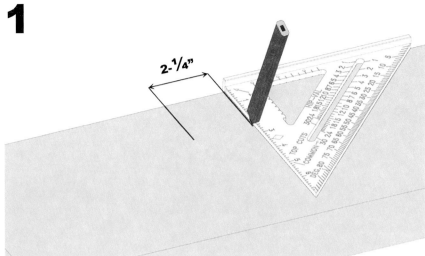

Scribe lines 2-¼" apart to denote width of notch. (The distance between notches will vary by the project so please refer to the instructions in that section for distance measurements.)

2

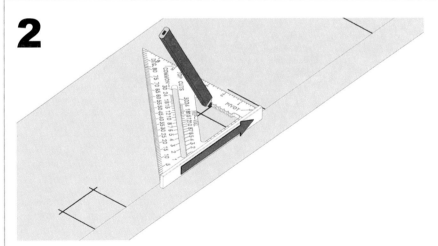

A quick way to use a speed square to mark lengthwise on the board is to orient the square as shown and place your pencil in the notch corresponding with the 2" mark. Push with your pencil while guiding the square along with your other hand. Notice we ran the lines long through each intersection. We did the "tails" on purpose; that way when you cut on a line, you can still see the mark afterwards.

3

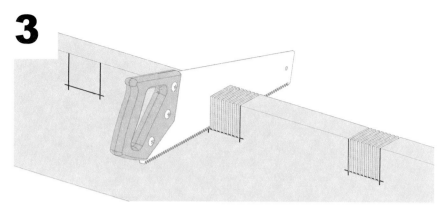

Once all of your lines are marked, it is time to start cutting. While this can all be accomplished with a handsaw, a circular saw is really the way to go. With a circular saw you can simply set the depth of the blade to correspond with the notch and go! Whether cutting by hand or with a circular saw, it won't hurt to over cut by a little bit (≈⅛"). Cut sequential notches as close together as you can get them. The closer the cuts, the easier the following steps will be.

4

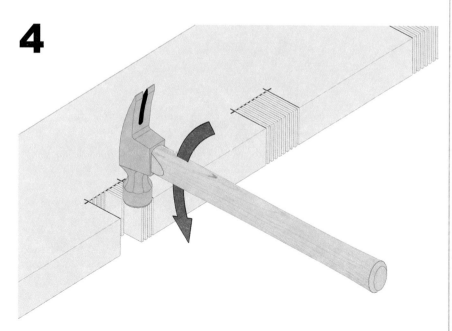

Using a hammer, proceed to knock out the slivers.

5

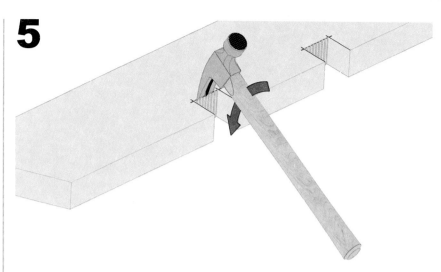

The base of the notch will most likely be a little rough and uneven. Using the claw of your hammer, you can chip at it to smooth things out. You can also try a chisel and/or a rasp. A circular saw run carefully side to side in the notch works wonderfully well at cleaning it up.

6

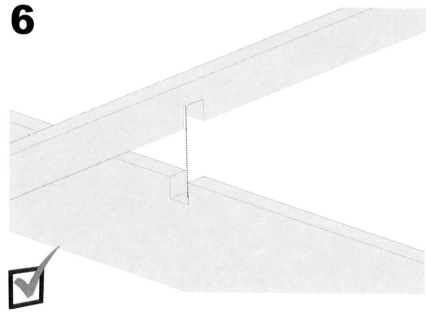

The notch is complete!

APPENDIX

Millennium Feathernet L-Bracket, Step 28, Page 129

Copy this page and cut this template out. **Scale 1:1**

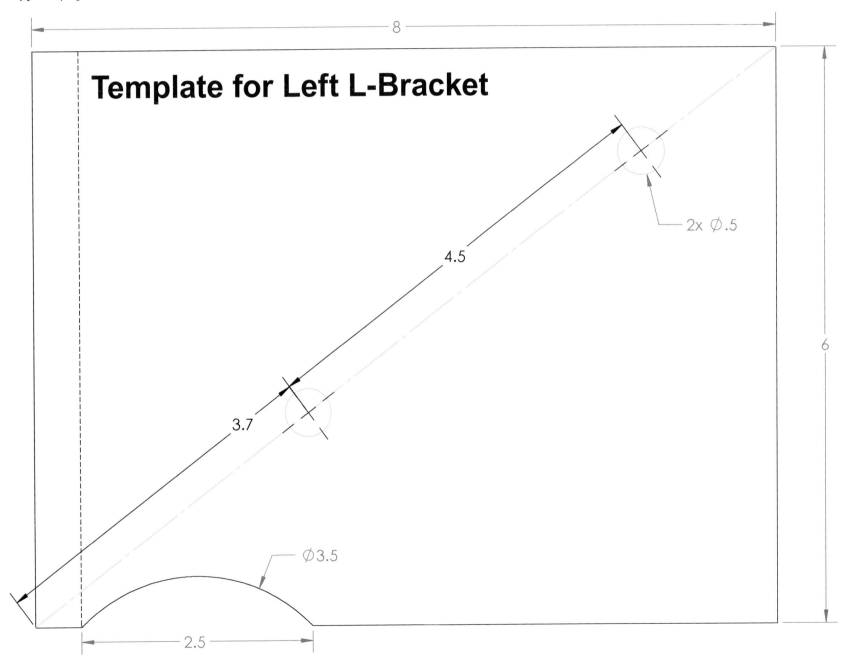

Template for Left L-Bracket

Millennium Feathernet L-Bracket, Step 28, Page 129

Copy this page and cut this template out. **Scale 1:1**

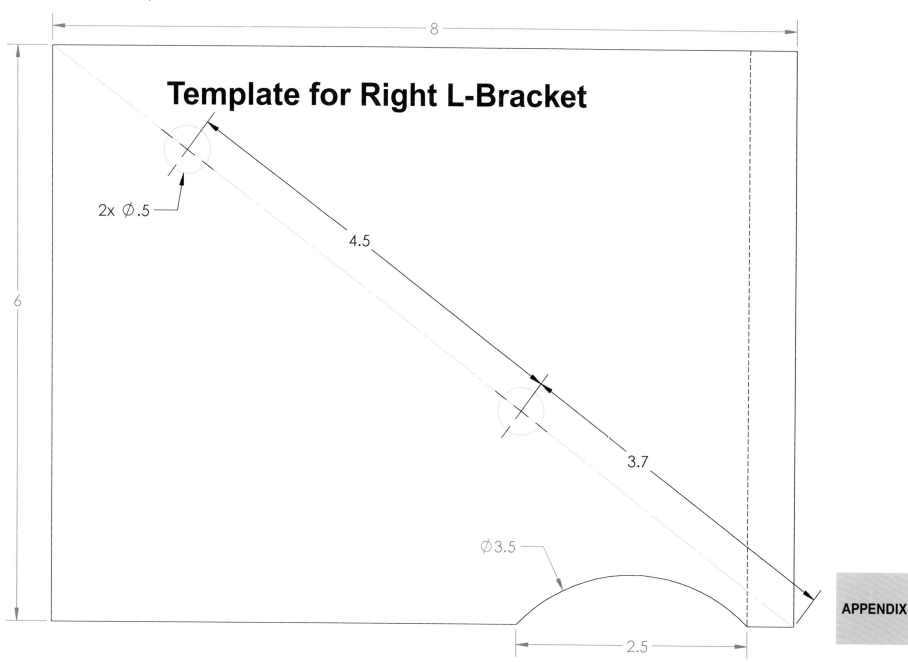

Template for Right L-Bracket

8

6

2x ⌀.5

4.5

3.7

⌀3.5

2.5

Broiler Shelter Dolly, Steps 1-4, Pages 30-34

Copy this page and cut these templates out. **Scale 1:1**

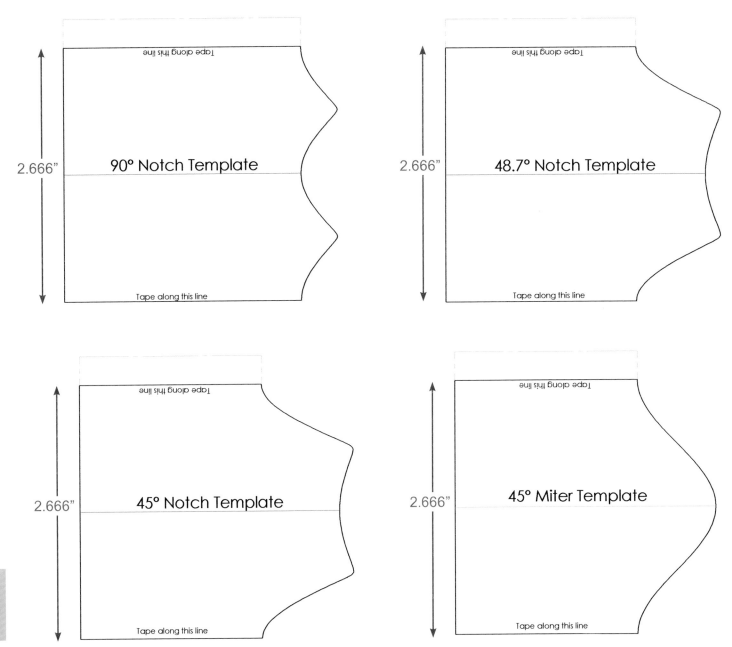

Broiler Shelter Dolly

Step 9, Page 36

This diagram depicts how the 3 templates (pages 556-558) should be laid out and taped together. Grayscale shows where the templates will overlap.

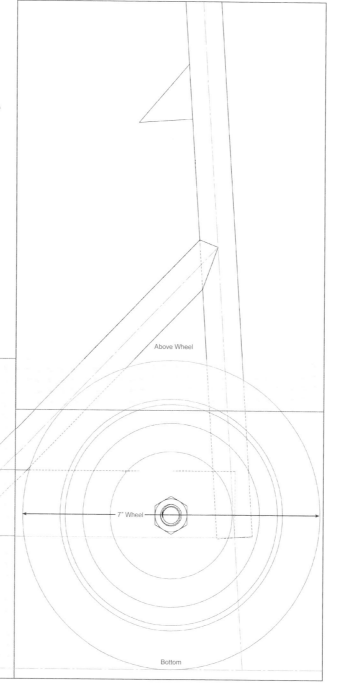

Above Wheel

7" Wheel

Bottom

Bottom

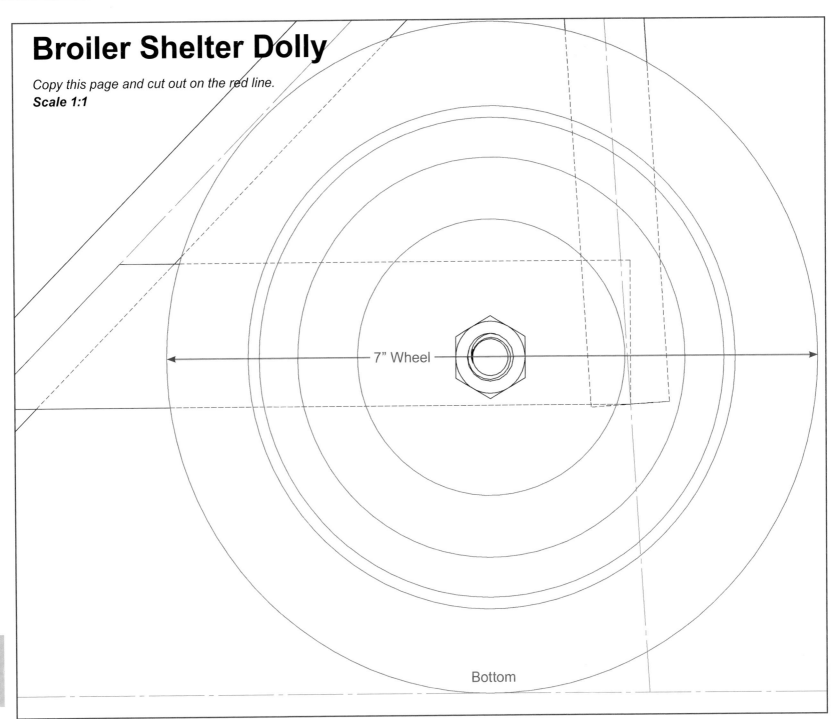

Broiler Shelter Dolly

Copy this page and cut out on the red line.
Scale 1:1

7" Wheel

Bottom

Broiler Shelter Dolly

Copy this page and cut out on the red line. **Scale 1:1**
Grayscale shows where the template will overlap.

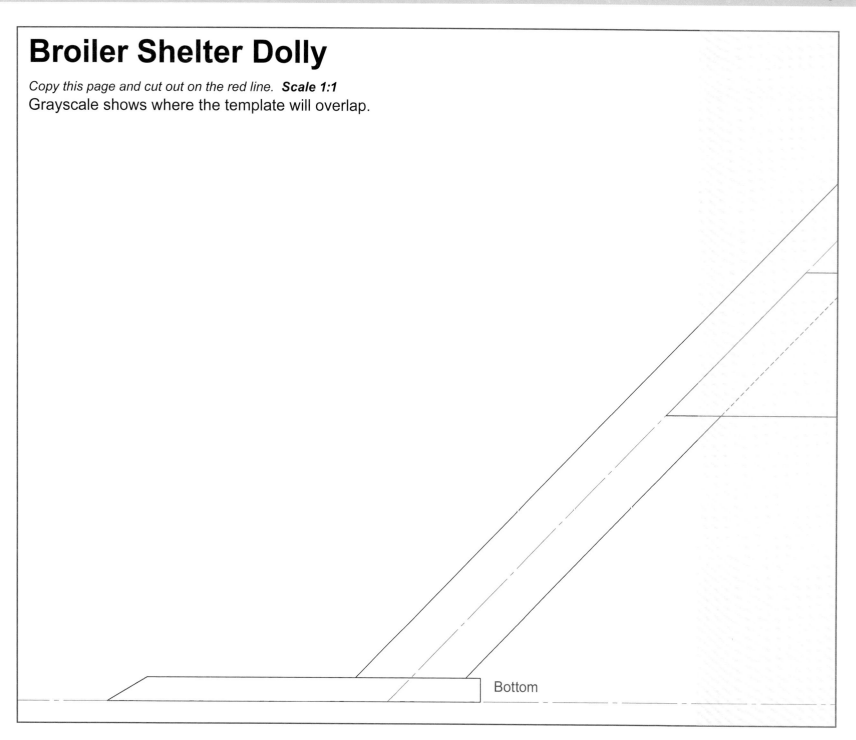

Bottom

Broiler Shelter Dolly

Copy this page and cut out on the red line. **Scale 1:1**
Grayscale shows where the template will overlap.

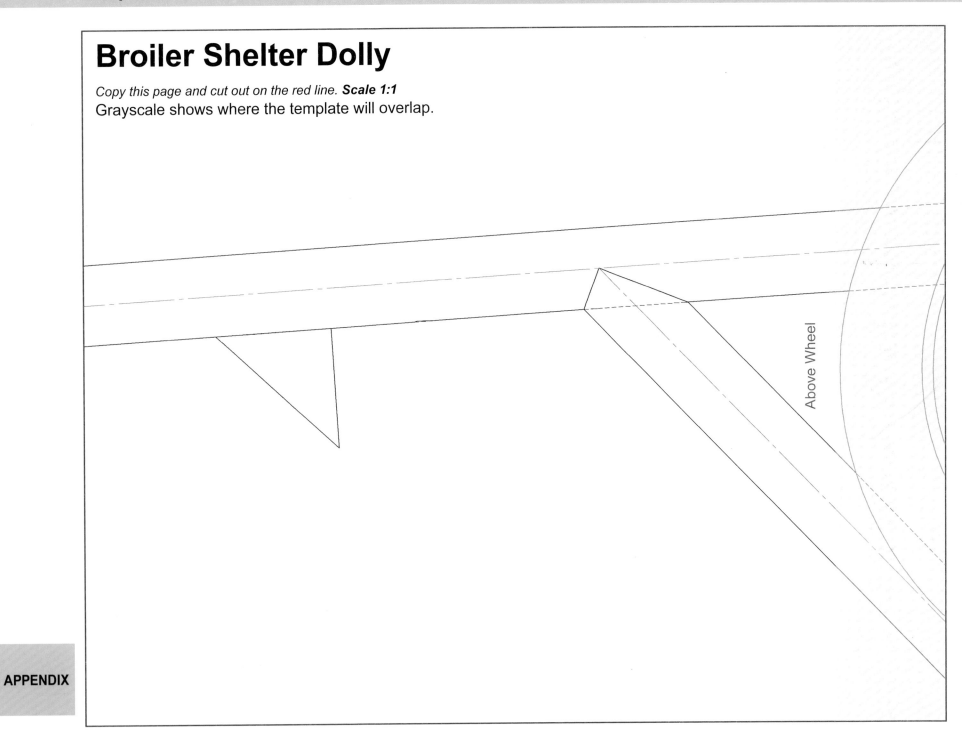

Above Wheel